Savanna Ecology and Management

AUSTRALIAN PERSPECTIVES AND INTERCONTINENTAL COMPARISONS

EDITED BY

PATRICIA A. WERNER PhD
CSIRO, Division of Wildlife and Ecology,
Tropical Ecosystems Research Centre,
Darwin, NT 0821, Australia

REPRINTED FROM THE
JOURNAL OF BIOGEOGRAPHY
VOLUME 17, NUMBER 4/5

OXFORD
BLACKWELL SCIENTIFIC PUBLICATIONS
LONDON EDINBURGH BOSTON
MELBOURNE PARIS BERLIN VIENNA

Editorial offices:
Osney Mead, Oxford OX2 0EL
25 John Street, London WC1N 2BL
23 Ainslie Place, Edinburgh EH3 6AJ
3 Cambridge Center, Cambridge
Massachusetts 02142, USA
54 University Street, Carlton
Victoria 3053, Australia

Other Editorial Offices:
Arnette SA
2, rue Casimir-Delavigne
75006 Paris
France

Blackwell Wissenschaft
Meinekestrasse 4
D–1000 Berlin 15
Germany

Blackwell MZV
Feldgasse 13
A–1238 Wien
Austria

First published 1991

Set by Burgess & Son (Abingdon) Ltd, Thames View, Abingdon, Oxfordshire
Printed and bound in Great Britain at the University Press, Cambridge

DISTRIBUTORS

Marston Book Services Ltd
PO Box 87
Oxford OX2 0DT
(*Orders*: Tel: 0865 791155
Fax: 0865 791927
Telex: 837515)

USA
Blackwell Scientific Publications, Inc.
3 Cambridge Center
Cambridge, MA 02142
(*Orders*: Tel: 800 759–6102)

Canada
Oxford University Press
70 Wynford Drive
Don Mills
Ontario M3C 1J9
(*Orders*: Tel: 416 441–2941)

Australia
Blackwell Scientific Publications
(Australia) Pty Ltd
54 University Street
Carlton, Victoria 3053
(*Orders*: Tel: 03 347–0300)

British Library
Cataloguing in Publication Data

Savanna ecology and management.
1. Grasslands. Management
I. Werner, Patricia A.
574.52643

ISBN 0-632-03199-9

Library of Congress
Cataloging-in-Publication Data

Savanna ecology and management:
Australian perspectives and intercontinental comparisons/
reprinted from the Journal of biogeography, volume 17, number 4/5, and edited by Patricia A. Werner.
p. cm.
Includes bibliographical references.
ISBN 0-632-03199-9
1. Savanna ecology—Australia.
2. Range management—Australia.
3. Savannas—Australia.
I. Werner, Patricia, 1941– .
II. Journal of biogeography.
QH197.S28 1991
574.5'2643'0994—dc20

Savanna Ecology and Management

Dedicated to Dr Michael G. Ridpath,
whose unwavering belief
in the scientific and practical value of the
wet-dry tropics of Australia
continues to inspire and change lives

Contents

Contributors

J. M. ACERO *GESER, Universidad de Buenos Aires, Buenos Aires 1248, Argentina*

J. ADAMOLI *GESER, Universidad de Buenos Aires, Buenos Aires 1248, Argentina*

A. N. ANDERSEN *Division of Wildlife and Ecology, CSIRO Tropical Ecosystems Research Center, Darwin, NT 0821, Australia*

E. R. ANDERSON *Queensland Department of Primary Industries, Rockhampton, Queensland 4702, Australia*

S. ARCHER *Department of Range Science, Texas A & M University, College Station, Texas 77843, USA*

A. J. BELSKY *The Cornell Plantations, Cornell University, Ithaca, NY 14853, USA*

B. BILBAO *Centro de Ecologia, IVIC, Caracas 1020-A, Venezuela*

A. C. BLACKMORE *Department of Botany, University of Witwatersrand, Witwatersrand 2050, Republic of South Africa*

R. W. BRAITHWAITE *Division of Wildlife and Ecology, CSIRO Terrestrial Ecosystems Research Center, Darwin, NT 0821, Australia*

W. H. BURROWS *Queensland Department of Primary Industries, Rockhampton, Queensland 4702, Australia*

J. O. CARTER *Queensland Department of Primary Industries, Longreach, Queensland 4730, Australia*

J. CLOBERT *Laboratoire d'Ecologie, URA CNRS 258, Ecole Normale Superieure, Paris Cedex 05, 75230, France*

J. R. CONNER *Department of Range Science, Texas A & M University, College Station, Texas 77843, USA*

M. J. COUCH *Department of Range Science, Texas A & M University, College Station, Texas 77843, USA*

R. J. COVENTRY *Division of Soils, CSIRO Davies Laboratory, Townsville, Queensland 4814, Australia*

K. A. DAY *Queensland Department of Primary Industries, Brisbane, Queensland 4001, Australia*

B. D. FORAN *Division of Wildlife and Ecology, CSIRO Center for Arid Zone Research, Alice Springs, NT 0871, Australia*

W. J. FREELAND *Conservation Commission of the Northern Territory, Darwin, NT 0831, Australia*

J. GIGNOUX *Laboratoire d'Ecologie, URA CNRS 258, Ecole Normale Superieure, Paris Cedex 05, 75230, France*

W. T. HAMILTON *Department of Range Science, Texas A & M University, College Station, Texas 77843, USA*

J. A. HOLT *Division of Soils, CSIRO Davies Laboratory, Townsville, Queensland 4814, Australia*

S. M. HOWDEN *Queensland Department of Primary Industries, Brisbane, Queensland 4001, Australia*

W. M. LONSDALE *Division of Entomology, CSIRO Tropical Ecosystems Research Centre, Darwin, NT 0821, Australia*

J. A. LUDWIG *Division of Wildlife and Ecology, CSIRO Rangelands Research Centre, Deniliquin, NSW 2710, Australia*

B. G. LYONS *Department of Range Science, Texas A & M University, College Station, Texas 77843, USA*

R. L. McCOWN *Division of Tropical Crops and Pastures, CSIRO Davies Laboratory, Townsville, Queensland 4814, Australia*

G. M. McKEON *Queensland Department of Primary Industries, Brisbane, Queensland 4001, Australia*

E. MEDINA *Centro de Ecologia, IVIC, Caracas 1020-A, Venezuela*

J. C. MENANT *Laboratoire d'Ecologie, URA CNRS 258, Ecole Normale Superieure, Paris Cedex 05, 75230, France*

M. T. MENTIS *Department of Botany, University of Witwatersrand, Witwatersrand 2050, Republic of South Africa*

F. A. MOOG *Department of Agriculture, Diliman, Quezon City, The Philippines*

J. J. MOTT *Division of Tropical Crops and Pastures, CSIRO Cunningham Laboratory, St Lucia, Queensland 4067, Australia*

B. R. MYRICK *Department of Range Science, Texas A & M University, College Station, Texas 77843, USA*

D. M. ORR *Queensland Department of Primary Industries, Brisbane, Queensland 4001, Australia*

M. A. PEMADASA *Department of Botany, University of Ruhuna, Matara, Sri Lanka*

C. PRADO *Laboratoire d'Ecologie, URA CNRS 258, Ecole Normale Superieure, Paris Cedex 05, 75230, France*

A. RESCIA *GESER, Universidad de Buenos Aires, Buenos Aires 1248, Argentina*

D. A. RIEGEL *Department of Range Science, Texas A & M University, College Station, Texas 77843, USA*

J. C. SCANLAN *Queensland Department of Primary Industries, Charters Towers, Queensland 4820, Australia*

W. J. SCATTINI *Queensland Department of Primary Industries, Brisbane, Queensland 4001, Australia*

R. J. SCHOLES *Department of Botany, University of Witwatersrand, Witwatersrand 2050, Republic of South Africa*

E. SENNHAUSER *GESER, Universidad de Buenos Aires, Buenos Aires 1248, Argentina*

J. F. SILVA *CIELAT, Universidad de los Andes, Merida, Venezuela*

D. M. STAFFORD SMITH *Division of Wildlife and Ecology, CSIRO Center for Arid Zone Research, Alice Springs, NT 0871, Australia*

P. STOTT *Department of Geography, School of Oriental and African Studies, University of London, London WC1H 0XG, UK*

J. W. STUTH *Department of Range Science, Texas A & M University, College Station, Texas 77843, USA*

B. H. WALKER *Division of Wildlife and Ecology, CSIRO, Canberra, ACT, Australia*

E. J. WESTON *Queensland Department of Primary Industries, Brisbane, Queensland 4001, Australia*

D. L. WIGSTON *Northern Territory University, Darwin, NT, Australia*

J. WILLIAMS *Division of Soils, CSIRO Davies Laboratory, Townsville, Queensland 4814, Australia*

W. H. WINTER *Division of Tropical Crops and Pastures, CSIRO Cunningham Laboratory, St Lucia, Queensland 4067, Australia*

P. S. YADAVA *Department of Life Sciences, Manipur University, Imphal 795003, India*

Preface

This volume is the fifth in a series of publications on the ecology of tropical savannas, over the past 5 years, arising from the work of the Responses of Savannas to Stress and Disturbance (RSSD) Program, under the auspices of the International Union of Biological Sciences (IUBS) Decade of the Tropics Programme, co-sponsored by the UNESCO Man and the Biosphere (MAB) Program.

Two of the first RSSD documents gave the international scientific community a framework of hypotheses, guidelines, and methods aimed at promoting various comparative, cooperative experiments, reaching across various savanna types and continents (Frost *et al.*, 1986; Walker & Menaut, 1988). Two other publications were proceedings of RSSD symposia in Harare, Zimbabwe (Walker, 1987) and in Guanare, Venezuela (Sarmiento, 1990). This fifth publication is comprised of research papers presented at the RSSD's international symposium, held in Darwin, Australia, in October 1988.

Over 100 scientists from six continents attended the Darwin meeting. They were asked to assess the status of current research in the context of intercontinental comparisons, especially considering the complex factors which interact to produce the savannas we know. Special attention was given to the Australian savannas. The collection was published initially as a double volume of the *Journal of Biogeography* (Vol. 17 (4/5), 1990). It is reprinted here with the addition of an extensive index which will aid the reader in cross-referencing the current state of knowledge about the important histories, processes, interactions, and species that make up the structure and function of the savannas of the world.

The Australian perspective is especially evident in the section dealing with management for large-scale livestock production, as well as in various comparisons of herbivory and fire in savannas on other continents. As for the overall intercontinental comparisons, it is not surprising that, at present, the breadth of questions, approaches, management problems, and research paradigms used by the various researchers make many comparisons difficult, albeit very interesting and informative. In total, savannas in over 46 countries are touched on by the 52 contributing authors.

This RSSD symposium, along with other international gatherings (e.g. Huntley & Walker, 1982; Tothill & Mott, 1985), and comprehensive volumes (Cole, 1986; NRC, 1990) have highlighted what research is required in regard to specific questions and needs.

All the authors, and many others cited by individual authors, made substantial contributions to the quality of this volume through their review of papers written by others. In addition to the authors and to anonymous reviewers, we thank the following for their assistance in manuscript review: Drs Martin Andrew, David Bowman, Elias Chacko, Garry Cook, Gordon Duff, Graham Harrington, Ian Noble, Tony Press, Michael Ridpath, John Taylor and John Woinarski; Professor David Wigston; Messrs Andy Chapman, Ian Cowie and Don Tulloch.

The book version was made possible by financial assistance from the International Union of Biological Sciences, Paris. Professor David L. Wigston was a co-compiler of the index; he and I gratefully acknowledge the editorial assistance of Mrs Janet Shepherd and Miss Laurel Moreland. Mr Philip Stott, as editor of the *Journal of Biogeography*, played a critical role in the review and production of the original volume. His photograph of a savanna open forest graces the cover of this book. Dr Brian H. Walker, Chief of the CSIRO Division of Wildlife and Ecology, Australia, was co-organizer of the Darwin symposium and provided guidance and assistance in innumerable ways throughout the meeting and publication. I am deeply grateful for the many discussions and practical experiences with colleagues and friends throughout the course of the past 10 years. They have helped to shape the editorial content of the introductory sections and conclusions of this volume. Any errors of fact or interpretation are my own.

PATRICIA A. WERNER
29 March 1991
Current Address:
Division of Biotic Systems and Resources;
National Science Foundation;
Washington, DC 20550; USA

REFERENCES

Cole, M.M. (1986) *The savannas: biogeography and geobotany*. Academic Press, London.

Frost, P., Medina, E., Menaut, J.-C., Solbrig, O., Swift, M. & Walker, B. (1986) *Responses of savannas to stress and disturbance: a proposal for a collaborative programme of research*. Biology International Special Issue 10. IUBS, Paris.

Huntley, B.J. & Walker, B.H. (Eds) (1982) *Ecology of tropical savannas*. Springer, Berlin.

National Research Council (1990) *The improvement of tropical and subtropical rangelands*. National Academy Press, Washington, DC, U.S.A.

Sarmiento, G. (Ed.) (1990) *The American savannas: biogeography, ecology, and management*. University of Los Almos Press, Mérida, Venezuela.

Tothill, J.C. & Mott, J.J. (Eds) (1985) *Ecology and management of the world's savannas*. Australian Academy of Science, Canberra.

Walker, B.H (Ed.) (1987) *Determinants of tropical savannas*. IRL Press, Oxford.

Walker, B.H. & Menaut, J.-C. (Eds) (1988) *Research procedure and experimental design for ecology and management*. RSSD Australia Publ. No. 1, CSIRO, Melbourne.

Introduction

Tropical savannas cover just under a third of the world's land surface, over half the area of Africa and Australia, and 45% of South America. About 10% of India and South-East Asia are classified as savanna. Overall, savannas contain a large and rapidly growing proportion of the world's human population, as well as the majority of its rangelands and livestock. Most savannas are experiencing increasing pressures from demographic and economic changes that have increased dramatically over the past few decades. In several regions there is now severe damage to vegetation and soils, and the greatest current human tragedy – the recurrent Sahel famines – is in large part due to the inability of the savannas concerned to withstand the pressures to which they have been subjected. In addition to the changing patterns in demography and economics, the forecasts of global warming (and concomitant changes in precipitation patterns) further alert us to a most important challenge – to conserve and manage wisely the savanna ecosystems of the world. To do this, we need to understand how they function. As a starting point, what are savannas, and how are they formed and how do they persist?

What are savannas?

Interpretation of the term 'savanna' varies greatly, but general accord has been reached using broad definitions, the common components of which may be summarized as the following: all tropical and subtropical ecosystems characterized by a continuous herbaceous cover of (heliophilous) C4 grasses that show seasonality related to water, and in which woody species are significant but do not form a closed canopy or a continuous cover (after Frost *et al.*, 1986). In general, savanna ecosystems are delineated both structurally (specifying a woody:grass composition), and climatically (seasonality of water availability).

How are they formed and maintained?

Indeed, the key determinant of savanna is the pattern of strong seasonality in available soil moisture, which sets a limit on the maximum plant productivity and a constraint on the types of plants that can survive the alternating periods of drought and favourable water relations. Total productivity and plant type are further constrained by availability of nutrients, which itself is partly interactive with precipitation and past vegetation history and partly independent (e.g. parent material). The core savannas of the world are determined by a set of water/nutrient interactions, including stresses that constrain the production of biomass. In general, the length of the wet season constrains the length of the growing season, and available nutrients constrain plant growth rate. Thus, it is not surprising that across the world's savannas, dry/eutrophic savannas have about equal total annual primary production as wet/dystrophic savannas. Given similar regimes of these two complex variables alone, in two different parts of the world, one could reasonably expect similar physiognomy and structure of the savannas with differences in species due to differences in evolutionary history. But, in fact, two other important factors, fire and herbivory, modify the patterns of primary production by removing biomass, usually shifting the balance of competitive interactions, and secondary relationships at higher trophic levels. These two factors can be either natural or human-controlled, but in either case they are not independent of climate and nutrients. The effects of fire and herbivory on a particular savanna's structure and function will depend upon the basic climatic/edaphic factors. For example, fire and herbivory in a dry/eutrophic savanna will have different probabilities of occurrences and degree of biomass removal than in a wet/dystrophic savanna (related to fuel load, humidity, nutritional value, migratory patterns of large mammals, etc.), and both short-term and long-term effects will differ accordingly.

On a continental scale of comparison, natural (non-human-based) fire and herbivore regimes have very different histories, both in geological time and in recent centuries. These differences undoubtedly account for some of the observed differences among savannas in structure, and (we must assume through lack of knowledge) in function.

Human influence

The human factor adds a great deal of complexity to most comparisons. Humans have modified natural fire and herbivore regimes in different ways on different continents (and in different areas within continents). In some savannas, basically determined by climate and edaphic characteristics, human-set fires have very long histories which have had a profound effect on structure and composition (e.g. Aboriginal fires in Australia beginning at least 50,000 years ago), whereas, in other savannas, humans have produced and maintained savannas out of what had been climatically-determined closed-canopy tropical forest (e.g. centuries of cutting and fires in parts of South-East Asia and the Indian subcontinent).

Aboriginal peoples in Australia and South America did not husband domesticated herbivores, yet now savannas on those two continents are subjected to intense herbivore pressure from ungulates, which were not part of their evolutionary history. Similar to fire, the effects of livestock have depended upon the basic climatic/edaphic regime; for example, drier savannas are especially sensitive to overgrazing in dry periods and can be shifted to shrubby thickets, devoid of grasses. (Most of the scientific disagreements and misunderstandings about savanna origins, structure, and function can be traced to this broad area of 'modifying determinants'.)

An integrative model

The integration of the four savanna determinants (climate, soil, fire and herbivory) allows for a single model of the structure and function of the world's savannas which, at the same time, permits an explanation of their differences in terms of the relative importance of these four factors. The model has been summarized diagrammatically in an earlier RSSD publication (Walker, 1987). (It is reproduced, with the essential connections and influences, as Fig. 4, in J. Belsky's paper, this volume p. 143.)

Thus, 'core' savannas are climatically and edaphically determined, and depending upon the particular water/nutrient regime, fire and herbivory (and the axe) may modify structure (such as woody:grass ratio) and function. At the 'margins' of savanna ecosystems they may increase savanna at the expense of closed forest (wet end) or lose savanna to shrub/thicket vegetation (dry end).

Future changes

Many savannas which are very similar in structure have vastly different histories. Their structures may have been developed and maintained by different compensating relative strengths of the four main factors determining savannas. Further, they may be changing in different ways, and their apparent convergence in physiognomy could be temporary, as similar points on their relative trajectories. Questions of stability and resilience require us to know both controlling factors and the trajectory of change. Unfortunately, as tempting as it is, one cannot infer common futures, or responses to particular management decisions, from similar-looking savannas, without understanding how and what processes shape those savannas. Fortunately, our continents possess abiotic and biotic 'experiments', both naturally and from exotic introductions and changes in land use, which, through collaborative studies, will give us insights into savanna function so that we might become better stewards of this basic resource.

Organization of this volume

The papers in this volume are organized into four sections. The first presents overviews of Australasian savannas. The Australian and Asian savannas differ significantly in their origins, most of the former being mainly climatically determined (and modified by non-agricultural human activities; Braithwaite) and much of the latter derived by intensive agriculture and human settlement (Stott). Both savannas are experiencing increased pressure from livestock grazing and other human activities (McKeon *et al.*, Stott, Yadava, Pemadasa) and require significant research in ecology, sociology, and resource economics.

Comparisons of Australia's geological and evolutionary history with other southern continents are very instructive and a rich source of hypotheses about savanna origins, function and stability under current developments (Braithwaite).

Section II examines the abiotic and biotic determinants of savannas. Several of the papers deal with interactions between factors, such as water and nutrients, water and fire, or nutrients, fire and decomposers (Medina & Silva, Scholes, Bilbao & Medina, Holt & Coventry). The relative roles of insect herbivory and large mammal herbivory in Australia are presented, with surprising conclusions that question commonly held opinions (Andersen & Lonsdale, Freeland). There is no dedicated chapter on the role of fire; however, fire is so ubiquitous a feature of savannas (especially Australian) that almost every author has included it.

The essential descriptor of various savanna ecosystems is the relative amounts and spacing of woody plants and grass (i.e. the tree/grass ratio). Section III deals with the development and persistence of tree/grass ratios as well as patches of different vegetation types within individual savannas (Archer, Blackmore *et al.*, Menaut *et al.*, Adámoli *et al.*). Theoretical models are also compared (Belsky) and a computer model is presented which examines temporal and spatial aspects of tree population dynamics as a function of fire and competition with grass (Menaut *et al.*).

The final section presents papers which have direct bearing on the management of savannas for pastoral industries, aimed at large-scale ranching operations (as opposed to pastoral, subsistence-level operations). Again, factors influencing productivity are discussed, mainly factors that impinge on fodder quality, quantity and timing (Burrows *et al.*, McCown & Williams, Moog). A strong argument is made for a change in management philosophy of Australia's cattle industry which includes the control of animal movements and pasture quality *via* fire management, leading to sustainable (yet increased) yields and conservation (Winter). Decision-support models for land management are presented, aimed both at the advisor (Stuth *et al.*) and the land manager (Stafford Smith & Foran, Ludwig). Every paper in this section recognizes that under current economic pressure, the savanna lands are in a precarious position and without enlightened management they face a future of widespread biological and physical deterioration.

P. A. Werner
B. H. Walker
P. A. Stott

Editorial note

The inclusion of accent marks in Spanish and French words has not always been possible due to the limitations set by the word processing system used in preparation of original manuscripts.

Section I
Australasian Savannas: Overviews

This introductory section begins with Braithwaite's discussion of the geological and evolutionary history of Australia, setting in context some of the broad differences among the floras and faunas of the savannas of the Southern Continents – Australia, Africa and South America. Braithwaite presents hypotheses about the relative strengths of various processes operating in these ecosystems, and makes the point that 'ghosts of historical events' will continue to affect predictions of structure and function made from global models of available nutrients and water (and hence, serve partly as tests of these models).

The savanna forests and open grasslands of mainland South-East Asia are described biogeographically and ecologically by Stott. He discusses possible origins of these ecosystems, and contrasts their stability and recent changes with respect to climatic fluctuations and the impact of human populations. There were most likely edaphic or topographic 'cores' from which savannas have been extended by fire and the axe; this process of conversion to savanna will undoubtedly continue under ever-increasing human population growth, although the effects of climate change are less certain.

Yadava describes the Indian savannas which have originated from woodland ecosystems through deforestation, abandoned cultivation and burning, producing a wide variety of savannas depending upon original forest types and seral stage of succession. Similarly, Pemadasa surveys the grasslands of Sri Lanka and India, and states that most of them occupy a 'tension-zone of biotic and anthropogenic forces working on potential forest lands.'

McKeon *et al.* provide a comprehensive review of savanna lands within Australia, from the perspective of management for pastoral production. They present evidence that El Niño/Southern Oscillation events account for over half the major ecologically significant droughts in Northern Australia over the past 120 years, and they have incorporated such climate variability into an integrated set of decision-support models to assist in management decisions. (See also Section IV.)

P.A.W.

1/Australia's unique biota: implications for ecological processes

R. W. BRAITHWAITE *CSIRO Division of Wildlife and Ecology, Tropical Ecosystems Research Centre, PMB 44, Winnellie, Darwin, Northern Territory 0821, Australia*

Abstract. The infertility of the soil and the climatic variability in Australia are the most extreme of all the continents and, after a long period of relative isolation, have resulted in a highly characteristic biota. Like those of the tropics of Asia, Africa and South America, the biota of tropical Australia has its major biogeographic affinities with the biota of its own continent. The eucalypts, the phyllodinous acacias and many of the vertebrate groups are distinctively Australian. The radiation of some biotic groups are clearly consequences of the evolution of particular resources in abundance (e.g. parrots and honeyeaters), but others are probably largely historical (e.g. marsupials).

The low productivity of native Australian ecosystems has favoured the occurrence of mutualistic relationships. It has also meant that spatial heterogeneity has great functional importance for spatial heterogeneity as small areas of higher productivity are crucial refuges and are sites of intense competition and predation. Further, great temporal variability in primary productivity has favoured invertebrates and ectothermic vertebrates. It has also resulted in high prevalence of fire and the evolution of a fire-adapted biota.

Within the wet–dry tropics of Australia, the greatest diversity of plants and animals is located in the extensive, but structurally simple and somewhat unproductive, eucalypt forests and woodlands (savannas). The wetlands and wet forests, being more productive, have more vertebrate herbivores and larger vertebrates predators but shorter food chains. Social insects dominate in the savannas, having substantial impacts on all aspects of the functioning of the savannas. Frequent fires of low intensity have resulted in a range of patterns of animal utilization of fire with consequences for community structure and function. The Australian savanna biota is clearly a distinctive eucalypt woodland biota rather than an attenuated rainforest biota as in South America or a grassland biota as in Africa.

Key words. Ecological processes, biota, biogeography, evolution, Australia.

INTRODUCTION

To understand the peculiarities of the Australian savannas, it is necessary to consider the biota and the environment of the Australian continent as a whole. These are the key to the structure of savanna communities of this continent, and this in turn influences the nature and relative importance of different ecological processes.

Although I consider all major components of the biota, I give greater emphasis to the native vertebrates than the other papers in this section. This is especially because in this part of the wet–dry tropics, we share the concern of most throughout our continent for the preservation of our unusual vertebrates. Australia has already lost many species. For example, even in sparsely populated central Australia, about 40% of mammal species have disappeared (Morton & Baynes, 1985). However, unlike the rest of the continent, the high-rainfall parts of the wet–dry tropics do not appear to have lost species. In spite of the introduction of various exotic plants and animals, the ecosystems have survived fairly intact (Braithwaite & Werner, 1987). The lack of success in agriculture due to isolation, poor soils and a hot, unreliable climate (Davidson, 1972), has been the protector of the natural systems of far north-west Australia.

Consequently this part of the world offers ecologists a special opportunity to study an intact landscape in order to investigate large-scale processes. Such a landscape also has tremendous appeal to tourists and is the basis of an important local industry.

THE ENVIRONMENTAL CONTEXT

Australia is located between 10 and 43°S, with the widest part of the continent in a zone of climatic uncertainty around 30°. The year-to-year variability in precipitation is high (Leeper, 1970; Gentilli, 1971). Seasonality in precipitation is also high (Nix, 1983). Australia is the driest continent (Brown, 1983).

The fertility of the soils is generally low (Nix, 1981). There are no substantial areas of geologically recent volcanic or tectonic activity or glaciation. The areas of higher fertility are small in area and patchily distributed. In general, the continent is characterized by a long history of weathering cycles. In fact, Cole (1986, pp. 301–302) argues that the actual distribution of major savanna categories is more complex than in South America and Africa because of the greater dissection of the Tertiary plantation surfaces which has produced a greater variety of habitat conditions over relatively short distances. Australian soils are particularly deficient in phosphorus. They have an average total phosphate content of 0.03%, compared with 0.06% for those of the United States and 0.04% for England (Blair, 1982). The continent also has deficiencies in a vast array of trace elements in a wide range of its soils (Taylor, 1983).

The continent is relatively flat with few mountainous areas and no peaks above 2250 m. This largely precludes the high habitat diversity associated with topographic variability which has long been regarded as a precursor to high density of species (e.g. Simpson, 1964). Altitudinal variation provides refuges from warmer and drier climates (Brown, 1971) which are a source of colonists when cooler and more mesic conditions return.

The latitudinal range of the continent is smaller than those of Asia, Africa and South America and has only a narrow zone of greater production around the northern, eastern and southern coasts. The extensive central arid zone limits seasonal migrations for many mobile birds and larger mammals in most years (see Nix, 1976).

Around 45 million years ago Australia broke off from Antarctica, completing the break-up of Gondwanaland. This followed the earlier separation of India, Africa and South America from the southern super-continent and marked the beginning of a long period of isolation for Australia. It was not until about 15 million years ago that Australia proceeded far enough north to collide with the Asian plate and end the isolation by exchanging parts of the biota with Asia (Kemp, 1981).

BIOGEOGRAPHIC AFFINITIES

The present terrestrial biota in Australia can be divided into six categories. The oldest groups are relics of the former large continents, Pangaea or the subsequent Gondwana. Blattodea (cockroaches), cycads and diplurans are examples of this (1) Pangaean or Archaic Element, while *Acacia*, proteacaeous plants, ratite birds, parrots, chelid turtles, diplodactyline geckoes and the marsupials are examples of the (2) Gondwanan or Old Southern element. An (3) Autochthonous element developed in the time gap of 20 million years between the Gondwanan and Tertiary Asian elements. Examples include eucalypts and honeyeaters. The (4) Asian Tertiary element is derived from the time the Australian plate approached the Asian plate sufficiently closely to receive immigrants. Examples include most families of snakes and lizards, the conilurine rodents and many birds and insects (Heatwole, 1987). Although it was also believed that much of the Australian tropical flora originated in Asia, Webb, Tracey & Jessup (1986) now argue that the tropical flora is Gondwanan in origin and was derived from the sclerophyllous element of the Australian flora following radiation from authochthonous rainforest species. The (5) Modern element of the biota are recent arrivals from Asia which have changed little since arrival. Examples include the ranid frog *Rana daemeli* (Steindachner), the turtle *Carettochelys insculpta* Ramsay, native rats *Rattus* and the Rainbow Bee-eater (*Merops ornatus*) Latham, and again many tropical plants. The (6) Introduced element is the very recent anthropochorous component of the biota which began with the dingo (*Canis familiaris dingo* (Meyer)) and the Tamarind tree (*Tamarindus indica* L.) and is continuing to increase at a substantial rate (Specht, 1981).

The long period of isolation has resulted in a high level of endemism for plants (Specht, 1981), mammals (Keast, 1972) and birds (Keast, 1981). While the radiation of some lizard groups has centred on the arid zone (e.g. skinks, geckoes), most animal groups have speciated as a result of the 'pulsating dry heart' of the Australian arid zone during the alternating mesic and arid periods through geological time. Much of the speciation has resulted from such vicariant separation of the forest refuges of the periphery of the continent. Horton (1984) has argued that during arid periods the savanna disappeared entirely wiping out its fauna including most of the megafauna of the continent. He further argues that much of the speciation took place in the eucalypt forests and woodlands. During mesic periods, mesic barriers which developed in the semi-arid zone facilitated radiation in that area. Consequently the savanna fauna of today is largely a eucalypt forest and woodland fauna with some intrusion of species from the semi-arid zone. I argue that this has resulted in a biota which has species which are essentially wet season specialists breeding in the wet season (e.g. *Carlia* skinks) and dry season specialists breeding in the dry season (Fig. 1) (e.g. *Ctenotus* skinks; see James & Shine, 1985). The result has been the great phenological range seen throughout the year in both plants (e.g. Kerle, 1985) and animals (e.g. James & Shine, 1985), with both wet season and dry season breeders rather than all phenological events being centred on the wet season.

In contrast with the eucalypt forest character of the Australian savannas, the South American savannas might be characterized as attenuated rainforest (cf. Prance, 1982; Redford & de Fonseca, 1986). On the other hand, the character of African savannas is one of vast, fertile grasslands with *Acacia* trees and large mammals (cf. Bell, 1982). Obviously this is simplistic but aspects of these dominant

FIG. 1. *Ctenotus storii* is one of the many insectivorous vertebrates of Australian savannas. (Photo: J. Wombey.)

themes affect all ecosystems of each continent to some extent. It is the consequences of these continental-scale evolutionary influences on ecological processes which I will now describe in greater detail for Australia.

BROAD FUNCTIONAL CONSEQUENCES

The low nutrient status of Australian soils is likely to have contributed to the success of legumes, particularly *Acacias*. Their ability to fix atmospheric nitrogen through mutualistic relations with bacteria housed in nodules associated with the root system enables them to compensate for low nitrogen availability in the soil. Nodulation has been recorded on about 300 species in fifty-six legume genera in Australia (Bowen, 1981). Nitrogen-fixing actinomycete associations with *Casuarina* and blue-green algal associations with cycads are also important contributors of nitrogen to Australian ecosystems (Bowen, 1981). Mycorrhizal symbioses between roots and various fungi are major means of increasing nutrient uptake, particularly phosphorus, from the soil. They are described from eight to nine tree families in Australia (Bowen, 1981) and have been found in all eucalypt species examined (Pryor, 1976).

The eucalypts have a strong tendency to develop hollows in their wood (e.g. Saunders, Smith & Rowley, 1982). 23% of Australian vertebrates are recorded as using tree holes in some way (Ambrose, 1980). Australia has more bird species which are obligate hole-nesters than southern Africa and North America (Saunders *et al.*, 1982). Australia has no primary excavators (species which produce their own holes, e.g. Picidae) among its birds. Excluding primary excavators, southern Africa has only 7% hole-nesters among its non-passerines while Australia has 21% (Saunders *et al.*, 1982). Micro-organisms and termites, in particular *Coptotermes acinaciformes* (Froggatt), are the main primary excavators in Australia (Perry, Lenz & Watson, 1985). Because of both local and regional associations with low soil fertility, it has been suggested that there may be a mutualistic relationship between some eucalypt species and *C. acinaciformis* (Braithwaite *et al.*, 1989). That is, termites supply increased nutrient availability to the tree and gain food and protection. This is a specific version of Janzen's (1976) nutrient hollow hypothesis and the relationship appears to be most strongly developed in the Top End region of Australia (Braithwaite *et al.*, 1989).

Eucalypts and other Australian plants have an impressive array of nutrient-conserving mechanisms (e.g. Florence, 1981). Included is the ability to extract almost all nutrients from dying leaves. However, this causes litter with a very high C:N ratio. Rogers & Westman (1977) found total litter fall of sub-tropical eucalypt forests predictable from climate and unaffected by soil, but the decomposition rate was rather low. Lee & Correll (1978) suggest that levels of activity of the litter fauna may often be limited by harsh climatic conditions in drier eucalypt forests. Consequently decomposition rates may be slow due to both adaptations by the flora to low fertility (e.g. high C:N ratio of litter) and the influence of the harsh climate on the litter fauna.

Most plant dispersal is by invertebrates rather than by vertebrates (Clifford & Drake, 1981). Fruit, a vertebrate-adapted dispersal mechanism is only common in the rainforest (Clifford & Drake, 1981) and is generally uncommon relative to other continents (e.g. Milewski, 1986; Keast, 1981). Presumably the availability of vertebrate dispersers is generally less certain in Australia overall and only reliable in rainforest. Some plants with fruit increase the probability of dispersal by having adaptations for dispersal by more than one type of agent. For example, *Petalostigma* spp. have a fruit dispersed by emus (*Dromaius novaehollandiae* (Latham)) but the seed is dehiscent and is also subsequently dispersed by ants (Monteith & Clifford, 1986). This type of generality decreases the risk in an uncertain environment and may be operating in pollination as well (see below).

Dispersal of seed by ants (Myrmecochory) is greatly enhanced by the presence of ant-attractant food bodies called elaiosomes attached to inedible seed. That elaiosomes comprise 5–10% of diaspore weight but only 0.6–3% of diaspore phosphorus content has led to the hypothesis that such dispersal adaptations are related to low availability of phosphorus in the soil (Westoby *et al.*, 1982). The prevalence of elaiosomes is greater in Australia than on any other continent (Berg, 1975). Removal of seeds from trays is most commonly by ants in Australia, but generally by birds and mammals in American deserts (Morton, 1985). In fact, 10% of the world's ant species are Australian (Riek, 1970) which may be a factor in the co-evolution of elaiosomes and ants.

Other Australian plant species produce tiny wind-dispersed seed which is protected in hard woody capsules (e.g. *Eucalyptus*). The prevalence of this seed protection mechanism has led to the magnificent radiation of Australian parrots. One sixth of the world's parrots are Australian and the majority feed on seed from species of *Eucalyptus, Acacia, Casuarina, Hakea, Grevillea, Banksia* and other dicots with woody capsules (Forshaw, 1969). It is not known if the parrots sometimes function as seed dispersers as well as seed predators. Evidence to the contrary has been presented by Bridgewater (1934).

The woody protection of seeds by some plant species, and in others the development of elaiosomes facilitating the burial of seeds in ant nests, are also adaptations in a fire-prone environment. The prevalence of the lignotuber, an underground woody storage organ from which resprouting occurs, is another adaptation to the harsh environment which is often characterized as fire-prone. It is useless to argue about whether such plant adaptations are adaptations in response to aridity or to fire (see also Gill, 1981). The two factors co-occur. Where the climate is extremely variable, seasonally or annually, there are periods of substantial vegetative growth followed by dry conditions which cure the vegetation to produce a substantial fuel load and weather conditions suitable for an intense fire. The frequency of fires appears to be largely determined by the frequency and distinctness of the pulses of productivity. Over the continent as a whole, if plants generally experience fires many years apart many plants are killed and the species present tend to regenerate from seed. If fires are more frequent each tends to be less intense and kill fewer plants and this fire environment favours species which resprout

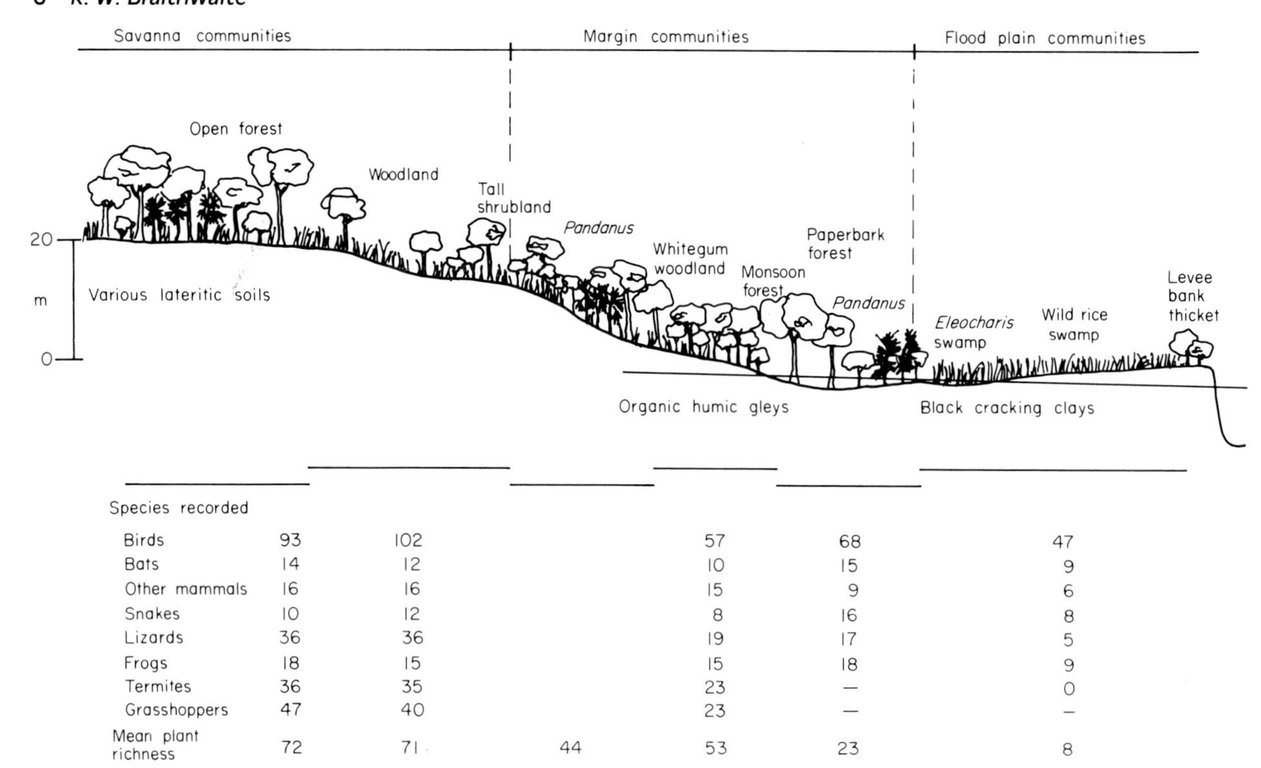

Species recorded						
Birds	93	102		57	68	47
Bats	14	12		10	15	9
Other mammals	16	16		15	9	6
Snakes	10	12		8	16	8
Lizards	36	36		19	17	5
Frogs	18	15		15	18	9
Termites	36	35		23	—	0
Grasshoppers	47	40		23	—	—
Mean plant richness	72	71	44	53	23	8

FIG. 2. A generalized gradient (3–10 km in length) from wetlands to upland eucalypt communities in Kakadu National Park showing numbers of plant and animal species recorded from different habitat types. The animal and plant data are from Braithwaite (1985) and Taylor & Dunlop (1985) respectively. The plant data are mean values per 0.16 ha.

FIG. 3. Termites are dominant herbivores in the Australian savannas. (Photo: R. Braithwaite.)

from above or below ground living tissue (Gill, Groves & Noble, 1981, *inter alia*).

Australian woody plants produce copious quantities of nectar. Ford, Paton & Forde (1979) suggest that as sunlight is rarely limiting in Australia, sugar is very cheap for a plant to produce. This has had a dramatic effect on the bird fauna. Sixty-seven of the 159 honeyeaters of the Family Meliphagidae occur in Australia and most of the rest are in New Guinea (Keast, 1959). The nectarivorous lorikeet parrots (Loriinae) also are a New Guinea–Australia radiation

(Keast, 1959). While nectar is the major source of food for twenty species of birds and a minor source for another twenty in southern Australia (34–40°S), there are no nectarivorous birds in Europe or North Africa, two in Asia, eight in North America and five in South America at equivalent latitudes. Indeed, the prevalence of nectarivorous bird species in Australasia and southern Africa seems to be associated with the prevalence of plants of the Families Myrtaceae and Proteaceae (Ford *et al*., 1979).

Although there is substantial floral variation among the plant species, there is a remarkable lack of specialization among Australian nectarivores. The honeyeaters are generalists and opportunists and individual plants attract a wide range of pollinators. The unspecialized flowers of *Eucalyptus* might be seen as part of a strategy for coping with erratic rainfall and unreliable flowering. Eucalypts appears to keep all their options open: they can be pollinated first by birds, later in the day by insects such as bees, flies and beetles, and at night by mammals and possibly moths (Ford *et al*., 1979; Turner, 1982). As the honeyeaters must take insects to obtain sufficient protein, it has been suggested that the insects attracted to the nectar might be part of the pollination attraction of the plant (Recher & Abbott, 1970). A further possibility is that the birds also act as protectors by removing many of the herbivorous insects (Ford *et al*., 1979) which can severely affect the growth of the trees (Morrow & LaMarche, 1978).

Stafford Smith & Morton (1989) also suggest that energy is not limiting and this is the reason some Australian tree species can tolerate high loads of sap-sucking insects. These insects attract animals such as ants which in turn probably act against herbivores grazing the trees (Buckley, 1982).

Invertebrates are often prominent in an infertile landscape. Braithwaite, Miller & Wood (1988) found termites and grasshoppers most diverse in the infertile parts of the landscape in the Kakadu National Park (Fig. 2). The invertebrates are better able to cope with infertile soils and harsh dry periods and quickly take advantage of the wet periods. Termites are able to fix nitrogen using symbiotic microorganisms in their hindguts and store grass in termitaria; grasshoppers have a high rate of population increase (Braithwaite *et al*., 1988); a variety of insects aestivate in mesic locations during the dry season (Monteith, 1982) and ants store seeds in nests. This importance of invertebrate herbivores (Fig. 3) in a system means that there is much food for insectivorous vertebrates. Morton & James (1988) have suggested this is the reason the arid zone of Australia supports the highest diversity of lizards in the world. The ectothermic vertebrates with their low metabolic rates and insectivorous diet are able to survive the harsh periods of little food better than endothermic birds and mammals.

Birds generally have the advantage of high mobility and are able to move to take advantage of remaining waterholes during drought and of patchy rainfall. Consequently it is no surprise that Keast (1959) classified a high 26% of Australian birds as nomads, with the remainder being 66% sedentary and only 8% north–south migrants. Nix (1976) describes the limited pattern of migration exhibited by Australian birds. A few of the larger mammals, like the Red Kangaroo (*Macropus rufus* (Desmarest)), also have sufficient mobility to track resources over a large area, but most mammals and many birds require local refuges to persist.

It has been argued that there is a marsupial strategy which includes the ability to terminate reproduction readily and that this is well adapted to the unpredictable climate of the continent (Low, 1978). Morton *et al*. (1982) have countered that most marsupials occur in the more predictable parts of the continent and the group exhibits a very wide range of life histories. However, the slow birth to weaning times of the smaller marsupials relative to similar-sized eutherians (Braithwaite & Lee, 1979) and 15% lower basal metabolic rate of marsupials (Dawson & Hulbert, 1970) may be of some adaptive significance, rather than merely being consequences of contraints of a different reproductive anatomy/physiology (cf. Lillegraven *et al*., 1987).

The clutch size of birds is generally lower in Australia than in other parts of the world (Woinarski, 1986). Yom Tov (1985, 1987) suggested that recently arrived (e.g. Pleistocene) groups of both birds and mammals have higher clutch/litter sizes than long-established groups. However, it could be argued that the recently arrived groups colonized the mesic and more fertile periphery of the continent. My own work on savanna mammals suggests a strong positive relationship between modal litter size of individual species and the proportion of insect material in the diet (Fig. 4). That is, in an infertile habitat, the litter size of herbivorous species is likely to be limited by protein availability. Alternatively, there are evolutionary constraints on diet for animals with a large litter size.

There have also been suggestions about the adaptive advantage of co-operative breeding in birds in the unreliable Australian climate (Rowley, 1975). Ford *et al*. (1988) demonstrate that highest species density of co-operative

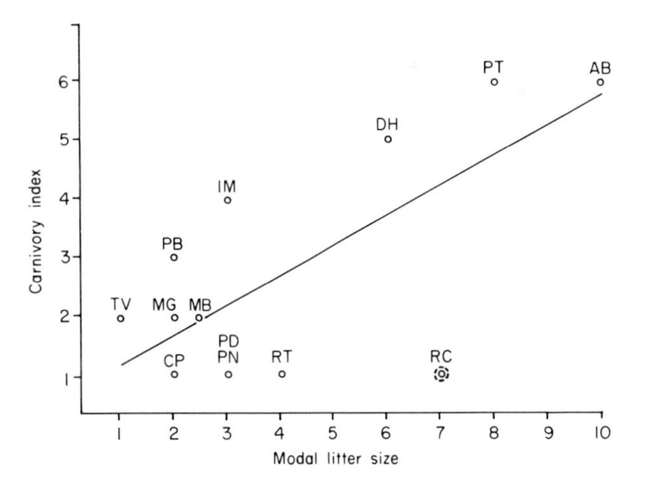

FIG. 4. The relationship between modal litter size and degree of carnivory/insectivory of the diet of an assemblage of small mammals in savanna in Kakadu National Park. The data are from Strahan (1983). The species codes are: AB, *Antechinus bellus* (Thomas); DH, *Dasyurus hallucatus* (Gould); PT, *Phascogale tapoatafa* Meyer; IM, *Isoodon macrourus* Gould; PB, *Petaurus breviceps* Waterhouse; TV, *Trichosurus vulpecula arnhemensis* Collett; CP, *Conilurus penicillatus* (Gould); MB, *Melomys burtoni* (Ramsay); MG, *Mesembriomys gouldii* Gray; PD, *Pseudomys delicatulus* (Gould); PN, *Pseudomys nanus* (Gould); RC, *Rattus colletti* Thomas; RT, *Rattus tunneyi* Thomas. For all species, $r = 0.71$, $P < 0.01$. If the circled *Rattus colletti* (mainly a species of the fertile wetlands) is excluded, $r = 0.84$, $P < 0.001$.

FIG. 5. A typical seep in savanna at Kapalga in Kakadu National Park. Such areas are important for the diversity and abundance of mammals and birds. (Photo: R. Braithwaite.)

breeders correlates closely with lack of annual seasonality but also suggest high risk of predation is important. The issue is complex. Perhaps the evolution of co-operative breeding in the nectarivorous species was a response to the lack of specialized pollinators amongst nectarivorous Australian species. Pugnaciousness amongst honeyeaters reaches its zenith in effectiveness in the Noisy and Bell Miners (Dow, 1970; Loyn *et al.*, 1983), where aggressive groups lay claim to a broad range of food resources by excluding most other bird species from a group territory. So

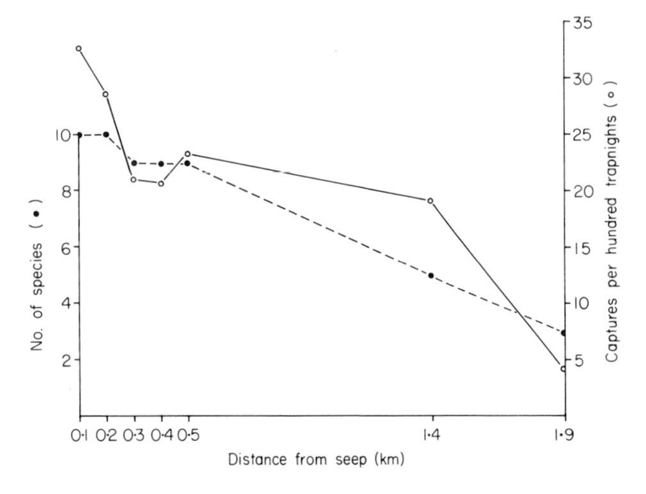

FIG. 6. Relative abundance and species richness of non-volant mammals less than 5 kg for a 3-year period along a 2 km gradient from seep to ridge at Kapalga in Kakadu National Park. The data for 0.1–0.5 km are from Braithwaite (unpublished) and for 1.4 and 1.9 km from Friend & Taylor (1985).

the adaptation to the broad context of the Australian environment may be an indirect one, i.e. an adaptation to the generalized pollination syndrome.

The predominance of infertile soil in the landscape means that small fertile patches, places in the landscape where nutrients and moisture accumulate, are very important to the productivity of the whole ecosystem (Noy Meir, 1973, 1983). Research we have done on the seepage zones (Fig. 5) of the Australian savanna suggests these areas are centres of diversity and abundance of the non-volant mammals less than 5 kg (Fig. 6), although they occupy less than 5% of the total savanna area. They seem to be sites of strong competitive interaction with relative species composition frequently changing (Braithwaite, 1985). Predation pressure also seems to be intense in the seepage zones (Braithwaite, 1985). They are also important areas for granivorous and nectarivorous birds (Braithwaite, 1985).

CONCLUSION

In Table 1 I have attempted to summarize what I perceive to be the main differences between ecological processes in Australia, Africa and South America. Of course, each continent is home for a range of biomes. However, the geological and recent history of each whole continent influences the evolution of the whole biota and this moulds the structure of the biota which occupies the savannas of each continent. This continent-specific biota with its radiations of different species groups influences characteristics of ecological processes in the savannas of each continent. The characteristics of many processes are merely consequences of present conditions (e.g. primary productivity)

TABLE 1. A preliminary synopsis of dominant features and the relative strength of processes in savannas of three continents.

Process	Africa	South America	Australia
Primary productivity	High	Intermediate	Low
Herbivory	Large vertebrates	Smaller vertebrates	Invertebrates
Predation	Large vertebrates	Medium vertebrates	Small verebrates
Nutrient acquisition	Easy	Intermediate	Difficult
Mutualism	Uncommon	Common	Common
Decomposition	Rapid	Intermediate	Slow
Pollination	Specialized	Very specialized	Unspecialized
Plant dispersal	Large vertebrates	Smaller vertebrates	Invertebrates
Animal migration	Major	Intermediate	Minor
Individual animal metabolism	Medium	High	Low
Animal population productivity	High	Medium	Low
Competition	Widespread	Patchy	Patchy
Fire accommodation	Medium	Low	High

while others are the results of long-term continental persistence of such conditions (e.g. pollination syndromes). As we continue to attempt to understand tropical savannas, we will find it possible to predict some aspects of ecological structure and function from global models based on plant-available moisture and nutrients, but the ghosts of historical events on different continents will continue to haunt us in the form of unexplained variation.

ACKNOWLEDGMENT

I thank W. M. Lonsdale and S. R. Morton for their critical reviews of a draft of the manuscript.

REFERENCES

Ambrose, G. (1980) Studying tree-holes as habitats: strategies and techniques. *Victorian Amateur Research Group Notes*, **16**, 7–14.

Bell, R.H.V. (1982) The effect of soil nutrient availability on community structure in African ecosystems. *Ecology of tropical savannas* (ed. by B. J. Huntley and B. H. Walker), pp. 193–216. Springer, Berlin.

Berg, R.Y. (1975) Myrmecochorous plants in Australia and their dispersal by ants. *Aust. J. Bot.* **23**, 475–508.

Blair, G.J. (1982) The phosphorus cycle in Australian agriculture. *Phosphorus in Australia* (ed. by A. B. Costin and C. H. Williams), pp. 92–111. C.R.E.S., Aust. Nat. Univ., Canberra.

Bowen, G.D. (1981) Coping with low nutrients. *The biology of Australian plants* (ed. by J. S. Pate and A. J. McComb), pp. 33–64. Univ. of West. Aust. Press, Perth.

Braithwaite, R.W. (1985) *The Kakadu fauna survey: an ecological survey of Kakadu National Park.* Unpublished report to Australian National Parks and Wildlife Service, Canberra.

Braithwaite, R.W. & Lee, A.K. (1979) A mammalian example of semelparity. *Amer. Nat.* **113**, 151–155.

Braithwaite, R.W., Miller, L. & Wood, J.T. (1988) The structure of termite communities in the Australian tropics. *Aust. J. Ecol.* **13**, 375–391.

Braithwaite, R.W., Parker, B.S., Wood, J.T. & Estbergs, J.A. (1989) Piping of tropical eucalypts by termites: predation or mutualism. (Unpublished manuscript).

Braithwaite, R.W. & Werner, P.A. (1987) The biological value of Kakadu National Park. *Search*, **18**, 296–301.

Bridgewater, A.E. (1934) The food of *Platycercus eximius* and *P. elegans*. *Emu*, **73**, 186–187.

Brown, J.A.H. (1983) *Australia's surface water resources.* Water 2000: Consultants Report No. 1. Dept. of Resources and Energy, Canberra.

Brown, J.H. (1971) Mammals on mountaintops: non-equilibrium insular biogeography. *Amer. Nat.* **105**, 467–478.

Buckley, R.C. (1982) Ant–plant interactions: a world review. *Ant–plant interactions in Australia* (ed. by R. C. Buckley), pp. 111–162. Junk, The Hague.

Clifford, H.T. & Drake, W. (1981) Pollination and dispersal in eastern Australian heathlands. *Ecosystems of the world. 9B: Heathlands and related shrublands: analytical studies* (ed. by R. L. Specht), pp. 39–50. Elsevier, Amsterdam.

Cole, M.M. (1986) *The savannas: biogeography and geobotany.* Academic Press, London.

Davidson, B.R. (1972) *The northern myth: limits to agricultural and pastoral development in tropical Australia*, 3rd edn. Melbourne University Press.

Dawson, T.J. & Hulbert, A.J. (1970) Standard metabolism, body temperature and surface areas of Australian marsupials. *Amer. J. Physiol.* **218**, 1233–1238.

Dow, D.D. (1970) Communal behaviour of nesting Noisy Miners. *Emu*, **70**, 131–134.

Florence, R.G. (1981) The biology of the eucalypt forest. *The biology of Australian plants* (ed. by J. S. Pate and A. J. McComb), pp. 147–180. Univ. of W. Aust. Press, Perth.

Ford, H.A., Bell, H., Nias, R. & Noske, R. (1988) The relationship between ecology and the incidence of cooperative breeding in Australian birds. *Behav. Ecol. Sociobiol.* **22**, 239–249.

Ford, H.A., Paton, D.C. & Forde, N. (1979) Birds as pollinators of Australian plants. *N.Z. Jl Bot.* **17**, 509–529.

Forshaw, J.M. (1969) *Australian Parrots.* Lansdowne Press, Melbourne.

Friend, G.R. & Taylor, J.A. (1985) Habitat preferences of small mammals in tropical open forest of the Northern Territory. *Aust. J. Ecol.* **10**, 173–185.

Gentilli, J. (1971) The main climatical elements. *Climates of Australia and New Zealand* (ed. by J. Gentilli), pp. 119–188. Elsevier, Amsterdam.

Gill, A.M. (1981) Adaptive responses of Australian vascular plant species to fires. *Fire and the Australian biota* (ed. by A. M. Gill, R. H. Groves and I. R. Noble), pp. 243–271. Aust. Acad. Sci., Canberra.

Gill, A.M., Groves, R.H. & Noble, I.R. (1981) *Fire and the Australian biota.* Aust. Acad. Sci., Canberra.

Heatwole, H. (1987) Major components and distribution of the terrestrial fauna. *Fauna of Australia*, Vol. 1A. *General articles* (ed.

by G. D. Dyne and D. W. Walton), pp. 136–155. Aust. Govt. Pub. Serv., Canberra.

Horton, D. (1984) Dispersal and speciation: Pleistocene biogeography and the modern Australian biota. *Vertebrate zoogeography and evolution in Australasia* (ed. by M. Archer and G. Clayton), pp. 113–118. Hesperian Press, Perth.

James, C.D. & Shine, R. (1985) The seasonal timing of reproduction: a tropical–temperate comparison in Australian lizards. *Oecologia (Berl.)*, **67**, 464–474.

Janzen, D.H. (1976) Why tropical trees have rotten cores. *Biotropica*, **8**, 110.

Keast, A. (1959) Australian birds: their zoogeography and adaptations to an arid continent. *Biogeography and ecology in Australia* (ed. by A. Keast, R. L. Crocker and C. S. Christian), pp. 115–135. Junk, The Hague.

Keast, A. (1972) Comparisons of contemporary mammal faunas of southern continents. *Evolution, mammals and southern continents* (ed. by A. Keast, F. C. Erk and B. Glass), pp. 433–502. State University of New York Press, Albany.

Keast, A. (1981) The evolutionary biogeography of Australian birds. *Ecological biogeography of Australia* (ed. by A. Keast), pp. 1586–1635. Junk, The Hague.

Kemp, E.M. (1981) Tertiary palaeogeography and the evolution of the Australian climate. *Ecological biogeography of Australia* (ed. by A. Keast), pp. 31–50. Junk, The Hague.

Kerle, J.A. (1985) Habitat preference and diet of the northern brushtail possum *Trichosurus arnhemensis* in the Alligator Rivers region, Northern Territory. *Proc. Ecol. Soc. Aust.* **13**, 161–176.

Lee, K.E. & Correll, R.L. (1978) Litter fall and its relationship to nutrient cycling in a South Australian dry sclerophyll forest. *Aust. J. Ecol.* **3**, 243–252.

Leeper, G.W. (1970) Climates. *The Australian environment* (ed. by G. W. Leeper), pp. 12–20. CSIRO, Melbourne.

Lillegraven, J.A., Thompson, S.D., McNab, B.K. & Patton, J.L. (1987) The origin of eutherian mammals. *Biol. J. Linn. Soc.* **32**, 281–336.

Low, B.A. (1978) Environmental uncertainty and the parental strategies of marsupials and placentals. *Amer. Nat.* **112**, 197–213.

Loyn, R.H., Runnalls, R.G., Forward, G.Y. & Tyers, J. (1983) Territorial Bell Miners and other birds affecting populations of insect prey. *Science*, **221**, 1411–1413.

Milewski, A.V. (1986) A comparison of bird-plant relationships in southern Australia and southern Africa. *The dynamic partnership: birds and plants in southern Australia* (ed. by H. A. Ford and D. C. Paton), pp. 111–118. D. J. Woolman, Govt. Printer, Adelaide.

Monteith, G.B. (1982) Dry season aggregations of insects in Australian monsoon forests. *Mem. Qd Mus.* **20**, 533–543.

Monteith, G. & Clifford, H.T. (1986) Seed dispersal in *Petalostigma*. Poster, Ecological Society of Australia Symposium, 25–27 August 1986, Brisbane.

Morrow, P.A. & LaMarche, V.C. (1978) Tree ring evidence for chronic insect suppression of productivity in subalpine *Eucalyptus*. *Science*, **201**, 1244–1245.

Morton, S.R. (1985) Granivory in arid regions: comparison of Australia with North and South America. *Ecology*, **66**, 1859–1866.

Morton, S.R. & Baynes, A. (1985) Small mammal assemblages in arid Australia: a reappraisal. *Aust. Mammal.* **8**, 159–169.

Morton, S.R. & James, C. (1988) The diversity and abundance of arid Australian lizards: a new hypothesis. *Amer. Nat.* **132**, 237–256.

Morton, S.R., Recher, H.F., Thompson, S.D. & Braithwaite, R.W. (1982) Comments on the relative advantages of marsupial and eutherian reproduction. *Amer. Nat.* **120**, 128–134.

Nix, H.A. (1976) Environmental control of breeding, post-breeding dispersal and migration of birds in the Australian region. *Proc. 16th Int. Ornith. Congr. Canberra*, pp. 272–305.

Nix, H.A. (1981) The environment of Terra Australis. *Ecological biogeography of Australia* (ed. by A. Keast), pp. 103–133. Junk, The Hague.

Nix, H.A. (1983) Climate of tropical savannas. *Ecosystems of the world, 13: Tropical savannas* (ed. by F. Bourliere), pp. 37–62. Elsevier, Amsterdam.

Noy Meir, I. (1973) Desert ecosystems: environment and producers. *Ann. Rev. Ecol. Syst.* **4**, 25–51.

Noy Meir, I. (1983) Desert ecosystem structure and function. *Ecosystems of the World, 12A: Hot deserts and arid shrublands* (ed. by M. Evanair, I. Noy-Meir and D. Goodall), pp. 93–103. Elsevier, Amsterdam.

Perry, D.H., Lenz, M. & Watson, J.A.L. (1985) Relationships between fire, fungal rots and termite damage in Australian forest trees. *Aust. For.* **48**, 46–53.

Prance, G.T. (1982) *Biological diversification in the tropics.* Columbia Univ. Press, New York.

Pryor, L.D. (1976) *The biology of Eucalypts.* Arnold, London.

Recher, H.F. & Abbott, I.J. (1970) The possible ecological significance of hawking by honeyeaters and its relation to nectar feeding. *Emu*, **70**, 90.

Redford, K.H. & da Fonseca, G.A.B. (1986) The role of gallery forests in the zoogeography of the Cerrado's non-volant mammalian fauna. *Biotropica*, **18**, 126–135.

Riek, E.F. (1970) Hymenoptera. *The insects of Australia (CSIRO)*, pp. 867–959. Univ. of Melbourne, Melbourne.

Rogers, R.W. & Westman, W.E. (1977) Seasonal nutrient dynamics of litter in a subtropical eucalypt forest, North Stradbroke Island. *Aust. J. Bot.* **25**, 47–58.

Rowley, I. (1975) *Bird life.* Australian Naturalist Library, Collins Sydney.

Saunders, D.A., Smith, J.T. & Rowley, I. (1982) The availability and dimensions of tree hollows that provide nest sites for cockatoos (Psittaciformes) in Western Australia *Aust. Wildl. Res.* **9**, 541–556.

Simpson, G.G. (1964) Species density of North American recent mammals. *Syst. Zool.* **13**, 57–73.

Specht, R.L. (1981) Major vegetation formations in Australia. *Ecological biogeography of Australia* (ed. by A. Keast), pp. 163–298. Junk, The Hague.

Stafford Smith, D.D. & Morton, S.R. (1989) A framework for the ecology of arid Australia. (Unpublished manuscript.)

Strahan, R. (1981) *A dictionary of Australian mammal names.* Angus and Robertson, Sydney.

Strahan, R. (1983) *The complete book of Australian mammals.* Angus and Robertson, Sydney.

Taylor, J.A. & Dunlop, C.R. (1985) Plant communities of the wet–dry tropics: the Alligator Rivers Region. *Proc. Ecol. Soc. Aust.* **13**, 83–127.

Taylor, J.K. (1983) The Australian environment. *Soils: An Australian viewpoint* (Division of Soils, CSIRO), pp. 1–2. CSIRO, Melbourne/Academic Press, London.

Turner, V. (1982) Marsupials as pollinators in Australia. *Pollination and evolution* (ed. by J. A. Armstrong, J. M. Powell and A. J. Richards), pp. 55–66. Roy. Bot. Gardens, Sydney.

Webb, L.J., Tracey, J.G. & Jessup, L.W. (1986) Recent evidence for autochthony of Australian tropical and subtropical rainforest floristic elements. *Telopea*, **2**, 575–589.

Westoby, M., Rice, B., Shelley, J.M., Haig, D. & Kohen, J.L. (1982) Plants' use of ants for dispersal at West Head, New South Wales. *Ant–plant interactions in Australia* (ed. by R. C. Buckley), pp. 75–88. Junk, The Hague.

Woinarski, J.C.Z. (1986) Breeding biology and life history of small insectivorous birds in Australian forests: response to a stable environment. *Proc. Ecol. Soc. Aust.* **14**, 159–168.

Yom Tov, Y. (1985) The reproductive rates of Australian rodents. *Oecologia (Berl.)*, **66**, 250–255.

Yom Tov, Y. (1987) The reproductive rates of Australian passerines. *Aust. Wildl. Res.* **14**, 319–330.

2/Northern Australian savannas: management for pastoral production

G. M. McKEON,[1] K. A. DAY,[1] S. M. HOWDEN,[1] J. J. MOTT,[2] D. M. ORR,[1] W. J. SCATTINI[3] and E. J. WESTON[4] [1]*Pasture Management Branch, Queensland Department of Primary Industries, Brian Pastures Research Station, Gayndah 4625,* [2]*Division of Tropical Crops and Pastures, CSIRO, 306 Carmody Rd, St Lucia 4067,* [3]*Pasture Management Branch, Queensland Department of Primary Industries, GPO Box 46, Brisbane 4001, and* [4]*Pasture Management Branch, Queensland Department of Primary Industries, PO Box 102, Toowoomba 4350, Australia*

Abstract. Our paper examines recent developments in climatology, systems analysis and decision support which are relevant to the management of northern Australian savannas. The structure, function and use of these communities have been well described in previous reviews which show the importance of pastoralism as the major economic activity.

Annual variability of rainfall is high, resulting in uncertainty in management decisions. Systems analysis models of pastoral enterprises are being constructed which predict the response of savannas to management alternatives against a background of annual climatic variation as well as expected long-term global climate change.

In northern Australia, El Niño/Southern Oscillation events account for over half the major ecologically significant droughts. The seasonal persistence of the Southern Oscillation phase allows forecasts to be made before the onset of summer rains. The potential of such forecasts is examined with respect to savanna management in northern Australia.

Models of soil water budgeting, grass production, pasture utilization, animal production and financial analysis are being developed for each savanna community in northern Australia. The key processes in these models are plant growth as a function of climate inputs and the effect of grazing on plant survival and production. We describe a general experimental methodology to apply existing models to specific grassland/soil combinations.

Examples of the application of these models show that: (1) periods of overgrazing can be identified when model output is combined with regional animal number statistics; and (2) management decisions such as burning can be improved when ENSO based forecasts are used.

The challenge of future savanna studies is to influence individual managers to make better management decisions based on reliable models of savanna processes. The uncertainty of future climate change suggests more flexible strategies will be required for the evolution of sustainable and economic savanna use.

Key words. Northern Australian savannas, ENSO, El Niño, Southern Oscillation, simulation, systems analysis, decision support.

INTRODUCTION

Recent definitive reviews have described northern Australian savannas from comparative structural and functional perspectives (Moore, 1970; Mott, Tothill & Weston, 1981; Mott *et al.*, 1985). Land use and comparative production potential have been described in detail at both a continental level (Tothill *et al.*, 1985) and at a regional level (Weston *et al.*, 1981). However, broad-scale geographical analyses lack the capability to address the management issues of the individual landowners and resource managers (for example, stocking rate, frequency and timing of burning). If the individual manager is to implement the principles of savanna management, general management rules (for example, desired level of pasture utilization) need to be made specific to a particular land resource. In savannas, management decisions need to be evaluated against a background of uncertainty in seasonal climate and price for product.

Recent major developments in savanna management throughout the world are: (1) a better understanding of global and regional climatology (Walker, 1989) with increasing accuracy of seasonal rainfall forecasts; (2) anticipated climate change as a continuing process as a result of the increases in greenhouse gases; and (3) the application of computer simulation models in decision-support packages (Stafford-Smith & Foran, 1988; Rickert, 1988) and Geographical Information Systems (Walker & Cocks, 1984).

In this paper we develop an understanding of the relationships between climate, soil and vegetation factors and savanna management for optimal pasture productivity. We

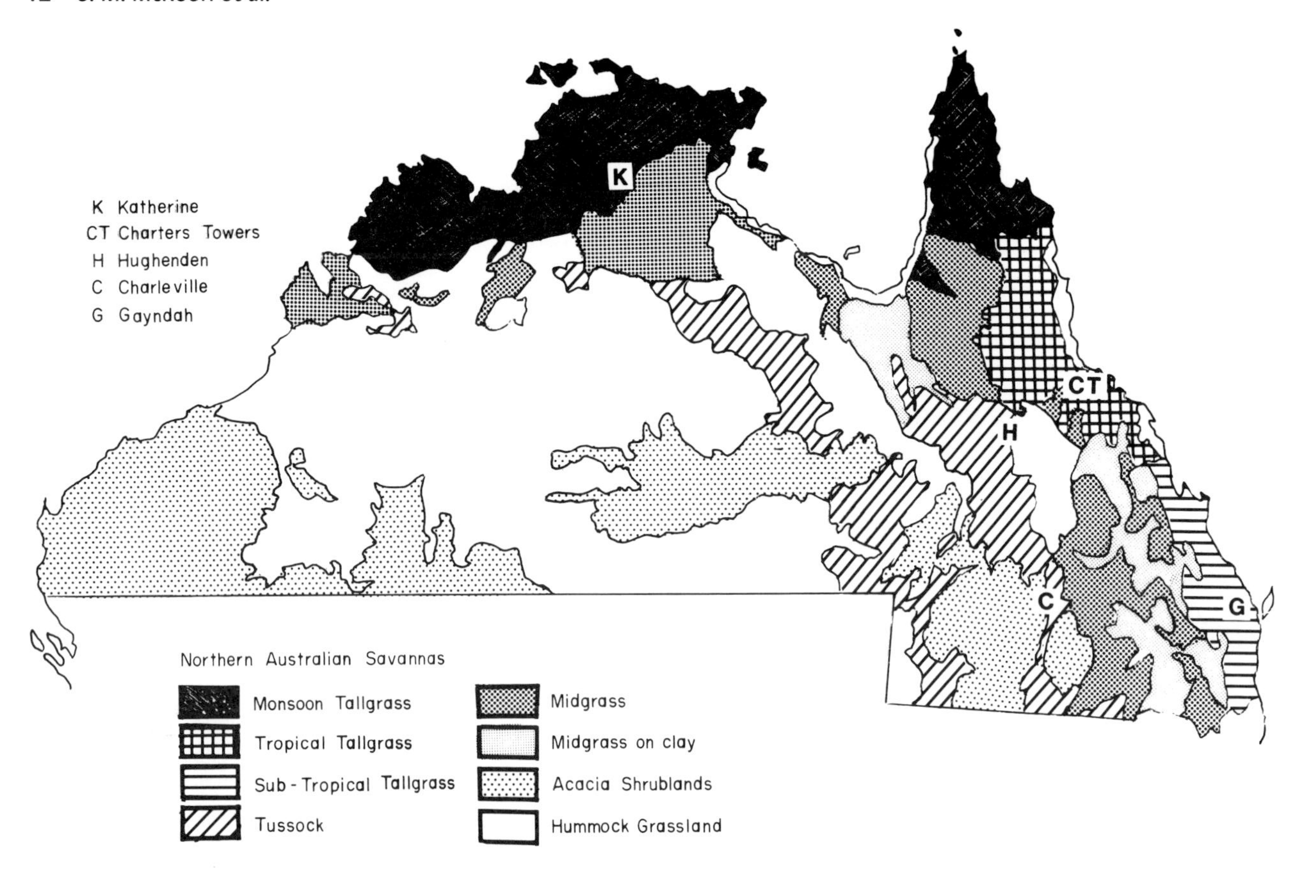

FIG. 1. Savanna communities of northern Australia (after Mott *et al.*, 1985) and locations mentioned in the text.

will consider examples from five main savanna types (Fig. 1, after Mott *et al.*, 1985). Monsoon Tallgrass (Katherine), Tropical Tallgrass (Charters Towers), Subtropical Tallgrass (Gayndah), Tussock Grasslands (Hughenden) and Acacia Shrublands (Charleville).

RESOURCE DESCRIPTION OF NORTHERN AUSTRALIAN SAVANNAS

Savanna vegetation is characterized by a continuous graminoid stratum, more or less interrupted by trees or shrubs (Johnson & Tothill, 1985). Other criteria used to define this vegetation type are: (1) some unity of vegetation type determined by environmental controls such as climate, soil and fire (Huntley & Walker, 1982); (2) grasses are mainly tall to mid-height tussock or bunch grasses, usually with flat basal and cauline leaves (Conseil Scientifique pour l'Afrique, 1956) which die back to the base during the dry season; and (3) C_4 grasses potentially dominate (Huntley, 1982).

In an Australian context, the geographical range of savannas is limited by predominantly summer rainfall, soil characteristics and the presence of fire which maintains the balance between grass, tree and shrub. Under these constraints, rainforest and coastal communities in the east are excluded and brigalow, gidgee and mulga communities are included in the savanna classification.

Native pasture communities have been described for Queensland (Weston *et al.*, 1981), based on soil associations (Northcote *et al.*, 1975). With the addition of data for the Northern Territory and Western Australia, the woodlands and grasslands of northern Australia are described (Mott *et al.*, 1981) using the same methodology. With further minor modifications the main vegetation types in the northern Australian savanna are described (Mott *et al.*, 1985, Fig. 1). Estimates of the areas of savannas are presented (Table 1).

In general terms, low fertility soils predominate in the

TABLE 1. Area of savanna types in Northern Australia (M ha).

	Qld	N.T.	W.A.	Total
Monsoon tallgrass	10.4	38.0	11.0	59.4
Black spear grass (northern and southern)	25.0	—	—	25.0
Blue grass–browntop	5.6	—	—	5.6
Queensland blue grass	2.4			2.4
Aristida/Bothriochloa (northern and southern)	33.5	0.1	—	33.6
Brigalow pastures	8.7	—	—	8.7
Gidgee pastures	4.8	—	—	4.8
Mitchell grass	29.5	8.3	2.0	39.8
Mulga pastures	19.1	11.3	40.0	70.4
Channel pastures	5.4	—	—	5.4
Spinifex	21.2	77.0	90.0	188.2
Other non-savanna	7.1	—	—	7.1
	172.8	134.7	143.0	450.5

TABLE 2. Surface chemical and physical characteristics of savannas.*

	pH	OC%	N%	Av. P (ppm)	Av. water (cm cm-1)
Cracking clays	7.5	2.4	0.22	214	0.17–0.21
Friable earths	6.2	4.6	0.46	47	0.05–0.13
Fertile duplex	6.2	2.1	0.13	25	0.05–0.17
Massive earths	6.1	1.7	0.08	8	0.04–0.05
Infertile duplex	6.3	1.2	0.07	10	0.04–0.08

*Data are averaged from published profile analysis, from Mott *et al.* (1985) and from Weston *et al.* (1981).

TABLE 3. Potential and current land use in Northern Australia (M ha).

	Crop	Sown pasture	Native pasture	Non-agric. land	Total
Queensland					
Potential	14.2	40.6	105.9	12.0	172.8
Current	3.0	4.0	153.8	12.0	172.8
Northern Territory					
Potential	0.3	8.1	63.3	63.0	134.7
Current	<0.1	0.1	71.6	63.0	134.7
Western Australia (north)					
Potential	0.1	1.8	67.0	74.0	143.0
Current	<0.1	<0.1	69.0	74.0	143.0
Northern Australia					
Potential	14.6	50.5	236.2	149.0	450.5
Current	3.0	4.1	294.4	149.0	450.5

north and west. A proportion of high fertility soils form a mosaic with soils of moderate to low fertility in the east. Some features of the major soils are shown (Table 2). Because of extensive areas of low fertility soils in the northern and western parts of the continent, 80% of sown pasture potential and over 95% of crop potential is in the eastern part of the area (Table 3).

This type of description of the land resource, when accompanied by area based estimates, has been used to determine relative productivity on the basis of animal production. Furthermore, such descriptions can be useful in determining the priorities for research funding.

Description of Savanna Lands

Monsoon Tallgrass

Schizachyrium pastures. Occurring across the northern parts of the continent, *Eucalyptus tetradonta* F. Muell and *E. dichromophloia* F. Muell low woodlands are common, while *Melaleuca* spp. occupy poorly drained sites. The dominant grass layer may contain some of the following: *Themeda triandra* Simon, *Heteropogon* spp., *Sorghum* spp., *Schizachyrium fragile* (R.Br.) A. Camus, *Chrysopogon fallax* S. T. Blake and *Eriachne* spp. Soils are predominantly earths and sands of low fertility.

Tropical and Subtropical Tallgrass

Black spear grass. Located along the eastern coast, it can be conveniently divided into northern and southern spear grass at approximately latitude 21°S and referred to as Tropical and Subtropical Tallgrass. The former contain *E. crebra* F. Muell, *E. alba* Blume and *E. dichromophloia* with an understorey of *Heteropogon contortus* (L.) Roem. and Schult, *H. triticeus* (R.Br.) Domin, *Themeda* and *Bothriochloa* spp. Southern spear grass communities, with *E. crebra, E. melanophloia* F. Muell. and *E. drepanophylla* Benth. have been extensively modified by man and animals. Timber thinning has been practised since settlement. The impact of animals has been expressed firstly in a change from *Themeda* pastures to *Heteropogon* and *Bothriochloa bladhii* (Retz.) S. T. Blake and more recently from *Heteropogon* to *Bothriochloa pertusa* (L.) A. Camus under continued heavy utilization. Soils are predominantly duplex, with low to moderate fertility.

Midgrass on clay

Brigalow pastures. Acacia harpophylla (Benth.) (brigalow) forests have been extensively cleared and the resulting grasslands now qualify as savanna. Native pastures which have developed contain *Dichanthium sericeum* (R.Br.) A. Camus *B. decipiens* (Hack.) C. E. Hubbard, *B. bladhii* and *Chloris* spp. Brigalow pastures and introduced grasses (*Cenchrus ciliaris* L., *Chloris gayana* Kunth, *Panicum maximum* var. *trichoglume* Robyns) are also important on cracking clay soils of moderate to high fertility with good water holding capacity.

Queensland blue grass. Grasslands and open woodlands

of *D. sericeum* occupy fertile clay soils in central and southern Queensland. Large areas are cultivated.

Blue grass–browntop. Grasslands of *D. fecundum* S. T. Blake, *Eulalia fulva* (R.Br.) Kuntze, *Astrebla squarrosa* C. E. Hubbard, *Sorghum* spp. and *Cyprus* spp. occur in north-west Queensland. Soils are clays, but of poor fertility and only moderate water holding capacity.

Midgrass

Aristida/Bothriochloa pastures. Major characteristics of these semi-arid woodlands are *E. populnea* F. Muell. in the south and *E. microneura* Maiden and Blakely in the north. Southern pastures are *Aristida* spp., *B. bladhii*, *B. decipiens* and *Chloris* spp. and they occur on duplex soils of moderate to low fertility. In the north, *Aristida* spp. and *Chrysopogon fallax* occur on low fertility duplex and sandy soils.

Acacia Shrublands

Mulga pastures. Pastures are sparse in dense mulga shrublands. In more open communities (referred to later as *Acacia* Shrublands), *Digitaria* spp., *Monochather paradoxa* Steud., *Eriachne* spp. and *Aristida* spp. are common. Soils are mainly massive earths and skeletal soils of low fertility and poor water holding capacity.

Gidgee pastures. Also extensively modified, these woodlands are replaced by *Cenchrus ciliaris* pastures. Soils are fertile clay and clay loams and *C. ciliaris* is well adapted to these soils conditions and has persisted.

Tussock Grasslands

Mitchell grass. Species of *Astrebla* are accompanied by the annual *Iseilema* spp. and *Dactyloctenium* spp. in extensive grasslands (referred to later as Tussock Grasslands). Soils are cracking clays of moderate to high fertility and good water holding capacity.

Channel pastures. The pastures of the inland river systems which are subject to flooding are characterized by Chenopods, *Trigonella* spp., *Atriplex nummularia* Lindl., *Echinochloa turnerana* (Domin) J. M. Black and *Muehlenbeckia cunninghamii* (Meisn.) F. Muell. Soils are fertile clays of good water holding capacity.

Hummock Grasslands

Spinifex. This most extensive hummock grassland has one of either *Triodia* spp. or *Plectrachne* spp. on infertile skeletal soils.

Savanna land use, present and future

The estimated potential for cropping and sown pastures in northern Australian savannas (14% of area, Table 3) suggests that there may be alternative land uses to that of the current pastoralism. Where cropping and pasture improvement have resulted in sustainable and productive agriculture, the 'loss' of savanna lands has been beneficial to economic growth and national living standards. However, the most recent period of development, beginning in the mid 1950s (Lloyd, 1980), was associated with periods of above-average rainfall and may have biased expectations of yields and returns. Furthermore, agroclimatic studies of long-term reliability of crop production at the cropping frontier (Clewett, 1985; Hammer, Woodruff & Robinson, 1987) caution against the enthusiasm of apparently successful attempts during periods of above-average rainfall. These considerations emphasize the importance of pastoralism as a sustainable and economic land use.

Current increases in greenhouse gases (Pearman, 1988) suggest that large changes in climate are likely to occur during (and beyond) the next 50 years (Tucker, 1988). Global circulation models are in general agreement that temperatures are likely to increase by 2–4°C in northern Australia but there is at present no agreement on likely precipitation changes (Coughlan, 1988). A scenario of 'plausible future climate' for the Australian region (Anon., 1988), made solely for the purpose of evaluating the impact of climatic change, suggested increases in summer rainfall (+30%) and decreases in winter rainfall (–20%) in eastern Australia. The large impact of such plausible scenarios on savannas (McKeon *et al.*, 1988; Graetz, Walker & Walker, 1988) indicate the importance of developing a capability to predict the savanna response to climate change. In northern Australian savannas the likely outcome of such a climate change may affect the areas used for cropping (McKeon *et al.*, 1988; Russell, 1988) and forestry (for preservation or timber). The global effect of climate change on cropping areas (existing and potential) elsewhere in the world could affect the economic viability of the Australian export grain and pastoral industries and hence the future of savanna lands. Comparative studies of savanna processes across wide geographical boundaries (for example, Mott *et al.*, 1985) would provide the basis for prediction of savanna response to climate change.

Geographic information systems will prove valuable in indicating the change in savanna land use with respect to climate change (Graetz *et al.*, 1988). Similarly Busby (1988) has shown how geographical information systems using vegetation/environment relationships can anticipate the changes in forest and wildlife habitats. However, our capability to plan for these changes to savannas is likely to be limited by the understanding of environment/vegetation/wildlife relationships (Braithwaite, 1985; Bowman & Minchin, 1987). The real value of such studies will be in the rapid evaluation of the impact of climate change as forecast by current global circulation models (Tucker, 1988).

CLIMATOLOGY OF NORTHERN AUSTRALIA

The many reviews of northern Australian Savannas have been in response to the need for resource description given the comparatively short (120 years) European occupation of the region. Similarly the recognition (Russell, 1981) of large temporal variability in rainfall on longer time scales (i.e. over decades and generations; Fig. 2a) suggests that a re-evaluation of appropriate land use and carrying capacity is required (Fig. 2b after Walker & Weston, 1990). In northern Australian savannas the evolution of agricultural

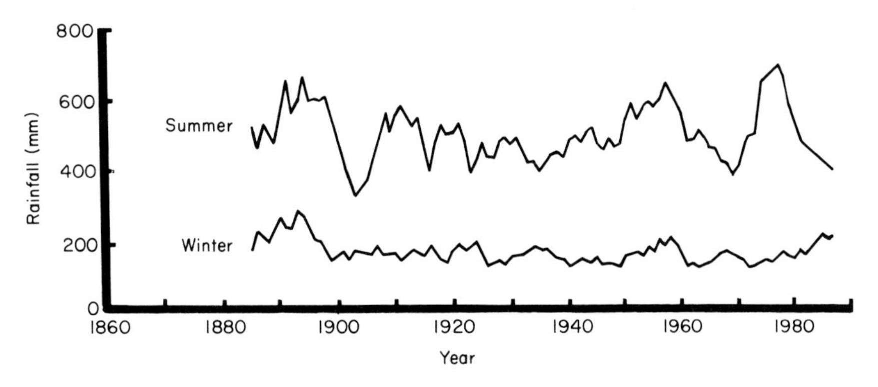

FIG. 2(a). Five-year moving averages of summer and winter rainfall for stations in the pastoral zone of Queensland with 100 years of records. Averages are plotted on year 5. Summer is October–March; winter is April–September.

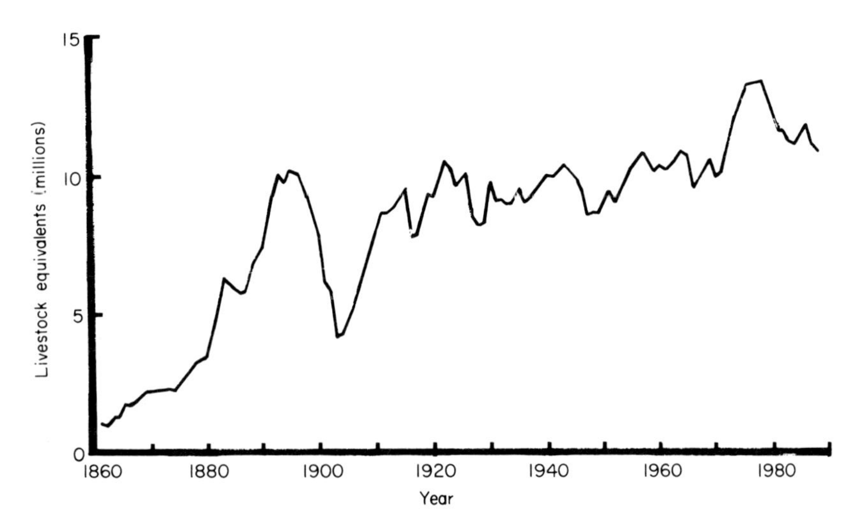

FIG. 2(b). Livestock equivalents in Queensland – annual totals (1 cattle equivalent = 8 sheep).

management 'rules' for sustainable cropping and pastoralism is a continuing process requiring understanding of the interaction of soil type, plant communities and climatic variability.

The major meteorological phenomena controlling seasonal rainfall in northern and eastern Australia are the movement of the intertropical convergence zone, the anticyclone (high) pressure belt, tropical cyclones and the Southern Oscillation phenomena (Brunt, 1961; Pittock, 1975; Russell, 1988). While Australian ecologists have long recognized the impact on savanna vegetation of high annual variability in rainfall (Dick, 1958) and temperature (Foley, 1945), the lack of understanding of the mechanisms causing this variability has led to a reactive rather than predictive approach to ecosystem management. It is hoped that a better understanding of the sources of temporal variability will allow the formulation of more appropriate models of savanna study and management (Walker, 1989; Graetz *et al.*, 1988).

The Southern Oscillation, the major source of climatic variation in eastern Australia outside seasonal variation, is a see-saw of atmospheric pressure between the South Pacific and Indian Oceans with a 'quasi-periodicity' in the range 2–10 years (Allan, 1988). Both extremes of the pressure fluctuations are associated with anomalous ocean and atmospheric conditions which may influence seasonal conditions in many parts of the world (Allan, 1988). The term ENSO even (*E* l *N* iño *S* outhern *O* scillation) has been used to describe one extreme of these global anomalies.

A commonly-used measure of the Southern Oscillation is the Troup Southern Oscillation Index (SOI), which is calculated as the normalized pressure difference between Tahiti and Darwin (Coughlan, 1988). The SOI 'cycles' (Fig. 3) from positive to negative by up to three standard deviations (i.e. from −30 to +30) with persistence varying with season (strongest in June–October, weakest in April; McBride & Nicholls, 1983). Negative indices of the SOI (below −5, i.e. half a standard deviation) may be termed ENSO events and are associated with: (1) a warming of sea surface temperature near the equator in the eastern Pacific (also referred to as El Niño); (2) *low* rainfall in eastern Australia, Indonesia, some areas of India, eastern and southern Africa, and northern China; and (3) *high* rainfall in Argentina, California and western South America. Opposite conditions are experienced at these locations during positive SOI periods (anti-ENSO events; Allan, 1988; Ropelewski & Halpert, 1989).

Studies of the influence of the Southern Oscillation on seasonal rainfall have provided a powerful understanding

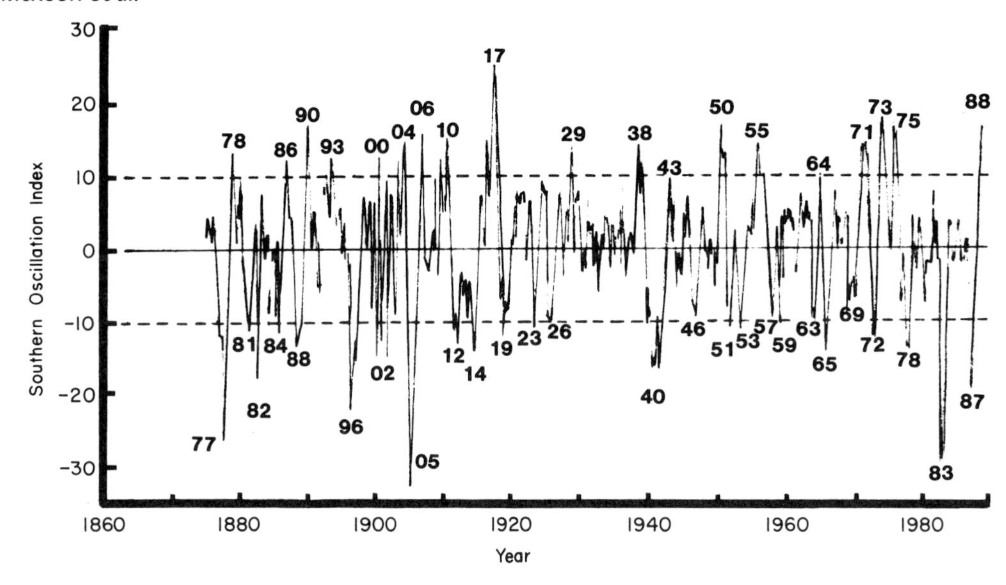

FIG. 3. Southern Oscillation Index supplied by T. Casey, Bureau of Meteorology (Melbourne). Selected dates of extreme positive and negative oscillations are indicated. Missing periods have been calculated from Wright (1975).

of comparative rainfall variability and frequency of extreme events (i.e. drought or flood). Nicholls (1988) has shown that for locations throughout the world where rainfall was affected by ENSO, relative variability was 30–50% higher than locations not affected by ENSO. Similarly the correlation of extreme flows (flood and drought) in river systems in Africa (the Nile), eastern India (the Krishna and Godavari Rivers) and eastern Australia (the Murray-Darling; Adamson, Williams & Baxter, 1987) indicates the global significance of the Southern Oscillation for comparative ecological studies.

Several studies have noted the larger annual variability of rainfall and stream flow in Australia when compared with other locations in the world (Dick, 1958; McMahon, 1982). The high Australian variability is likely to be a result of the stronger control of meteorological mechanism associated with the phases of the Southern Oscillation. Correlations with SOI are higher in eastern Australia ($r = 0.4$–0.7) than in many other areas of the world (Nicholls & Wong, 1990) and the magnitude of SOI anomalies influence the area of eastern Australia in drought or 'flood' (Pressland & McKeon, 1989). The phase of the Southern Oscillation is best regarded as a second dimension to the traditional seasons of summer, autumn, etc. (Table 4).

The main differences between representative locations in northern Australia are: (1) the extreme seasonality of rainfall but lower between-year variation in the Monsoon Tallgrass compared to the other savanna communities; and (2) the high summer rainfall variability of Tropical Tallgrass, Tussock Grasslands, Mulga Shrublands resulting in a greater frequency of 'drought', i.e. <60% of summer rainfall. However, simple comparison of climatic averages and variation have proved misleading (Taylor & Tulloch, 1985). For example, many climatological analyses have been carried out for the Monsoon Tallgrass communities (Slatyer, 1960; McAlpine, 1976; Nicholls, McBridge & Ornerod, 1972; Taylor & Tulloch, 1985; Ridpath, 1985) and each study has emphasized the high variability of rainfall from ecological or agricultural perspectives. This is best summarized by Taylor & Tulloch (1985) who, after an analysis of Darwin rainfall 1870–1983, stated that; 'As approximately 77% of years have significant departures from the two most frequent rainfall patterns (respectively consisting of sixteen and eleven years), notions of a normal or average rainfall year are of dubious value in this environment.'

INTERACTION BETWEEN SOUTHERN OSCILLATION AND SAVANNA DYNAMICS

Australian ecologists are just beginning to examine the importance of extremes of the Southern Oscillation in influencing community composition and demography. There was no reference to the Southern Oscillation in the 1984 Proceedings of the Australian Ecological Society devoted to the topic 'Are Australian ecosystems different?' However, recent papers have shown the importance of ENSO and anti-ENSO events in plant population dynamics (Taylor & Tulloch, 1985; Austin & Williams, 1988; Graetz *et al*., 1988). Taylor & Tulloch (1985) showed that a range of biological phenomena associated with recruitment, distribution and survival, were strongly affected by extreme rainfall events, approximately half of which were associated with extremes of the Southern Oscillation. At all representative locations in northern Australian savannas relatively wetter (>decile 7) or drier (<decile 3) conditions have a two- to six-fold difference in frequency of occurrence depending on the phase of the Southern Oscillation (Table 4).

Given the large control of the Southern Oscillation on eastern Australian rainfall, the ecological impact of historical variability in the SOI is worthy of evaluation. Climatologists comment on the uniqueness of each ENSO event explaining to some extent the apparent randomness of year-to-year rainfall variability (Allan, 1988). The understanding of the mechanisms for rainfall extremes should allow savanna managers to extrapolate short-term experience

TABLE 4. Relationship between summer and winter rainfall and phase of the Southern Oscillation Index at representative locations in northern Australia based on >100 years of records.

	Katherine		Charters Towers		Hughenden		Gayndah		Charleville	
	Sum.	Win.	Sum.	Win.	Sum.	Win.	Sum.	Win.	Sum.	Win.
Mean rainfall	929	48	519	133	401	89	552	223	327	163
Decile 3 rainfall	763	8	372	76	295	41	474	154	224	110
% years < Decile 3 SOI<−5	42	43	70	42	50	43	50	54	48	46
% years < Decile 3 SOI>+5	17	14	4	11	8	25	8	19	8	7
Decile 7 rainfall	1052	69	603	165	465	122	617	278	386	189
% years > Decile 7 SOI<−5	8	18	4	25	23	22	27	11	17	15
% years > Decile 7 SOI>+5	48	38	63	39	54	50	54	48	46	38
Coefficient of variation	28	53	45	64	46	82	33	47	53	49
% of years <60% mean	6		16		20		10		23	

Two 'phases' of the SOI are considered: (1) Years when the seasonal SOI was less than −5 (22% of years from 1871 to 1988). (2) Years when the seasonal SOI was greater than +5 (22% of years from 1871 to 1988).

The chance of 'dry' years (i.e. less than Decile 3 rainfall) and 'wet' years (i.e. rainfall greater than Decile 7 rainfall) are shown for each 'phase'. By definition the expected chance of these conditions is 30%.

(5–10 years) over longer periods. Management decision rules derived from short-term experiments (for example, stocking rate and burning policies formulated in the 1950s compared with the 1960s, Fig. 2) should be evaluated in the context of historical ENSO and anti-ENSO experience. Furthermore, 'there is evidence in the historical record for a "doubling" or "tripling up" of ENSO (1899–1902, 1939–42) and anti-ENSO (1915–17, 1954–56, 1973–76) events' (Allan, 1988). These events have had a major impact on grazing lands in eastern Australia.

Following legislation for closer land settlement in western Queensland in 1883, a rapid build-up in animal numbers (sheep) occurred during the above-average rainfall of the early 1890s (a period of strongly positive SOI, Fig. 3). The triple ENSO (1899–1902) resulted in very low rainfall and a 60% decrease in animal numbers (for example, Fig. 2b). Heavy utilization of edible grasses and shrubs resulted in increasing spread of inedible shrubs and grasses (Heathcote, 1985, p. 177). Heathcote (1965), in a detailed history of western NSW and Queensland, found that carrying capacities did not return to 1890 values despite subsequent property improvements and further land subdivision.

Similarly in the Subtropical grasslands the apparently rapid change in composition from *Themeda triandra* to *Heteropogon contortus* during 1881–82 (Shaw, 1957) appears to have occurred as a result of overgrazing with sheep during ENSO-related droughts in these years. Other ENSO-related droughts in Queensland have caused large stock and crop losses. Gibbs & Maher (1967) documented fifteen major drought periods in Queensland from 1882 to 1966. The droughts were mainly described as 'severe' or 'widespread', and losses of stock and crop as 'heavy' or 'devastating'. Thirteen of these droughts included periods of large negative SOI values (<−10, Fig. 3).

The above-average rainfall during the anti-ENSO periods was associated with: (1) the initial large increase in the conversion of grassland to cropping (1954–56; Lloyd, 1980); (2) the successful establishment of sown pastures (1970–76; Eyles & Cameron, 1985); and (3) the loss of the naturalized pasture legume *Stylosanthes humilis* Kunth. in northern spear grass pastures from a weather related fungal disease (1973–76; P. L. Lloyd, pers. comm.). The low beef prices of 1974–76 (Daly, 1983) resulted in a large increase in stocking rate as graziers held cattle on their properties in expectation of higher prices. Fortunately above-average rainfall occurred during this anti-ENSO period but attempts to maintain these stocking rates during the drier 1980s has led to pasture degradation and species changes (invasion of *Aristida* species) in southern spear grass (M. Quirk and D. M. Orr, pers. comm.) and soil erosion and invasion of *Bothriochloa pertusa* in northern spear grass (Gardener, McIvor & Williams, 1988; Howden, 1988).

The Southern Oscillation is one of several global meteorological phenomena affecting rainfall variability. However, its apparent influence on northern Australian savannas is strong enough to suggest that the Southern Oscillation should be included in any prescription for savanna management. Given the high annual variability in rainfall, such management recommendations are difficult to formulate from short-term experience (Taylor & Tulloch, 1985) and hence methodologies that allow extrapolation of research and experience are required. The next sections describe the analysis of savanna production from a pastoral perspective and the examination of how improved climatological understanding can be used for better management.

SAVANNAS AS A PASTORAL SYSTEM

The adoption of a systems approach to savanna management provides a framework to examine the flow of dry matter, energy and nutrients in the savanna ecosystem. A systems approach attempts to describe the dominating processes and relationships such that the savannas' response to environmental or management induced 'stress' can be predicted. Given the dominant land use of pastoralism, we concentrate on examining savannas as a grazing system considering productivity, reliability and sustainability (Walker, 1989).

Management models and decision support packages

The development of decision-support packages requires the collection of appropriate data on different soil types that allow extrapolation to other locations and times using simple resource data banks (geological landscapes, soil depth, climate surfaces; Nix, 1987). The recognition of soil–landform–vegetation interrelationships has been regarded as a fundamental step in deriving management rules for cropping potential (Christian & Stewart, 1953), conservation (Bowman & Minchin, 1987; Taylor & Dunlop, 1985) and pastoralism (Taylor, 1989). We present here a case study of the data collection, model development, validation and application to savanna management.

Where understanding/prediction of the interaction of management (stocking rate, burning, tree and shrub clearing) and environment is required, then a more refined system description is needed to examine the range of species composition and soil types occurring in savannas. Such models (for example, McKeon & Scattini, 1980) have a general structure simulating productivity and reliability by describing the flow of plant dry matter (growth, detachment, litter disappearance and animal intake). The sustainability of production can be simulated by combining population dynamic models of changes in species composition (Torssell & McKeon, 1976; Mott & Andrew, 1985; Howden, 1988; Watkinson, Lonsdale & Andrew, 1989) and plant density (Johnston & Carter, 1986; Orr, 1988).

The final 'product' of these models is a decision-support package which allows research information/knowledge to be used by individual ecosystem managers. Analysis of pastoral enterprises requires models of the major savanna relationships: the effects of environment, tree/shrub density and stocking rate on grass and animal production. We will present examples of models of three of these relationships: (1) climate and grass growth; (2) stocking rate and grass growth; and (3) stocking rate and animal growth.

The effect of climate on grass growth

A variety of approaches to modelling grass growth are available from 'top-down' approaches based on empirical analysis of field data to 'bottom-up' approaches based on physiological analysis. We will examine how both approaches can contribute to the understanding of grass growth and model development at a suitable level for decision support.

Empirical geographic models of grass growth

Models of soil water balance and grass growth have been developed to allow geographical analyses of the impact of climate on the spatial variability of grassland production (Fitzpatrick & Nix, 1970; McCown, 1973; McKeon *et al.*, 1980; Mott *et al.*, 1985, Williams & Probert, 1984; Ridpath, 1985). Fitzpatrick & Nix (1970) formulated a plant growth index as the multiplication of the separate effects of temperature, solar radiation and soil moisture and plant growth. Simple soil water balance models (that is, one soil layer, weekly time step) allow simulation of the seasonal plant growth patterns from relatively simple climatic inputs. These simulation models have shown the general effects of temperature and moisture, and available-soil-water range on plant growth. With the assumption of one general temperature response for C_4 species, many studies have shown that moisture is the major climatic limitation to productivity in Monsoon Tallgrass, Tropical Tallgrass communities while temperature limits plant production in Subtropical Grasslands and *Acacia* Shrublands (Fitzpatrick & Nix, 1970; McCown, 1981; Mott *et al.*, 1985).

Such geographical approaches necessarily use generalized plant responses and soil description to derive average seasonal plant and animal growth patterns (for example, Fitzpatrick & Nix, 1970). These approaches are somewhat simplistic in that they fail to recognize that savanna grass species have a wide range of growth responses to temperature (Christie, 1975; Ivory & Whiteman, 1978). Grass species co-existing in the same habitat may have different temperature responses: *Themeda triandra* and *Bothriochloa* spp. have greater growth at low temperatures (15°C) than *Heteropogon contortus* (S. J. Cook, pers. comm.); C_3 grass species in *Acacia* shrubland (*Thyridolepis*) are more productive during winter than C_4 species such as *Aristida* spp. (Pressland & Lehane, 1980).

Many of the geographical analyses necessarily use long-term average temperature and evaporation rates. However, autumn/winter rainfall in the subtropics may be associated with temperatures above the seasonal mean (for example, 1983, 1988) and hence analysis using average conditions over-estimates the limiting effect of temperature on C_4 grass growth. In the Tussock Grasslands, winter rainfall produces growth from herbaceous C_3 forbs (Orr, 1986) in the tussock interspaces. Substantial autumn and winter rainfall has occurred in southern Queensland following ENSO-related summer droughts (for example, 1978, 1983, 1988; R. Stone, pers. comm.). The subsequent C_4 grass and forb growth supports high liveweight gains in cattle (McKeon *et al.*, 1980, 1986) and sheep (Orr *et al.*, 1988) respectively.

Where high seasonality in grass production occurs due to reliable seasonal drought (monsoonal) or low temperature restriction, grass production has been directly related to rainfall (Bremen & de Wit, 1983; Dye & Spear, 1982; Sala *et al.*, 1988; Singh, Hanxi Yang & Sajise, 1985). Such approaches have proved successful in comparing the interaction of rainfall and soil texture (Walker, 1985; Sala *et al.*, 1988) across savanna communities. Where run-off and deep drainage are variable but relatively minor components of the soil water budget, the calculation of evapotranspiration using simple soil water balance models provides a satisfactory interpretation of rainfall/growth relationships, especially when annual variation in rainfall is considered. For example, McCown, Gillard & Edye (1974) showed that the relationship between pasture growth and evapotranspiration (estimated from a simple soil-water balance model) was relatively constant over a 5-year period.

Thus while geographical models provide a general framework for comparative studies between communities, more detailed approaches are required for accurate analysis at specific locations.

Physiological modelling of grass productivity

One approach of analysing the complexity of soil × landform × community combinations is to develop a general model of primary production. This requires the definition of a minimum data set for field measurement to allow model calibration and validation as has been achieved in recent developments in crop modelling (Hammer *et al.*, 1989). These crop management models examine dry matter production in both terms of light use efficiency (LUE, i.e. growth per unit of intercepted light or radiation; Charles-Edwards, Doley & Rimmington, 1986) and transpiration efficiency (TE, growth per unit of transpired water; Tanner & Sinclair, 1983). Genetic differences in both LUE and TE occur between crop hybrids and have been used to explain genotype-by-environment interactions in a range of field environments (Hammer *et al.*, 1989). In savannas such an approach is likely to provide the soundest physiological basis for interpreting the limitations imposed by temperature and nutrient availability. However, perennial grasses are more difficult to study than crops since: (1) any disturbance to the soil environment is likely to provide different nutrient availability than that occurring under field conditions; and (2) the large root biomass (Singh *et al.*, 1985) makes measurement of total production (root + shoot growth) difficult under field conditions.

Interaction of nutrients, water and light-use efficiency

Above-ground grass growth is a function of plant available moisture and available nutrients. The relative effects of these variables are best measured where both fertilizer and irrigation are applied to existing vegetation. For example, in a Subtropical Grassland (Narayen) fertilizer and irrigation were applied both separately and together for 2 years. Swards were burnt annually in spring and yield was measured regularly throughout the growing season. Fertilizer treatments including combined and separate treatments of N (600 kg/ha/year) and P (300 kg/ha/year). At this site (granitic sand 10 ppm P) there was little plant growth response to applied P alone and only the results from combined fertilizer treatments are presented (Table 5). Even under severe water limitation (transpiration ratio <0.2) available nutrients limited growth. When production was expressed in terms of radiation use efficiency, nutrient limitation resulted in 80% of potential production at low levels of water availability and 60% of potential with water not limiting.

The apparent sensitivity of savanna grasses to available nutrients highlights a major problem in assessing production potential other than *in situ*. Any disturbance (cultivation, replanting in pots) is likely to provide a different nutritional supply, hence the determination of basic physiological parameters (water use efficiency) should be done in the field (Robbins, 1984). Many comparisons between grasses have actually confounded nutritional status and species differences by comparing 'young' sown pastures (for example, *Panicum maximum*) with existing undisturbed 'native' pastures (for example, *Heteropogon contortus*). Where both species have been sown at the same time the major effect on plant and animal production is age since

TABLE 5. Response of a Subtropical Spear Grass community to fertilizer and irrigation.

	No fertilizer		Added fertilizer	
	Growth	RUE	Growth	RUE
No irrigation				
Year 1	1376	1.66	2453	2.00
Year 2	597	1.62	1082	1.76
Irrigation				
Year 1	2916	2.62	9049	4.44
Year 2	2640	2.40	8760	3.91

Growth is maximum dry matter production (kg ha^{-1}) by April (1986 and 1987). RUE is radiation use efficiency (kg ha^{-1} MJ^{-1} m^2) where radiation is total solar radiation and % interception = 100–%transmission. Fertilizer is N (600 kg ha^{-1} $season^{-1}$) and P (300 kg ha^{-1} $season^{-1}$). Irrigation is >75% pan evaporation.

cultivation; demonstrating the overriding effect of declining N mineralization (Robbins & Bushell, 1985; Robbins, Bushell & Butler, 1987).

Few studies have been conducted on the light use efficiency of savanna grasses in the field. Sward structure, senescence and varying species composition and phenological development, variation in plant density, difficulty of root growth measurements and the scarcity of radiation data, all limit the application of conventional crop growth analysis. In annual-crop modelling a compromise between empirical and physiological approaches has been developed using a transpiration-based analysis of plant production (Tanner & Sinclair, 1983; Hammer *et al.*, 1989) and has been applied to annual grassland production (Van Keulen, 1975). The application of this approach to northern Australian savannas is described in the model GRASP (McKeon *et al.*, 1982).

'GRASP' model description

GRASP (GRASs Production) combines two successful approaches in modelling plant growth, viz those of McCown *et al.* (1974) and Fitzpatrick & Nix (1970). The soil water budget is simulated using three soil layers (0–10, 10–50, 50–100 cm). The processes of run-off, drainage, soil evaporation and transpiration are separately calculated on each day from inputs of rainfall and pan evaporation (Rickert & McKeon, 1982). The soil variables required include: available soil water range for each layer, maximum rate of bare soil evaporation and infiltration parameters. A plant growth index is calculated from a soil water index, plant growth response to average air temperature, and solar radiation. Following burning or mowing, plant growth is calculated as a function of growth index, plant density (% basal area), and potential regrowth. As green cover increases, plant growth is calculated from a combination of temperature response and transpiration, using a transpiration efficiency (kg/ha/mm of transpiration). The main plant variables (potential regrowth after mowing, and transpiration efficiency) are derived from calibrating the model against field data. Tanner & Sinclair (1983) showed that from both theoretical analysis and field work,

TABLE 6. Relationship of transpiration efficiency to vapour pressure deficit in subtropical tallgrass (Gayndah).

Season	Rainfall (mm)	Transpiration* (mm)	Measured growth (kg ha^{-1})	Transpiration efficiency (kg ha^{-1} mm^{-1})	Vapour pressure deficit (Dec.–Apr.) (hPa)	K (hPa kg ha^{-1} mm^{-1})
1961–62	622	419	3600	8.6	16.0	138
1962–63	410	250	2100	8.4	16.8	141
1963–64	381	215	1050	4.9	20.4	100
1964–65	592	296	1550	5.2	20.0	104
1965–66	456	354	2530	7.1	18.0	128
1966–67	527	363	3030	8.3	16.0	133
1986–87	642	359	2400	6.7	19.3	129

* Transpiration calculated by GRASP model.

transpiration efficiency, TE (kg/ha/mm) was inversely proportional to daytime vapour pressure deficit (VPD), i.e. TE = K/VPD where K is constant for a given species by nutrient combination. Studies in Subtropical Tallgrass where variation in VPD is high (12–40 hPa) indicate the importance of changes in humidity conditions (Table 6). Lower TE values were associated with drought conditions (1963/64, 1964/65) when soil water was likely to limit above-ground growth (for example, McCown, 1973, 1981).

In Queensland an informal group of pasture scientists is collaborating by measuring native pasture production at more than fifteen locations. By combining data collected under different environmental conditions, the group plans to develop a general model which is applicable to northern Australian grasslands. The project is popularly known as GUNSYND, an acronym for Grass Under Nutritional Stability: Yield Nitrogen and Development.

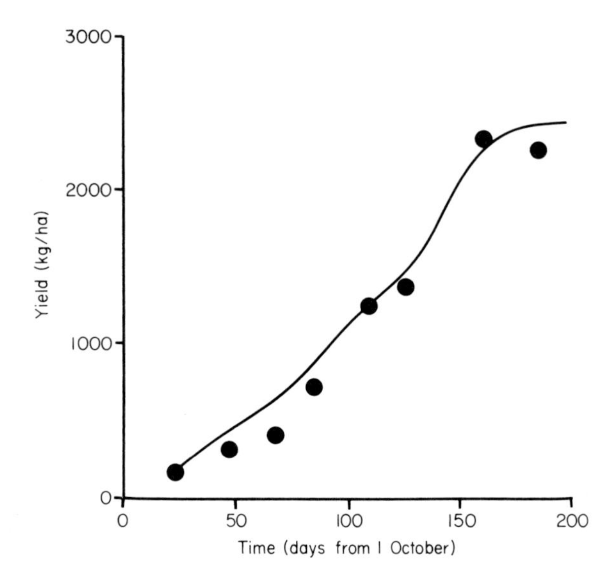

FIG. 4. Calibration of 'GRASP' model to 'Gunsynd' data in a Subtropical Tallgrass community (Gayndah: infertile clay with 70% *Bothriochloa bladhii*, 1986–87).

'Gunsynd' measurements

Sites (30 × 30 m) of uniform vegetation/soil were selected and fenced with netting to exclude all grazing animals. Within each site three plots (8 × 15 m) were chosen avoiding microsite variations. The sites were burnt or mown at the start of the growing season (October). At each harvest, estimates of species composition (by yield), % green cover and % bare soil were taken; above-ground plant yield and height were measured (four 1 × 0.5 m quadrats per plot or larger depending on variability). Sub-samples were taken to measure N concentration; and for the separation of green leaf, dead leaf, live and dead stem, inflorescence, dicots and litter. Soil moisture was measured (two holes per plot) in 10 cm intervals (as deep as possible). Harvests were taken every 3 weeks during the growing season and 6-weekly during the dry or winter season. Basal cover (%) was measured at each site by one of us (D. M. Orr) to avoid operator variability.

The major soil and plant parameters required for the model can be derived from these experimental measurements or general relationships across sites. An example of model calibration (Fig. 4) and validation against independent data for Subtropical Tallgrass (Fig. 5) indicated that the soil and plant parameters are relatively constant despite a wide range of management practices (burning, grazing, mowing) (McKeon *et al.*, 1982).

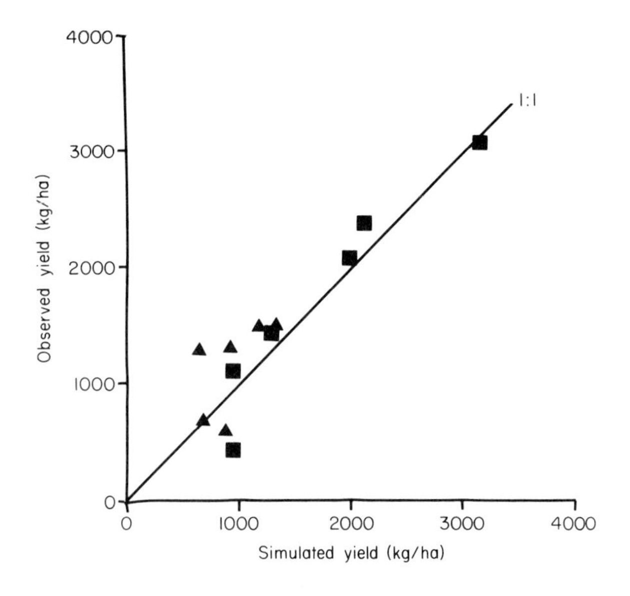

FIG. 5. Validation of 'GRASP' model against independent data: 1962–67 (Scattini, 1973) (▲, start of grazing; December; ■, end of season: May).

Effect of stocking rate on grass growth

Sustainable pasture production requires that desirable species be maintained despite the stresses of defoliation and drought. In Subtropical Grasslands a major pasture composition change occurred with the loss of *Themeda triandra* in the severe droughts of the 1880s (Shaw, 1957). In Tropical Tallgrass changes of dominance from *Heteropogon contortus* to *Bothriochloa pertusa* are occurring in response to greater grazing pressure (due to better adapted cattle, use of animal supplements, lower beef prices; Gardener *et al.*, 1988; Howden, 1988). In the Monsoon Tallgrass, loss of existing species has occurred with greater grazing pressure due to patch grazing and the introduction of pasture legumes (Mott, 1987). Where species are replaced by stoloniferous grasses there may in fact be no decrease in animal production per hectare although the risks of feed shortages and invasion of undesirable species are greater. However, a loss in production occurs where existing grass species are replaced by woody weeds such as *Sida retusa* in Monsoon Tallgrass and unpalatable grasses such as *Aristida* spp. in Subtropical Tallgrass, *Acacia* Shrubland and Tussock Grassland.

Derivation of sustainable stocking requires experimentation on the mechanisms of species loss and pasture composition change. In Subtropical Tallgrass Mott (1986) compared the tolerance of *Themeda triandra*, *Heteropogon contortus* and the introduced *Panicum maximum* to frequency of defoliation with two pre-growing season treatments: (1) simulating a very dry winter period; and (2) 'natural' winter period with winter rainfall. Frequent defoliation during the growing season following a dry winter period resulted in high death rates in *Themeda triandra*. Such dry winter periods occur frequently in the tropical Monsoon Tallgrass and Tropical Tallgrass and approximately one in five years in subtropical savannas (usually in association with ENSO events). The experiment provides a possible hypothesis for the disappearance of *Themeda triandra* from the subtropical grasslands following the winter drought of 1881 (Shaw, 1957).

In Tropical Tallgrass, Howden (1988) examined the effect of frequency of defoliation on a sward of *Themeda triandra*, *Heteropogon contortus* and *Bothriochloa pertusa*. At low frequency of defoliation (for example, one cut per growing season) *Themeda triandra* increased in density while with frequent defoliation (weekly) *Bothriochloa pertusa* increased to dominate the sward. The combination of defoliation and moisture stress resulted in the greatest death of *Heteropon contortus* (Howden, 1988).

A model of the effect of stocking rate on basal cover

The effects of very high grazing pressure (>60% utilization) have been studied in a long-term grazing experiment (20 years) on subtropical grasslands at Gayndah (Scattini, 1973). Cleared pastures dominated by *Heteropogon contortus*, *Bothriochloa bladhii* and *Dichanthium sericeum* were grazed at three stocking rates (0.74, 1.24, 2.47 weaners/ha) during summer and autumn.

Analysis of changes in basal cover of the tussock (or bunch) grasses showed that large declines occurred where high utilization rates (>50%) occurred during 'drought' type (transpiration ratio <0.3) growing seasons. Rapid increases occurred during good seasons. The changes in basal cover for these species could be described as a function of utilization and pasture growth (Fig. 6) and provided a good representation ($r^2 = 0.84$) of observed changes over 6 years of drought and wet conditions (Fig. 7). Pasture production from October to May can be: (1) measured in an exclosure; or (2) estimated from pasture yields and animal intake; or (3) simulated from the 'GRASP' model. Utilization is calculated from animal intake per unit area expressed as a per cent of above-ground pasture production. Animal intake is calculated as a function of liveweight gain and animal size. No independent data are available for validation but the relationship is consistent with other observations of pasture composition changes and basal cover changes in other communities (for example, *Acacia* Shrublands, Carter & Johnston, 1986; Tussock Grasslands, Orr, Bowly & Evenson, 1986) and the more detailed studies of Mott (1986) and Howden (1988).

In *Acacia* Shrublands, annual pasture production was linearly related to basal cover (Christie, 1978), but in subtropical savannas stoloniferous grasses (*Chloris* spp.) or *Aristida* spp. invaded at the highest stocking rate and plant production was reduced to 60% of production at lower stocking rates.

The calculation of sustainable stocking rates involves not only consideration of the level of utilization of summer growth but also different methods of herbivore grazing, for example, hedge grazing by sheep in Tussock Grasslands (Orr, 1986) and patch grazing by cattle in Monsoonal Tallgrass (Mott, 1987). Studies on unburnt *Themeda triandra* dominant pastures at Katherine showed that cattle concentrated on the same area of the paddock during the growing season while the remaining areas was ungrazed and rapidly became rank. In the grazed areas, high utilization rates (50–100%) exceeded the capacity for plant survival and resulted in bare areas. Over 4 years, the area overgrazed increased by 0.3 ha/head/year. Burning was required to redistribute

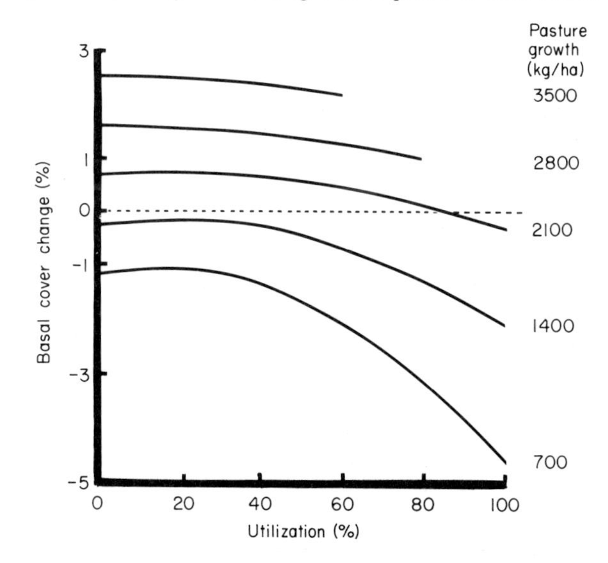

FIG. 6. Change in basal cover as a function of utilization and summer pasture growth in a subtropical grassland (after Scattini, 1973).

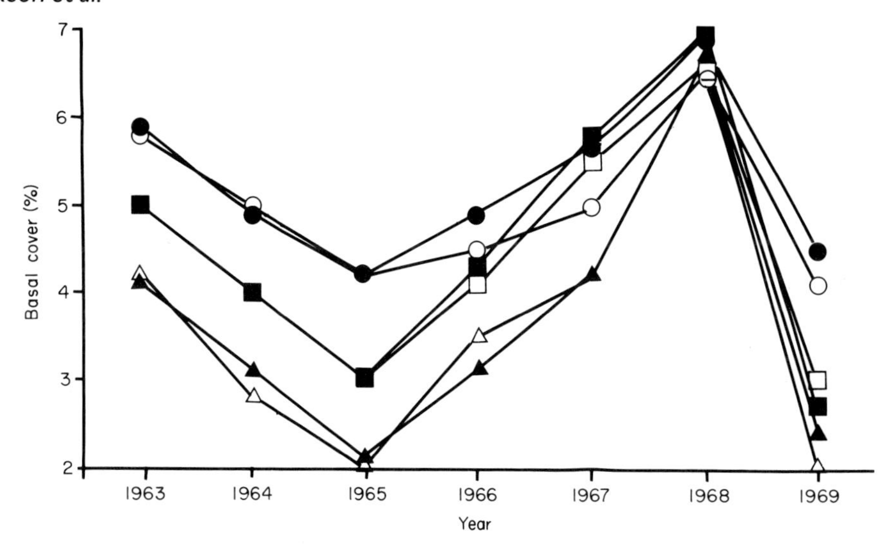

FIG. 7. Observed and predicted basal cover in a Subtropical Tallgrass community (Gayndah) at three stocking rates: 0.74 (obs., ●; pred., ○), 1.24 (obs., ■; pred., □) and 2.47 (obs., ▲; pred., △) weaners ha^{-1}. Basal cover is that of *Heteropogon contortus, Bothriochloa bladhii* and *Dicanthium sericeum* (Scattini, 1973).

grazing pressure to previously ungrazed areas and this rested overgrazed areas which had insufficient yield to carry a fire (Mott, 1987; Andrew, 1986; Winter, 1987). Similarly Hassal (1979) devised a system of rotational grazing and burning for Tropical Tallgrass (*Themeda triandra* dominant) which involved grazing an area (25%) of pasture during the growing season only once in 4 years.

The development of general models of changes in plant density and pasture composition is limited by the lack of data for the combinations of species × soil × rainfall. The above results (Figs. 6 and 7) show that simple experiments (Mott, 1986; Howden, 1988) and longer-term grazing trials can be linked. The development of an overall model for other savanna communities is a high priority if better management rules are to be derived.

Effect of stocking rate on animal production

Annual variability in rainfall and temperature are the major climatic factors influencing both the reliability and sustainability of animal production in northern Australian savannas. Variation in plant production (kg DM/ha/year) determines carrying capacity, or with constant animal numbers results in year-to-year variation in utilization (consumption expressed as a proportion of above-ground production). Diet quality (and hence animal production) is mainly determined by the availability of green leaf (Ash *et al.*, 1982; Robbins *et al.*, 1987) which is a result of the length of the growing season (McCown, 1981) and frequency of frosts (Wilson & t'Mannetje, 1975). Thus carrying capacity and individual animal production are not necessarily correlated as they are determined by different climatic factors.

In examining the effect of climatic variation on animal production from savannas both quality and quantity of feed must be considered. For beef cattle production, liveweight gain per animal is linearly related to stocking rate (i.e. animals per unit area; Jones & Sandland, 1974) such that $y = a - bs$ where y is the liveweight gain (LWG) per animal; a is potential liveweight gain (i.e. at a very low stocking rate); b is the slope coefficient; s is the stocking rate (beast per ha).

Both potential LWG and the slope coefficient vary greatly between pastures (Walker, Hodge & O'Rourke, 1987). However, in subtropical pastures (*Heteropogon contortus, Panicum maximum*) a general model has been developed (McKeon & Rickert, 1984) which covers a range of potential LWG (100–224 kg/head/year) and plant production (1500–7000 kg DM/ha/year). Potential LWG can be calculated for each season from daily climate data and model-derived growth indices (for example, McCown, 1981). In Monsoon Tallgrass and Tropical Tallgrass the length of the growing season or 'green' season and 'green' weeks in the dry season (McCown, 1981) allow calculation of potential LWG. In Subtropical Grassland both soil moisture indices and frost incidence must be considered (McKeon *et al.*, 1986).

As utilization increases in pastures with strongly seasonal growth, the proportion of leaf in the diet declines resulting in a reduction in rate of passage through the animal and corresponding reductions in intake (Poppi, Minson & Ternmouth, 1981). Hence, the decline in LWG with increasing stocking rate can be generalized by expressing LWG as relative intake (proportion of potential, RI) calculated from utilization (proportion of plant production consumed since the start of the growing season, U):

$$\mathrm{RI} = a' - b'.U$$

While the above equation has been validated for a wide range of stocking rates and pasture types ($a' = 1.05$ $b' = 0.46$, native and sown grass; McKeon & Rickert, 1984) it is likely to be inappropriate for pastures which have substantially differing leaf/stem ratios of morphology such as *Themeda triandra*, *Aristida* spp. and *Bothriochloa pertusa*. The further refinement of the above model will require measurement of the production and quality (nitrogen and digestibility) of leaf and stem components and dietary composition. Where such data have been collected, successful models of diet selection and animal production have been developed (White, 1978, for sheep in Tussock Grasslands; Hendricksen *et al.*, 1982, for beef cattle in Subtropical Grasslands).

Limitation of simplistic modelling approach

However, a major limitation to the application of models such as GRASP to savannas other than the Subtropical Grasslands, is the lack of data on cover/run-off relationships. High pasture utilization can reduce surface infiltration rates by reducing plant cover and degrading surface soil structure (Mott, Bridge & Arndt, 1979). Run-off losses can be a large proportion of the rainfall in Monsoon Tallgrass (Ive *et al.*, 1976), Tussock Grasslands (Clewett, 1985), Tropical Tallgrass and *Acacia* Shrublands (Pressland & Lehane, 1982). Similarly the redistribution of water either from run-off to run-on areas or through lateral drainage on shallow sloping duplex landscapes are important components of the hydrological cycle and may limit the application of the single landscape unit approach.

Further limitations are the lack of knowledge of the direct climatic and mineral nutrition effects on animal production especially for domestic animals of European origin which are at the limits of their adaptation (for example, sheep in northern Tussock grasslands; Entwistle, 1974). In Subtropical Grasslands comparison of Hereford with Sahiwal cross (first and second generation) steers showed substantial benefits to adapted cattle (Robbins & Esdale, 1982; Laing *et al.*, 1984). In Monsoon Tallgrass large liveweight gains could be achieved in steers with the addition of salt (Winter, 1987). Similarly in both Tropical and Tallgrass Subtropical communities direct dietary supplement with phosphorus, and fertilizer P increased steer liveweight gain on low P soils (Kerridge & McLean, 1988). Thus, considerable care should be taken in comparative studies of animal production between savannas as comparisons may be confounded by: (1) breed differences; (2) response to nutrient supplements; (3) overriding soil nutrient limitations (for example, P); (4) incidence of weather-related diseases (for example, ephemeral fever) and parasites (ticks); and (5) direct climatic effect on animal functioning.

THE APPLICATION OF RESOURCE DESCRIPTION, CLIMATE ANALYSIS AND DECISION-SUPPORT TO SAVANNA MANAGEMENT

The association of major cultural and ecological events with rainfall extremes prompts an evaluation of rainfall variability and pasture management. For the past 120 years graziers and land administrators have been forced to operate without any knowledge of the major mechanisms causing rainfall variability. Management recommendations have been based on neither strategic or tactical approaches. Strategic recommendations include constant stocking rates set to an average 'safe' level to avoid pasture degradation in low rainfall years and allow necessary management tools such as burning following above-average years. Tactical recommendations have been formulated to change stocking rates in autumn to achieve a desired utilization (for example, 30%) of known pasture yield. Such a policy has been demonstrated to maintain desirable pasture species and avoid large changes in plant density and production (Orr *et al.*, 1986).

The use of simulation models of pastures and crops allows a better interpretation of spatial and temporal climatic variability (for example, pasture growth, McCown *et al.*, 1974; beef cattle production, Gillard, 1979, McCown, 1981; stocking rate management, Rickert, McKeon & Prinsen, 1981; sorghum production, Clewett, 1985; wheat production, Hammer *et al.*, 1987). Historical analysis of management decisions (frequency of burning, stocking rate) indicate that different generations of graziers should expect to adopt different management strategies. For example, in *Acacia* shrublands pasture burning is recommended to control invasion of woody weed species (*Eremophila* spp.; Pressland *et al.*, 1986). The duration required for fuel accumulation represents a period of financial loss as destocking is usually required. Carter & Johnston (1986) used the GRASP model with historical rainfall data to calculate frequency of burning. Even for a cleared pasture in good condition, the lower rainfall in the 1917–47 period (Russell, 1981) would have led to fewer opportunities to use fire and greater reluctance of graziers to practise burning.

The persistence of the SOI from season to season is being used as the basis for rainfall forecasts in eastern Australia (McBride & Nicholls, 1983; Coughlan, 1988) where correlations between seasonal SOI values and next season's rainfall exceed 0.4 (i.e. 15–50% of variation depending on season). Alternative approaches (Clewett, Young & Willocks, 1988) calculate different probability distributions of rainfall based on three classes of the SOI. The recent development of these forecasts means that their utility in pasture management is yet to be assessed. Initial surveys of farmers suggests that their probabilistic nature is acceptable to managers.

Application to burning and stocking rate policy

In Subtropical Tallgrass, considerable debate has occurred regarding the practice of spring burning (Tothill, 1971) to improve diet quality and reduce patch grazing. If dry conditions follow burning there is economic loss due to feed restrictions reducing intake, reductions in pasture growth, and greater risk of soil erosion. The model GRASP has been used to simulate spring pasture growth following burning at Gayndah (114 years of historical rainfall). At conventional regional stocking rates (0.2 beasts/ha), the spring (September–November) pasture growth requirement to allow 30% utilization, is 600 kg DM/ha. This level of growth was *not* achieved in 36% of years. The winter SOI value (June–August) was significantly correlated with subsequent spring pasture growth ($r = 0.36$, $n = 114$). The percentage of years when spring pasture growth was less than 600 kg/ha was 53% when the winter SOI was less than –5 and 23% when the winter SOI was greater than +5. Thus seasonal SOI forecasts could be used to make better

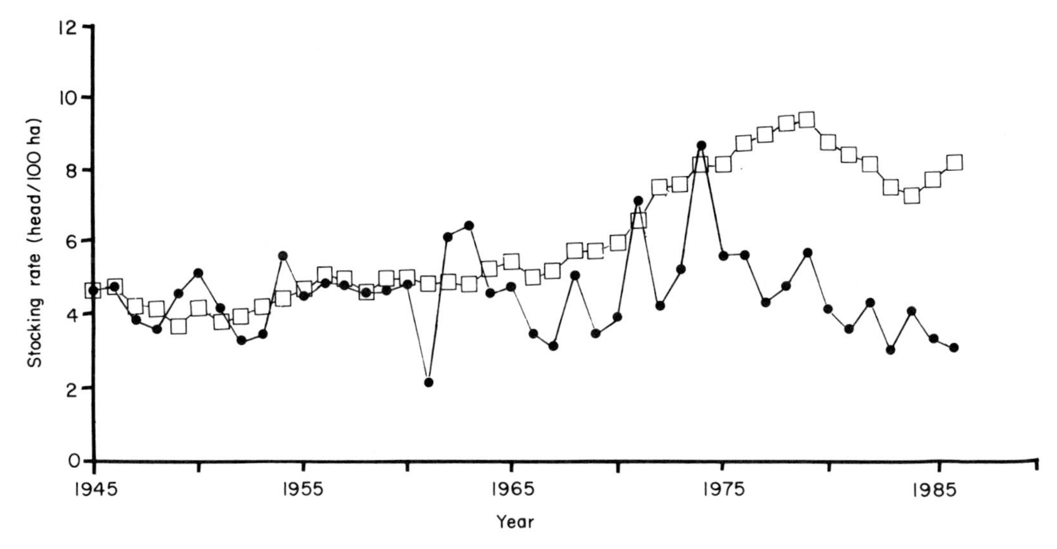

FIG. 8. Comparison of potential stocking rate (●) and actual stocking rate (□) in the Dalrymple Shire, 1945–86. Potential stocking rate = pasture growth (kg/ha Dec.–May) × safe utilization in summer (30%) ÷ animal intake for 6 months (1800 kg per cattle equivalent) × Shire Index (effect of trees and other land-uses). Pasture growth was calculated using the GRASP model with pasture and soil data from the Springmount GUNSYND site (L. Punter and R. Hendricksen, unpub. data). The Shire Index (0.304) was calculated for 1945–63 (a period of pasture stability) as the ratio of the actual shire stocking rate (4.54 cattle equivalents per 100 ha) to the calculated safe stocking rate for the cleared pasture (14.92 cattle equivalents per 100 ha). The actual shire stocking rate was calculated as the ratio of total beef cattle to the shire area.

decisions to reduce area or frequency of burning under conditions (winter SOI <–5) when production losses and erosion risks are greatest.

Pressland & McKeon (1989) used the GRASP model to calculate sustainable stocking rates at Charters Towers using historical rainfall records and regional cattle number statistics (Dalrymple Shire). Safe carrying capacities were calculated using simulated pasture growth and recorded regional stocking rates for a period of known botanical stability (1945–63).

Animal numbers increased after 1966 as a result of improved animal husbandry, change in breed to better adapted cattle, and the export beef price collapse in 1974–76 when graziers were unwilling to sell excess stock (Fig. 8). Rainfall and simulated pasture growth were above average from 1971 to 1976 in association with a period of anti-ENSO conditions (Fig. 3). However, average and below average conditions occurred from 1982 to 1988 and actual stocking rates were double the simulated safe stocking rates. Without pasture improvement (timber clearing, pasture legumes) these stocking rates were excessive and pasture degradation in the shire occurred during this period (Gardener *et al.*, 1988; A. J. Pressland, pers. comm.)

Pressland & McKeon (1989) suggested that monitoring of regional animal numbers combined with simulated pasture growth may provide the basis for recommendations which avoid overstocking. Graziers were likely to have had their expectations of property carrying capacity biased by the long period of favourable rainfall conditions (1971–81). Simulation analysis using long-term climatic records provides a basis indicating sustainable stocking rates and providing warning of periods when stock numbers and pasture growth are out of balance.

From 1982 to 1988 very low rainfall (<60% of average) occurred in four wet seasons (1982/83, 1984/85, 1985/86, 1987/88). Of these four wet seasons, both the 1982/83 and 1987/88 droughts were predictable given their association with very negative SOI values (Fig. 9). Hence, resource damage due to overgrazing could have been avoided in these seasons.

However, the summer droughts of 1984/85 and 1985/86 were not associated with ENSO variation. Damage to the savanna resource as a result of overgrazing during these droughts could only be avoided by more conservative stocking rate policy. The impact of regional savanna management would greatly benefit from regional rainfall forecasts although such a capability is yet to be demonstrated.

The relationship between the SOI and pasture growth was examined for Charters Towers by using the GRASP

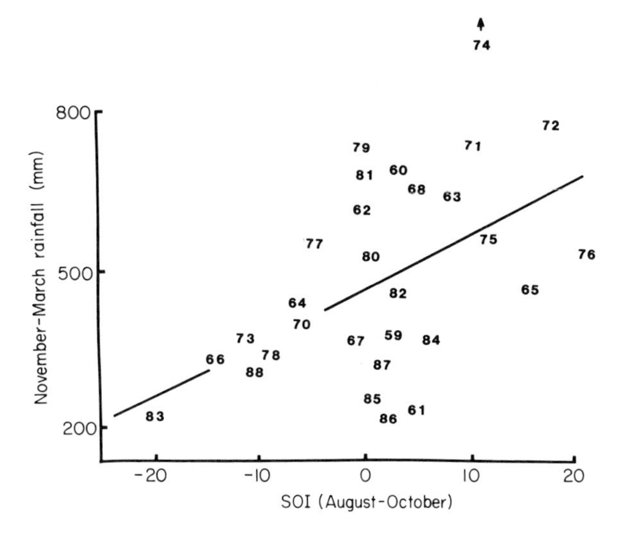

FIG. 9. Relationship between the mean Southern Oscillation Index (SOI) recorded from August to October and the actual rainfall (*R*) received at Charters Towers between the following November and March. The regression shown ($R = 472 + 9.6$ SOI : $r = 0.49$, $n = 29$) does not include the 1973–74 summer.

model and the historical daily rainfall record (102 years). Summer (November–April) pasture growth was significantly (>0.01) correlated with both spring ($r = 0.46$) and summer SOI ($r = 0.57$). When spring (August–October) SOI values were less than –5 (30% of all years) the chance of conditions when large decreases in basal area would occur due to over-utilization doubled (Fig. 6).

Thus, models such as GRASP allow the calculation of probability distributions of pasture growth which can be used to indicate safe long-term stocking rates both for optimizing pasture condition (30% utilization of summer growth) and animal production (McKeon *et al.*, 1986). Moreover the use of current seasonal forecasting capabilities provides early warning of conditions when overgrazing is likely to occur. Decision support packages (Stafford-Smith & Foran, 1988) and GRASSMAN (J. Scanlan, pers. comm.) allow graziers and advisors to examine alternatives for herd structure and stocking rate management.

CONCLUSIONS

Comparative studies across northern savannas are leading to the development of general models of pastoral systems. The specification of minimum data sets will ensure adequate data collection to support the models and allow extrapolation to individual pasture enterprises. When the models are incorporated in decision support packages, they will allow graziers and their advisers to quickly evaluate economic performance and sustainable production.

ACKNOWLEDGMENTS

Some of the research described in this paper was supported by the Australian Wool Corporation and the Australian Meat and Livestock Research and Development Corporation.

REFERENCES

Adamson, D., Williams, W.A.J. & Baxter, J.T. (1987) Complex late Quaternary alluvial history in the Nile, Murray-Darling and Ganges basins: three rivers presently linked to the Southern Oscillation. *International geomorphology* (ed. by V. Gardiner), Part II, pp. 875–887. John Wiley and Sons, London.

Allan, R.J. (1988) El Nino Southern Oscillation influences in the Australasian region. *Progr. Phys. Geogr.* **12**, 313–348.

Andrew, M.H. (1986) Use of fire for spelling monsoon tallgrass pasture grazed by cattle. *Trop. Grassl.* **20**, 69–78.

Anon. (1988) Climate change in Australia to the year 2030 AD. *Greenhouse: planning for climate change* (ed. by G. I. Pearman), pp. 737–740. CSIRO, Melbourne.

Ash, A.J., Prinsen, J.H., Myles, D.J. & Hendricksen, R.E. (1982) Short-term effects of burning native pasture in spring on herbage and animal production in south-east Queensland. *Proc. Aust. Soc. Anim. Prod.* **14**, 377–380.

Austin, M.P. & Williams, O.B. (1988) Influence of climate and community composition on the population demography of pasture species in semi-arid Australia. *Vegetatio*, **77**, 43–49.

Bowman, D.M.J.S. & Minchin, P.R. (1987) Environmental relationships of woody vegetation patterns in Australia monsoon tropics. *Aust. J. Bot.* **35**, 151–169.

Braithwaite, R.W. (1985) Biological research for national park management in ecology of the wet-dry tropics. *Proc. Ecol. Soc. Aust.* **13**, 323–331.

Bremen, H. & de Wit, C.T. (1983) Rangeland productivity and exploitation in the Sahel. *Science*, **221**, 1341–1347.

Brunt, A.T. (1961) The climate of Queensland. *Introducing Queensland*, pp. 19–26. ANZAAS, Brisbane.

Busby, J.R. (1988) Potential impacts of climate change on Australian flora and fauna. *Greenhouse: planning for climatic change* (ed. by G. I. Pearman), pp. 387–398. CSIRO, Melbourne.

Carter, J.O. & Johnston, P.W. (1986) Modelling expected fuel loads for fire at Charleville in western Queensland. *Third Queensland Fire Research Workshop Proceedings* (ed. by B. R. Roberts), pp. 55–67. Darling Downs Institute of Advanced Education, Toowoomba.

Charles-Edwards, D.A., Doley, D. & Rimmington, G.M. (1986) *Modelling plant growth and development.* Academic Press, Sydney.

Christian, C.S. & Stewart, G.A. (1953) General report on survey of Katherine-Darwin region, 1946. *CSIRO Aust. Land Res. Ser.* No. 1.

Christie, E.K. (1975) Physiological responses of semi-arid grasses. III. Growth in relation to temperature and soil water deficit. *Aust. J. agric. Res.* **26**, 447–457.

Christie, E.K. (1978) Herbage condition assessment of an infertile range grassland based on site production potential. *Aust. Rangel. J.* **1**, 87–94.

Clewett, J.F. (1985) Shallow storage irrigation for sorphum production in north-west Queensland. *Queensland Department of Primary Industries Bulletin*, QB85002.

Clewett, J.F., Young, P.D. & Willcocks, J.R. (1988) Effect of climate change on agriculture in Central Queensland: rainfall variability analysis. *The changing climate and central Queensland agriculture* (ed. by E. R. Anderson), pp. 43–52. Australian Institute of Agricultural Science, Rockhampton, Qld.

Conseil Scientifique pour l'Afrique (C.S.A.) (1956) Reunion de speciales du C.S.a. en matiere de Phytogeographie (CAS Specialist Meeting on Phytogeography). Yangambi CCTA, London, Publ. No. 53.

Coughlan, M.J. (1988) Season climate outlooks. *The changing climate and central Queensland agriculture* (ed. by E. R. Anderson), pp. 17–26. Australian Institute of Agricultural Science, Rockhampton, Qld.

Daly, J.J. (1983) Queensland Beef Industry 1930 to 1980: lessons from the past. *Qd Agric. J.* **109**, 61–97.

Dick, R.S. (1958) Variability of rainfall in Queensland. *J. Geogr.* **11**, 32–42.

Dye, P.J. & Spear, P.T. (1982) The effects of bush clearing and rainfall variability on grass yield and composition in south west Zimbabwe. *Zimbabwe J. agric. Res.* **20**, 103–118.

Entwistle, K.W. (1974) Reproduction in sheep and cattle in Australian Arid zone. Studies of Australian Arid Zone 2. *Animal production* (ed. by A. D. Wilson), pp. 85–97. CSIRO, Melbourne.

Eyles, A.G. & Cameron, D.G. (1985) The contribution of science to Australian tropical agriculture. 3. Tropical Pasture Research. *J. Aust. Inst. agric. Sci.* **51**, 17–28.

Fitzpatrick, E.A. & Nix, H.A. (1970) The climatic factor in Australian grassland ecology. *Australian grasslands* (ed. by R. M. Moore), pp. 1–26. ANU Press, Canberra.

Foley, J.C. (1945) Frost in the Australian region. *Commonwealth Meteorological Bureau Bulletin*, No. 32.

Gardener, C.J., McIvor, J.G. & Williams, J.C. (1988) Dry tropical rangelands: solving one problem and creating another. *Proc. Ecol. Soc. Aust.* **16** (in press).

Gibbs, W.J. & Maher, J.V. (1967) Rainfall deciles as drought indicators, Bulletin 48, Bureau of Meteorology, Melbourne, pp. 118.

Gillard, P. (1979) Improvement of native pasture with Townsville stylo in the dry tropics of sub-coastal northern Queensland. *Aust. J. exp. Agric. Anim. Husb.* **19**, 325–336.

Graetz, R.D., Walker, B.H. & Walker, P.A. (1988) The consequences of climatic change for seventy per cent of Australia. *Greenhouse: planning for climate change* (ed. by G. I. Pearman), pp. 399–420. CSIRO, Melbourne.

Hammer, G.L., Donatelli, M., Farquhar, G.D., Hubick, K.T. & Nade, L.J. (1989) Radiation use efficiency, water use efficiency and crop improvement in grain sorghum. *Proceedings Australian Sorghum Workshop*, Toowoomba, February, 1989 (in press).

Hammer, G.L., Woodruff, D.R. & Robinson, J.B. (1987) Effect of climatic variability and possible climatic change on reliability of wheat cropping – a modelling approach. *Agric. For. Met.* **41**, 123–142.

Hassall, A.C. (1979) Native pasture management. *Trop. Grassl.* **10**, 53–54.

Heathcote, R.L. (1965) *Back of Bourke, study of land appraisal and settlement in semi-arid Australia.* Melbourne University Press.

Hendricksen, R.E., Rickert, K.G., Ash, A.J. & McKeon, G.M. (1982) Beef production model. *Proc. Aust. Soc. Anim. Prod.* **14**, 204–208.

Howden, S.M. (1988) Some aspects of the ecology of four tropical grasses with special emphasis on *Bothriochloa pertusa.* Ph.D. thesis, Griffith University.

Huntley, B.J. (1982) Southern African savannas. *Ecology of tropical savanna* (ed. by B. J. Huntley and B. H. Walker), pp. 101–119. Springer, Berlin.

Huntley, B.J. & Walker, B.H. (1982) *Ecology of tropical savannas* (ed. by B. J. Huntley and B. H. Walker), pp. 1–4. Springer, Berlin.

Ive, J.R., Rose, C.W., Wall, B.H. & Torssell, B.W. (1976) Estimation and simulation of sheet run-off. *Aust. J. Soil Res.* **14**, 129–138.

Ivory, D.A. & Whiteman, P.C. (1978) Effect of temperature on growth of five subtropical grasses. I. *Aust. J. Plant Physiol.* **5**, 86–99.

Johnson, R.W. & Tothill, J.C. (1985) Definition and broad geographic outline of savanna lands. *Ecology and management of the world's savannas* (ed. by J. C. Tothill and J. J. Mott), pp. 1–13. Australian Academy of Science, Canberra.

Johnston, P.W. & Carter, J.O. (1986) The role of fire in production systems in western Queensland: a simulation approach. *Third Queensland Fire Research Workshop Proceedings* (ed. by B. R. Roberts). Darling Downs Institute of Advanced Edcuation, Toowoomba, Qld.

Jones, R.J. & Sandland, R.L. (1974) The relation between animal gain and stocking rate. Derivation of the relation from the results of grazing trials. *J. agric. Sci., Camb.* **83**, 335–342.

Kerridge, P.C. & McLean, R.W. (1988) Fertiliser and supplementary phosphorus responses by cattle on legume pastures in south-east Queensland. *Proc. Aust. Soc. Anim. Prod.* **17**, 426.

Laing, A.R., Taylor, W.J., Robbins, G.B. & Bushell, J.J. (1984) The influence of filial generation of Sahiwal-Hereford steers on live weight, liveweight gain, carcass weight and fat thickness. *Proc. Aust. Soc. Anim. Prod.* **15**, 416–419.

Lloyd, P.L. (1980) *Perspectives in productivity.* Queensland Department of Primary Industries, Brisbane.

McAlpine, J.R. (1976) Climate and water balance. Lands of the alligator Rivers Area, Northern Territory. CSIRO Land Res. Ser. No. 38, pp. 35–49.

McBride, J.L. & Nicholls, N. (1983) Seasonal relationships between Australian rainfall and the Southern Oscillation. *Mon. Weath. Rev.* **110**, 14–17.

McCown, R.L. (1973) An evaluation of the influence of available soil water storage capacity on growing season length and yield of tropical pastures using simple water balance models. *Agric. Meteorol.* **11**, 53–63.

McCown R.L. (1981) The climatic potential for beef cattle production in tropical Australia: Part I. Simulating the annual cycle of liveweight change. *Agric. Systems*, **6**, 303–318.

McCown, R.L., Gillard, P. & Edye, L.A. (1974) The annual variation in yield of pastures in the seasonally dry tropics of Queensland. *Aust. J. exp. Agric. Anim. Husb.* **14**, 328–333.

McKeon, G.M., Howden, S.M., Silburn, D.M., Carter, J.O., Clewett, J.F., Hammer, G.L., Johnston, P.W., Lloyd, P.L., Mott, J.J., Walker, B., Weston, E.J. & Willcocks, J.R. (1988) The effect of climate change on crop and pastoral production in Queensland. *Greenhouse: planning for climate change* (ed. by G. I. Pearman), pp. 546–563. CSIRO, Melbourne.

McKeon, G.M. & Rickert, K.G. (1984) A computer model of the integration of forage options for beef production. *Proc. Aust. Soc. Anim. Prod.* **15**, 15–19.

McKeon, G.M., Rickert, K.G., Ash, A.J., Cooksley, D.G. & Scattini, W.J. (1982) Pasture production model. *Proc. Aust. Soc. Anim. Prod.* **14**, 202–204.

McKeon, G.M., Rickert, K.G., Robbins, G.B., Scattini, W.J. & Ivory, D.A. (1980) Prediction of animal performance from simple environmental variables. *Fourth Biennial Conference, Simulation Society of Australia*, pp. 9–16. Department of Engineering, University of Queensland, Brisbane.

McKeon, G.M., Rickert, K.B. & Scattini, W.J. (1986) Tropical pastures in the farming system: case studies of modelling integration through simulation. *Proceedings 3rd Australian Conference Tropical Pastures*, pp. 92–100. Tropical Grassland Society, Brisbane.

McKeon, G.M. & Scattini, W.J. (1980) Integration of feed sources in property management: modelling approach. *Trop. Grassl.* **14**, 246–252.

McMahon, T.A. (1982) Hydrological characteristics of selected rivers of the world. *Technical Documents in Hydrology.* UNESCO.

Miller, C.P. & Webb, C.D. (1988) Phosphorus supply to cattle grazing stylo pastures. *Proc. Aust. Soc. Prod.* **17**, 441.

Moore, R.M. (1970) Australian Grasslands. *Australian grasslands* (ed. by R. M. Moore), pp. 85–100. ANU Press, Canberra.

Mott, J.J. (1986) Pastoralism and Australian Savannas: a question of stability. *Proc. 4th Aust. Rangel. Soc. Conference, Armidale*, pp. 61–64.

Mott, J.J. (1987) Patch grazing and degradation in native pastures of the tropical savannas in northern Australia. *Grazing lands research of the plant animal interface* (ed. by F. P. Horne, J. Hodgson, J. J. Mott and R. W. Brougham), pp. 153–162. Winrock International, Morrilton, Arkansas.

Mott, J.J. & Andrew, M.H. (1985) The effect of fire on population dynamics of native grasses of north-west Australia. *Proc. Ecol. Soc. Aust.* **13**, 231–239.

Mott, J.J., Bridge, B.J. & Arndt, W. (1979) Soil seals in tropical tallgrass pastures of northern Australia. *Aust. J. Soil Res.* **17**, 483–494.

Mott, J.J., Tothill, J.C. & Weston, E.J. (1981) Animal production from the native woodlands and grasslands of northern Australia. *J. Aust. Inst. agric. Sci.* **47**, 132–141.

Mott, J.J., Williams, J., Andrew, M.H. & Gillison, A.N. (1985) Australian savanna ecosystems. *Ecology and management of the world's savannas* (ed. by J. C. Tothill and J. J. Mott), pp. 56–82. Australian Academy of Science, Canberra.

Nicholls, N. (1988) El Nino – Southern Oscillation and Rainfall Variability. *J. Climate*, **1**, 418–421.

Nicholls, N., McBridge, J.L. & Ornerod, R.J. (1982) On predicting the onset of the Australian wet season at Darwin. *Mon. Weath. Rev.* **110**, 14–17.

Nicholls, N. & Wong, K.K. (1990) Dependence of rainfall variability on mean rainfall, latitude and the Southern Oscillation. *J. Climate* (in press).

Nix, H.A. (1987) The role of crop modelling, minimum data sets and geographic information systems in the transfer of agricultural technology. *Agricultural environment: characterisation, classification and mapping* (ed. by A. H. Bunting). CAB International, Wallingford, U.K.

Northcote, K.H., Hubble, G.D., Isbell, R.F., Thompson, C.H. & Bettenay, E. (1975) *A description of Australian soils.* CSIRO, Australia.

Orr, D.M. (1986) Factors affecting the vegetation dynamics of *Astrebla* grasslands. Ph.D. thesis, University of Queensland, Brisbane.

Orr, D.M. (1988) Interaction of rainfall and grazing on the demography of *Astrebla* spp. in north western Queensland. *Proc. 3rd Int. Rangel. Congr*, pp. 192–194. New Delhi.

Orr, D.M., Bowly, P.S., Evenson, C.J. (1986) Effects of grazing management on the basal area of perennial grasses in *Astrebla* grassland. *Rangelands: a resource under siege* (ed. by P. J. Joss, P. W. Lynch and O. B. Williams), pp. 56–57. Australian Academy of Science, Canberra.

Orr, D.M., Evenson, C.J., Jordan, D.J., Bowly, P.S., Lehane, K.J. & Cowan, D.C. (1988) Sheep productivity in an *Astrebla* grassland of south west Queensland. *Aust. Rangel. J.* **10**, 39–47.

Pearman, G.I. (1988) Greenhouse gases: evidence for atmospheric changes and anthropogenic causes. *Greenhouse: planning for climate change* (ed. by G. I. Pearman), pp. 3–21. CSIRO, Melbourne.

Pittock, A.B. (1975) Climatic change and the patterns of variation in Australia rainfall. *Search*, **6**, 498–504.

Poppi, D.P., Minson, D.J. & Ternouth, J.H. (1981) Studies of cattle and sheep eating leaf and stem fractions of grasses. I. The voluntary intake, digestibility and retention time in reticulorumen. *Aust. J. agric. Res.* **32**, 99–108.

Pressland, A.J., Cowan, D.C., Evenson, C.J. & Bowly, P.S. (1986) Benefits of infrequent fire in the mulga (*Acacia aneura*) rangelands of Queensland. *Rangelands: a resource under siege* (ed. by P. J. Joss, P. W. Lynch, and O. B. Williams), pp. 608–609. Australian Academy of Science, Canberra.

Pressland, A.J. & Lehane, K.J. (1980) Production and water use of a wiregrass (*Aristida* spp.) pasture in south western Queensland. *Aust. Rangel. J.* **2**, 217–221.

Pressland, A.J. & Lehane, K.J. (1982) Runoff and the ameliorating effect of plant cover in the mulga communities of south western Queensland. *Aust. Rangel. J.* **4**, 16–20.

Pressland, A.J. & McKeon, G.M. (1989) Monitoring animal numbers and pasture condition for drought administration – an approach. *Proc. 5th Australian Soil Conservation Conference, Perth* (in press).

Rickert, K.G. (1988) Computer models and the study of grazing systems. *Trop. Grassl.* **22**, 145–9.

Rickert, K.G. & McKeon, G.M. (1982) Soil water balance model: WATSUP. *Proc. Aust. Soc. Anim. Prod.* **14**, 198–200.

Rickert, K.G., McKeon, G.M. & Prinsen, J.H. (1981) Growing beef cattle on native pasture oversown with fine stem stylo in subtropical Australia. *Proc. XIV Int. Grassl. Congr.*, pp. 762–765.

Ridpath, M.G. (1985) Ecology in the wet-dry tropics: how different? *Proc. Ecol. Soc. Aust.* **13**, 3–20.

Robbins, G.R. (1984) Relationships between productivity and age since establishment of pastures of *Panicum maximum* var. *trichoglume*. Ph.D. thesis, University of Queensland, Brisbane.

Robbins, G.B. & Bushell, J.J. (1985) Productivity of grazed irrigated pastures of ryegrass, berseem clover, and ryegras/berseem clover in south-east Queensland. *Trop. Grassl.* **21**, 133–138.

Robbins, G.B., Bushell, J.J. & Butler, K.L. (1987) Decline in plant and animal production from ageing pastures of green panic (*Panicum maximum* var. *trichoglume*). *J. agric. Sci., Camb.* **108**, 407–417.

Robbins, G.B. & Esdale, C.R. (1982) Accounting for differences in growth rates between Hereford and Sahiwal-Hereford cattle used in pasture studies. *Proc. Aust. Soc. Anim. Prod.* **14**, 395–397.

Ropelewski, C.F. & Halpert, M.S. (1989) Precipitation patterns associated with the high index phase of the Southern Oscillation. *J. Climate*, **2**, 268–264.

Russell, J.S. (1981) Geographic variation in seasonal rainfall in Australia – an analysis of the 80 year period 1895–1974. *J. Aust. Inst. agric. Sci.* **47**, 59–66.

Russell, J.S. (1988) The effects of climatic change on the productivity of Australian agroecosystems. *Greenhouse: planning for climate change* (ed. by G. I. Pearman), pp. 491–505. CSIRO, Melbourne.

Sala, O.E., Parton, W.J., Joyce, L.A. & Lauenroth, W.K. (1988) Primary production of the central grassland region of the United States. *Ecology*, **69**, 40–45.

Scattini, W.J. (1973) A model for beef cattle production farm rangeland and sown pasture in south-eastern Queensland, Australia. Ph.D. thesis, University of California, Berkeley.

Shaw, W.H. (1957) Bunch spear grass dominance in burnt pastures in south eastern Queensland. *Aust. J. agric. Res.* **8**, 325–34.

Singh, J.S., Hanxi Yang, Sajise, P.C. (1985) Structural and functional aspects of India and southeast Asian savanna ecosystems. I. *Ecology and management of the world's savannas* (ed. by J. C. Tothill and J. J. Mott), pp. 34–51. Australia Academy of Science, Canberra.

Slatyer, R.O. (1960) Agricultural climatology of the Katherine Area, N.T. *CSIRO Aust. Division Land Resources Regional Survey Technical Paper* No. 13.

Stafford-Smith, D.M. & Foran, B.D. (1988) Strategic decisions in pastoral management. *Aust. Rangel. J.* **10**, 82–95.

Tanner, C.B. & Sinclair, T.R. (1983) Efficient water use in crop production: research or re-search? *Limitation to efficient water use in crop production* (ed. by H. M. Taylor, W. R. Jordan and T. R. Sinclair). ASA, CSSA, SSSA, Madison, Wisconsin.

Taylor, J.A. (1989) The role of animals in restoration processes. *Aust. Rangel. J.* (submitted).

Taylor, J.A. & Dunlop, C.R. (1985) Plant communities of the wet-dry tropics of Australia: the Alligator Rivers regions, Northern Territory. *Proc. Ecol. Soc. Aust.* **13**, 83–127.

Taylor, J.A. & Tulloch, D. (1985) Rainfall in the wet-dry tropics: extreme events at Darwin and similarities between years during the period 1870–1983 inclusive. *Aust. J. Ecol.* **10**, 281–295.

Torssell, B.W.R. & McKeon, G.M. (1976) Germination effects on pasture composition. *J. appl. Ecol.* **13**, 593–603.

Tothill, J.C. (1971) A review of fire in management of native pasture with particular reference to north-eastern Australia. *Trop. Grassl.* **5**, 1–10.

Tothill, J.C., Nix, H.A., Stanton, J.P. & Russell, M.J. (1985) Land use and productive potentials of Australian savanna lands. *Ecology and management of the world's savannas* (ed. by J. C. Tothill and J. J. Mott), pp. 125–141. Australian Academy of Science, Canberra.

Tucker, G.B. (1988) Climate modelling: How does it work? *Greenhouse: planning for climate change* (ed. by G. I. Pearman), pp. 22–34. CSIRO, Melbourne.

Van Keulen, H. (1975) *Simulation of water use and herbage growth in arid regions*. Pudoc, Wageningen.

Walker, B.H. (1985) Structure and function of savannas: an overview. *Ecology and management of the worlds savannas* (ed. by J. C. Tothill and J. J. Mott), pp. 83–92. Australian Academy of Science, Canberra.

Walker, B.H. (1989) Autecology, synecology, climate and livestock as agents of rangeland dynamics. *Aust. Rangel. J.* **10**, 69–75.

Walker, B., Hodge, P.B. & O'Rourke, P.K. (1987) Effects of stocking rate and grass species on pasture and cattle productivity of sown pastures on a fertile brigalow soil in central Queensland. *Trop. Grassl.* **21**, 14–23.

Walker, B. & Weston, E.J. (1990) Pasture development in Queensland. The success story. *Trop. Grassl.* (In preparation).

Walker, P.A. & Cocks, K.D. (1984) Computerised choropleth mapping of Australian resource data. *Cartography*, **13**, 243–252.

Watkinson, A.R., Lonsdale, U.M. & Andrew, M.H. (1989) Modelling the population dynamics of an annual plant *Sorghum intrans* in the wet-dry tropics. *J. Ecol.* **77**, 162–181.

Weston, E.J., Harbison, J., Leslie, J.K., Rosenthal, K.M. & Mayer, R.J. (1981) Assessment of the agricultural and pastoral potential of Queensland. *Agriculture Branch Technical Report* No. 27. Queensland Department of Primary Industries, Brisbane.

White, B.J. (1978) A simulation based evaluation of Queensland's northern sheep industry. *Geography Department, James Cook University Monograph Series* No. 10, pp. 109.

Williams, J. & Probert, M.E. (1984) Characterisation of the soil/climate constraints for predicting pasture production in the semi-arid tropics. ACIAR. *Proceedings Workshop on Soils*. Townsville, Australia.

Wilson, J.R. & t'Mannetje, L. (1978) Senescence, digestibility and carbohydrate content of buffel grass and green panic leaves in swards. *Aust. J. agric. Res.* **29**, 503–516.

Winter, W.H. (1987) Using fire and supplements to improve cattle production from monsoon tallgrass pastures. *Trop. Grassl.* **21**, 71–81.

Wright, P.B. (1975) An index of the Southern Oscillation. Climatic Research Unit Report 4, University of East Anglia, pp. 21.

3/Stability and stress in the savanna forests of mainland South-East Asia

PHILIP STOTT *Department of Geography, School of Oriental and African Studies (University of London), Thornhaugh Street, Russell Square, London WC1H OXG, U.K.*

Abstract. In this paper, the savanna forests and the much rarer open savanna grasslands of mainland South-East Asia, a region stretching from Manipur State in India, through Burma, Thailand, Laos, Kampuchea (Cambodia) and Viet-Nam, are described biogeographically and ecologically. These distinctive formations, the canopies of which are dominated by six leaf-shedding members of the Dipterocarpaceae, have been much neglected by ecologists, although they often comprise the most important single formation over much of the region, and play a significant role both in the ecology of the area's distinctive forest wildlife and in the economy of many local peoples. The 'forest' character of the formation is particularly noteworthy when compared with many similar or parallel formations, particularly those of Africa and Latin America.

The possible origins of the associations involved are considered, and an attempt is made to contrast stability and change in their recent ecological history, with particular respect to climatic fluctuations and to the impact of human populations. The formations are seen to possess edaphic or topographic 'cores', from which they have been spread by fire and the axe, but only within the general area of a monsoon forest (*Am*) or savanna forest (*Aw*) climate, where 'running' fire is characteristic in the dry season. Physiognomically, ecophysiologically and phenologically, the key taxa exhibit a wide range of adaptations to the main ecological stresses of such environments, and the natural limits of these adaptations are examined. The ecotonal characteristics of savanna forests and savanna are considered in relation to a wide range of parapatric formations. In drawing comparisons with savannas in other regions of the world, the essential homogeneity of the South-East Asian associations parallels most closely that of the dry eucalypt savanna woodlands of Northern Australia, both contrasting sharply with the remarkable heterogeneity of West African and Latin American savannas.

Present-day processes for change are then considered. In particular the following ecological stress factors are monitored, namely: agricultural extensification and development; war; the mismanagement of fire; flood; edaphic and geomorphological pressures; over-exploitation; and faunal changes. The importance of the formations is reassessed in terms of their overall biology, for the management of wildlife, as ecological controls, and as an economic resource with potential for further development. Finally, all this technical discussion is set against the development of new attitudes to the conservation and management of natural and semi-natural habitats in South-East Asia, and in particular in Thailand.

Key words. Savanna forest, savanna, mainland South-East Asia, ecological status, ecological stress, role of fire, conservation.

INTRODUCTION

Kurz (1876, 1877) was one of the first biologists to describe the savannas and savanna forests of South-East Asia. Writing of Burma, he recognized a wide range of 'open' or 'savanna(h)' forests which dominated much of the more seasonal and drier parts of the country. There were also open grasslands, but of these he wrote that 'the savannahs (*sic*) are the undergrowth of the savannah forests, and as such do not differ from these in any point except that they are void, or nearly void, of trees.' The essential 'forest' character of the savannas of mainland South-East Asia has since been acknowledged by many writers, including Schimper (1898, 1903), Stamp (1924, 1925), Champion (1936), Richards (1952), Davis (1960), Vidal (1960), Blasco (1983) and Stott (1984), among several others.

True open savannas certainly exist in mainland South-East Asia, as in Thailand on the so-called 'bald hills' of the Petchabun range, and at Khao Yai National Park; around Mondolkiri in Kampuchea (Cambodia); and at Ban Me Thuot in Viet-Nam (Schmid, 1962, 1974; Martin, 1971; Blasco, 1983), but in the main these isolated fragments represent only a very small fraction of the natural and semi-natural vegetation of the countries concerned. The origins of these open grassland communities appear to lie either in heavy inundation during the rainy season, as at Sakon Nakhon in North-East Thailand, although this site is now largely lost, or in intense cutting, burning and grazing, as on the Petchabun mountains. The total annual precipitation may also fall below 500 mm in the region of some of these savanna enclaves (Khemnark, 1979). In the Thai language, open savannas are called *pàa yâa*, meaning 'wild grass', 'grass wilderness' or even 'grass forest', and the most characteristic grasses, herbs and woody plants are *Imperata cylindrica* L., *Panicum repens* L., *Saccharum spontaneum* L., *Sorghum halepense* Perr., *Vitiveria zizanoides* Stapf., *Eupatorium odoratum* L., *Acacia catechu* Willd., *A. siamensis* Craib, *Careya arborea* Roxb., and *Pterocarpus macrocarpus* Kurz. Where such grasslands occur, they provide important habitats for large browsing and grazing mammals, such as barking deer (*Muntiacus muntjak* Zimmermann, 1780), sambar deer (*Cervus unicolor* Kerr, 1792), and banteng (*Bos javanicus* D'Alton, 1823).

By contrast, the true 'savanna forests' of mainland South-East Asia, in the sense first fully described by Schimper (1903: 260), are much more closed and wooded. In Stamp's (1925) analysis, repeated by Richards (1952: 330), a number of forest types fall under this wide general designation, including '*te*' forest with *Diospyros burmanica* Kurz, and groups with *Acacia catechu*, *Terminalia oliveri* Brandis, and *Tectona hamiltoniana* Wall. *ex* Ham. Unquestionably, however, the key savanna forest type throughout nearly all the seasonal parts of mainland South-East Asia is the dry deciduous dipterocarp forest or *forêt claire à diptérocarpacées* ('open dipterocarp forest': Burmese *indaìng*, Thai *pàa tengrang*), which is a highly distinctive formation extending from the remote Kabaw valley on the borders of Manipur State in India (Kaith, 1936; Sen Gupta, 1939), through Burma, Thailand, Laos and Kampuchea (Cambodia), to Viet-Nam (Stott, 1984).

In this paper, we shall therefore above all concern ourselves with the ecology and dynamics of the dry deciduous dipterocarp forests of the region, associations which perhaps find their closest parallels taxonomically in the celebrated *sāl* (*Shorea robusta* Gaertn. *f.*) forest and woodlands of the Indian sub-continent (Champion, 1936; Puri, 1960), but structurally more with the equally homogenous eucalypt savanna woodlands of Northern Australia (e.g. Braithwaite & Estbergs, 1985; Stott, 1989; Taylor & Dunlop, 1984) (Fig. 1a, b) and the *miombo* woodlands of southern Africa. To a large extent, these associations have been singularly neglected by scholars of savanna vegetation, and they remain very much the 'cinderellas' of the subject, even in some modern texts (e.g. Cole, 1986). To date, the main contributors to our understanding of their basic ecology have been French scholars working in Indochina (e.g. Rollet, 1953; Vidal, 1956, 1960; Boulbet, 1982), Japanese ecologists involved with biomass and productivity studies (e.g. Ogawa, Yoda & Kira, 1961), biologists and foresters of the Thai Royal Forest Department (e.g. Smitinand, 1968), and staff in the Faculty of Forestry at Kasetsart University, Bangkok. Under Professor Sangha Sabhasri and Dr Somsak Sukwong, both former Deans of the Faculty, work on savanna forests has developed apace following advice from the present author, with projects on classification and structure (e.g. Kutintara, 1975), productivity (e.g. Sabhasri *et al.*, 1968), phenology (e.g. Sukwong, Dhamanitayakul & Pongumphai, 1975), the role of fire, fauna and economic importance. Stott (1984, 1988a, b) reviews much of the earlier work, the 1984 paper providing a base-line for future studies.

DRY DECIDUOUS DIPTEROCARP FOREST

Dry deciduous dipterocarp forest (savanna forest *sensu* Schimper) may be broadly defined according to the following fundamental criteria:

Macroclimate. Köppen's *Am* and *Aw* climates, though mainly the latter (Mizukoshi, 1971); annual value of Angström's humidity coefficient (*H*) usually <240, i.e. semi-arid (Ogawa *et al.*, 1961: 78); evaporation may exceed rainfall for more than 9 months of the year; dry season of 5–7 months (October/November to March/April or May); most typical rainfall mean of 1000–1500 mm per annum; mean temperature of the coldest month rarely falls below 20°C; potential fires from November to May.

Microclimate. Usually sites which are topographically 'dry', either plateaux in the rain shadow of mountain ranges (e.g. the Korat plateau of North-East Thailand), or slopes in the shadow of hills; usually thin, dry, freely-draining soils of the Red-Yellow Podzolic or Grey Podzolic groups, with a high sand fraction.

Geology and soils. Sandstones, granite slopes and ridges, old alluvium, but above all the first; below 1000 m above sea-level (exceptionally 1300 m); soils mainly thin and stony, or with a laterite layer forming a hardpan, the latter sites waterlogging during the rainy season; humus content very low, <2% to 4%; C/N ratio *c.* 12.7; pH 5.0–6.2 (Sangtongpraow & Dhamanonda, 1973); total nitrogen content low at all soil depths; high degree of stoniness; very low clay fraction; mainly sandy loams; sometimes saline.

Structure. One to three tree storeys, normally the top storey from 9 m to 30 m height, the highest ever recorded by the author being an exceptional stand of 38 m in Huay Kha Khaeng Wildlife Sanctuary; median tree height 8–11 m, canopy cover 53–77%, percentage through-forest visibility $(-1/6D+10)^2$, where *D* is the distance in metres (Neal, 1967); nine to twelve trees per 10 × 20 m quadrat; slight shrub layer; thin to dense pygmy bamboo (*Arundinaria* spp.), grasses, and herbs; fungi after fire; some bryophytes, especially in burnt patches and by boulders; epilithic lichens common; epiphytes only abundant over 700 m above sea-level.

Dominants. Formation dominated by various mixtures of six leaf-shedding members of the Dipterocarpaceae, namely: *Dipterocarpus intricatus* Dyer, *D. obtusifolius* Teijsm., *D. tuberculatus* Roxb., *Shorea obtusa* Wall. *ex* Blume, *S. roxburghii* G. Don (syn. *S. talura* Roxb.), and *S. siamensis* Miq. (=*Pentacme suavis* and its varieties of Smitinand, 1968); other highly characteristic tree taxa include, among many: *Dillenia* spp., *Irvingia malayana* Oliv. *ex* A. Benn., *Melanorrhoea usitata* Wall., *Pinus merkusii* Jungh. & De Vriese, *Pterocarpus macrocarpus*, *Sindora siamensis* Teijsm. *ex* Miq., *Terminalia* spp., and *Xylia kerrii* Craib & Hutch.; ground flora: small shrubs and herbs, but above all

(a)

(b)

FIG. 1. (a) Savanna forest at Sakaerat Environmental Research Station, Pak Thong Chai, Nakhon Ratchasima, north-eastern Thailand. Note the open wooded character of the formation and the homogeneity of the ground cover. (b) For comparison, a dry open eucalypt forest in Kakadu National Park, Northern Territory, Australia. Physiognomically and structurally, these formations are quite close, although they are dominated by completely different taxa. Both are extremely homogenous when compared with West African and South American savanna formations. (Photos: the author.)

pygmy bamboo (*Arundinaria* spp.) and grasses, e.g. genera like *Arundinella, Capillipedium, Eulalia, Heteropogon, Hyparrhenia, Imperata, Polytoca* and *Themeda*; the dwarf palm complex, *Phoenix acaulis/humilis*, and the cycad, *Cycas siamensis* Miq., are also highly characteristic.

Physiognomy and phenology. Main tree dominants have thick, hard, rough, fire-resistant bark, with tap roots and lateral roots similarly protected; leaf-shedding between November and April; maximum diameter growth in September; ground flora dominated by geophytes, hemicryptophytes and therophytes; *Cycas siamensis* and *Phoenix* protected by overlapping leaf-bases; standing dead crop of ground flora seems to peak in February and March.

Fauna. Quite rich in herpetofauna with some highly characteristic species, e.g. *Leiolepis belliana* (Gray), 1827 and *Varanus bengalensis* (Daudin), 1802; poor bird fauna, with few typcial species; mammals mainly crepuscular and nocturnal browsers and grazers, with a small range of insectivores and rodents; top carnivores invasive from other formations; the habitat is too dry and too hot for many species, and the essential 'forest' character has not permitted the development of a great plains fauna; most species are only transgressives; classic savanna forest food-chains would be: (a) tiger (*Panthera tigris corbetti* Mazak, 1968) (in adjacent tropical semi-evergreen forest or monsoon forest) → sambar deer/barking deer/siamese hare (*Lepus peguensis siamensis* Bonhote, 1902) → ground flora and small salt licks; (b) hawks and snakes → *Leiolepis belliana* → insects and ground flora.

Human economy. Important source of valuable timbers, especially *Shorea obtusa* and *S. siamensis*, two species which in many characteristics are preferable even to teak (*Tectona grandis* L. *f.*); also fine woods such as *Pterocarpus macrocarpus* and *Xylia kerrii*; wood oil from the *Dipterocarpus* spp., Burma lacquer varnish (*Melanorrhoea usitata*), resins and turpentines, strychnine (*Strychnos nux-vomica* L.), forest foods (insects, wild meat, fungi, leaves and fruits), grazing and thatching materials, orchids and wild birds.

PAST AND PRESENT ECOLOGICAL STATUS

The absence of satisfactory pollen diagrams for mainland South-East Asia makes any interpretation of the ecological history and status of the dry deciduous dipterocarp forest somewhat difficult and dangerous. However, three simple facts are cogent to such an analysis, namely that the formation is clearly adapted first to dry seasonal climates and to 'dry' topographic and edaphic sites; secondly, to a moderate level of fire; and, finally, to the axe, the main dipterocarps all coppicing well and recovering quickly after cutting and extraction. Wacharakitti & Intrachandra (1970), in an experiment carried out at Lampang in Northern Thailand, showed that *Dipterocarpus obtusifolius* was the most productive after coppicing, followed by *Shorea siamensis, D. tuberculatus* and *S. obtusa*. On the other hand, Stott (1986, 1988a, b) has demonstrated a wide range of physiognomic and ecophysiological adaptations to fire levels below a fire-line intensity (I) of 400 kW m^{-1} following Byram's formula (Chandler *et al.*, 1983: 24–28), where the flame height is around a maximum of just over 1 m, and the speed of forward spread is less than 3.0 cms^{-1}. When fire occurs annually, or at least biennially, the build-up of leaf litter and the growth of ground cover will never be sufficient to produce fires above this intensity, and many will be very tame affairs, with fire-line intensities of only 170 kW m^{-1} and a forward spread of less than 1.0 cms^{-1}.

Traditionally villagers throughout mainland South-East Asia have burnt the forest to produce paths and 'drives' in hunting, fungi on the forest floor, new grazing, flushes of valuable leaves, such as those of *Milientha suavis* Pers., as well as to control pests and diseases, and to release nutrients which are washed down to their fields with the onset of the monsoon. Many fires have also escaped into the forest from shifting cultivation (swidden) burns or field fires, from roadsides, and from plantations, while purely accidental fires, ignited by dropped cigarettes, ill-quenched cooking fires, or sparks from a passing locomotive, are not uncommon. The mesmeric force of flames has also caused fires to be set for ritual purposes and for sheer devilment. Since the advent of Neolithic peoples in the region some 12,000 years ago, fire has thus been an increasingly potent force in the ecology of the savanna forests and, along with the axe, has clearly been a key factor in extending the formation into the less fire-resistant associations with which it shares an ecotone. Above all, sharp fire-honed boundaries are characteristic where savanna forest abuts on tropical semi-evergreen rain forest, regular burning causing the savanna forest to expand slowly but inexorably into the latter until the boundary is eventually stabilized by some fire-halting topographic feature, such as a sharp break of slope or a stream bed. Ecotones with other formations, as with monsoon forest, tend to be much wider and more complex, and controlled by a range of factors, mainly edaphic and microclimatic. Yet again, however, heavy and regular burning will tend to spread the savanna forest at the expense of the neighbouring formation.

But how important was fire before its intensive use by humans, and to what extent are the savanna forests a fire-climax? Natural fires certainly occur in the savanna forest, the chief causes being lightning during electrical storms, refraction and deflection. A classic example is known to occur annually in the west of Thailand, in Kanchanaburi province, where the sun's rays are deflected into the forest from a crystallized quartzite cliff (Miss Saranarat Kanjanavanit, pers. comm., 1988). The potential for natural fire is thus clearly present, and it must have occurred throughout the formation even before the more widespread and systematic use of fire by humans. However, what is also clear is that the distribution of savanna forest is not primarily governed by the presence or absence of annual burns. This point was made most cogently by Barrington (1931: 17) in his little-known, but excellent, survey of the Hlaing Forest Circle in Burma, when he observed that savanna forest was: '. . . the apparent climax on coarse sand almost everywhere in the lowlands of Burma. In the lower half of the province, fire protection for thirty or forty years had no appreciable effect on the vegetation, so *indaing* [the Burmese for savanna forest] is at least an edaphic, or soil-controlled climax. North of Mandalay it seems to be a preclimax stabilized by annual fires; protection encourages an evergreen undergrowth which prevents the reproduction of *in* [*Dipterocarpus tuberculatus*] and would obviously change the consociation to a moister type.'

Stott (1976) has also shown that savanna forests comprise a whole series of different associations in which the role of fire is markedly different, and the distribution of the 'core' savanna forest communities would appear, as Barrington inferred, to be essentially controlled by the presence of topographically and edaphically 'dry' sites, such as steep slopes with rocky outcrops and freely-draining coarse sands. Fire occurs at these 'core' sites, but it is not the essential determining ecological factor, although the forest is naturally adapted to the burns, the adaptations for drought proving equally effective against moderate fires. Elsewhere, however, away from these 'core' communities, it is fire and cutting which have caused the savanna forests to invade new territories at the expense of other forest types. In these locations, fire is therefore essential to the maintenance of savanna forest, and savanna forest will not persist without it.

The ecological status and history of savanna forest may thus be summarized as follows. Savanna forest as a formation clearly evolved to occupy topographically 'dry' sites in those regions of mainland South-East Asia which experienced a strongly seasonal and relatively low total annual rainfall pattern. During the last glaciation, when the tropics were drier than now (Flenley, 1979a, b, 1982), 'core' savanna forests may well have extended much more widely in South-East Asia, maybe even into peninsular Malaysia. With the full re-establishment of the monsoon paralleling the retreat of the ice, savanna forest probably contracted once more to its key core locations on coarse sands, steep rocky slopes, and poor soils with a laterite layer near the surface. With the advent of Neolithic peoples, however, and the increasing use of fire and the axe, savanna forest would again spread from its cores into areas where the climate still permitted fires to burn in the dry season. It is extremely interesting to note that many key Neolithic sites in mainland South-East Asia (such as Non Nok Tha and Ban Chiang in North-East Thailand) are located within savanna forest areas, although the age-old question of which came first is not easy to resolve. The process continues today where fire is regularly employed around savanna forest ecotones.

The savanna forests of mainland South-East Asia are therefore an edaphic climax *and* a fire climax *and* a climatic climax, all at the same time, the distribution of the 'core' formations being governed by topography and soil, with a secondary spread controlled by the use of fire and the axe, but in both instances only within the general area of a monsoon forest (*Am*) or savanna forest (*Aw*) climate. The same 'form' is thus produced by a complex series of processes operating in different ways at different sites at different times; the old arguments about the key determinant of savannas, which afflicted the study of the African and South American formations for so long, are clearly obsolete, and we must strive to understand the complex ecological dynamics of a wide range of associations in the specific localities where they occur.

STABILITY AND STRESS IN THE SAVANNA FORESTS OF MAINLAND SOUTH-EAST ASIA

What then are the present-day processes of change in the savanna forests of mainland South-East Asia?

First and foremost, as elsewhere throughout the world, there is the complete loss and destruction of the habitat because of a wide range of factors. In Thailand, the country about which we know the most in the region, the Government persists in claiming that 30% of the land remains forested (e.g. many Government statistics and *Thailand: Natural Resources Profile*. Thailand Development Research Institute, 1987), but the reality is sadly very different, the actual figure being around 14%. Many areas classed in legal terms as possessing forest of one type or another no longer carry a single tree. Since 1960, the rate of loss has markedly steepened, and now stands around 3.15% per annum (Allen & Barnes, 1985). It appears that in some regions of the country, savanna forest has been one of the worst sufferers. This is particularly the case on the Korat Plateau in the North-East, a region of poor, sandy soils with a classic savanna forest climate, where originally up to 80% of the vegetation comprised dry deciduous dipterocarp forest. Fig. 2 presents the current situation, and shows the loss of this and other formations in the region since 1975. It was this alarming state of affairs which prompted H.M. the King of Thailand to initiate *The Green E-Sarn Project* under General Chaovalit Yongchaiyudh in August 1987 (see Biwater, 1987).

Sukaesinee Subhadira *et al.* (1987) have carried out an excellent series of studies on the pressures affecting the savanna forests of the Korat Plateau. Fig. 3 illustrates the very recent pattern of forest loss around two villages, *Baan Phu Hang* and *Baan Non Amnuay* (*Baan* is the Thai for 'house' or 'village'), in the Dong Mun forest, a region once completely covered with dense savanna forest and tropical semi-evergreen rain forest. As one villager vividly said, 'Years ago the forest was so thick with large trees, vines, wild animals and filled with malaria that only a few hunters had the courage to enter it.' The area remained relatively untouched by human activity until the early 1970s when the first permanent settler (quoted above) arrived at the site of Phu Hang. He brought with him an extended family of nineteen people who had been forced to leave the Lam Pao region following the development of a dam there under the Mekong Project. I find it somewhat ironic that my first visit to Thailand, in 1972, was to help on this project. By 1986, the village had grown to 600 households. Non Amnuay was established later in 1979 by relatives of the founder of Phu Hang who were attracted to the large expanse of unsettled and unclaimed land. In 1985, Non Amnuay had 200 households with more than 1400 people. Immigration has been the main source of development and this has occurred despite the fact that the area was a battleground between leftist guerillas and the Border Patrol Police (BPP) in the late-1970s. This very thorough study provides a cameo of a process which is common on the forest margins throughout South-East Asia, and which is eating into the forest resources of the region, in the main unchecked and unmanaged. Squatter settlement in pioneer areas and agricultural

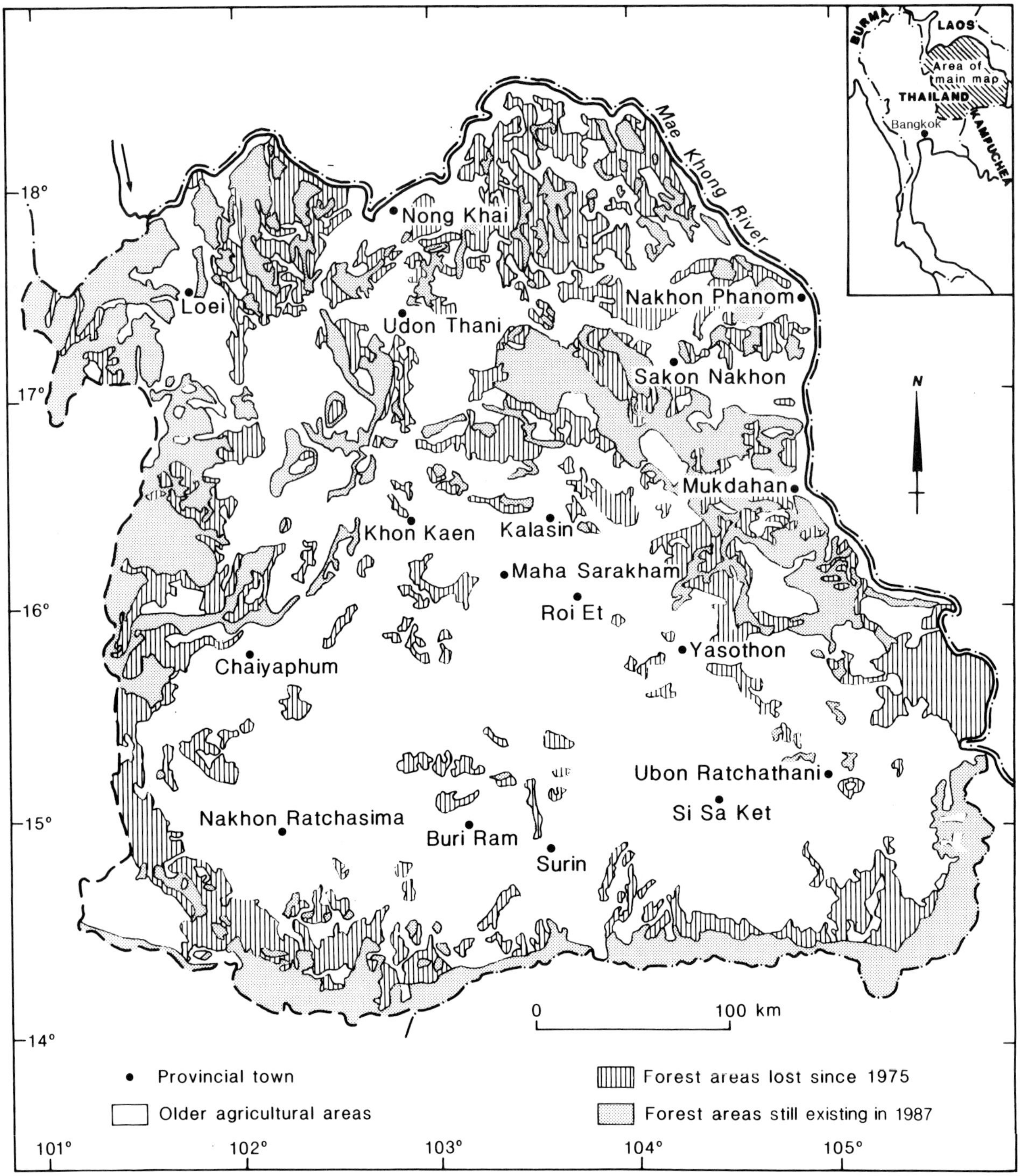

FIG. 2. Recent deforestation in the Korat Plateau of North-East Thailand. (*After* the *Final Report* on 'Green E-Sarn: Investigation and preparation of a water resource development programme for North East Thailand', 1987, Biwater, Dorking, U.K., Map 5.)

extensification are the two main causes of forest loss, a process then exacerbated by the development of roads and tracks, illegal logging and hunting, mining and army incursions. And once the settlements are established, the pressure on the forest resource will simply grow and become increasingly destructive, if not planned and managed. Fig. 4 illustrates graphically the continuing impact of Phu Hang and Non Amnuay on the surrounding forest areas.

This then is a very characteristic picture of the destruction of savanna and other forest types in Thailand; we know far less of the process in Burma, Laos, Kampuchea and Viet-Nam, although war has taken an inevitable toll of the forest resources in some of these states. The near extinction of the world's largest cattle, the kouprey (*Bos sauveli* Urbain, 1937), which was once abundant in the monsoon forests and savanna forests of Northern Cambodia, is a

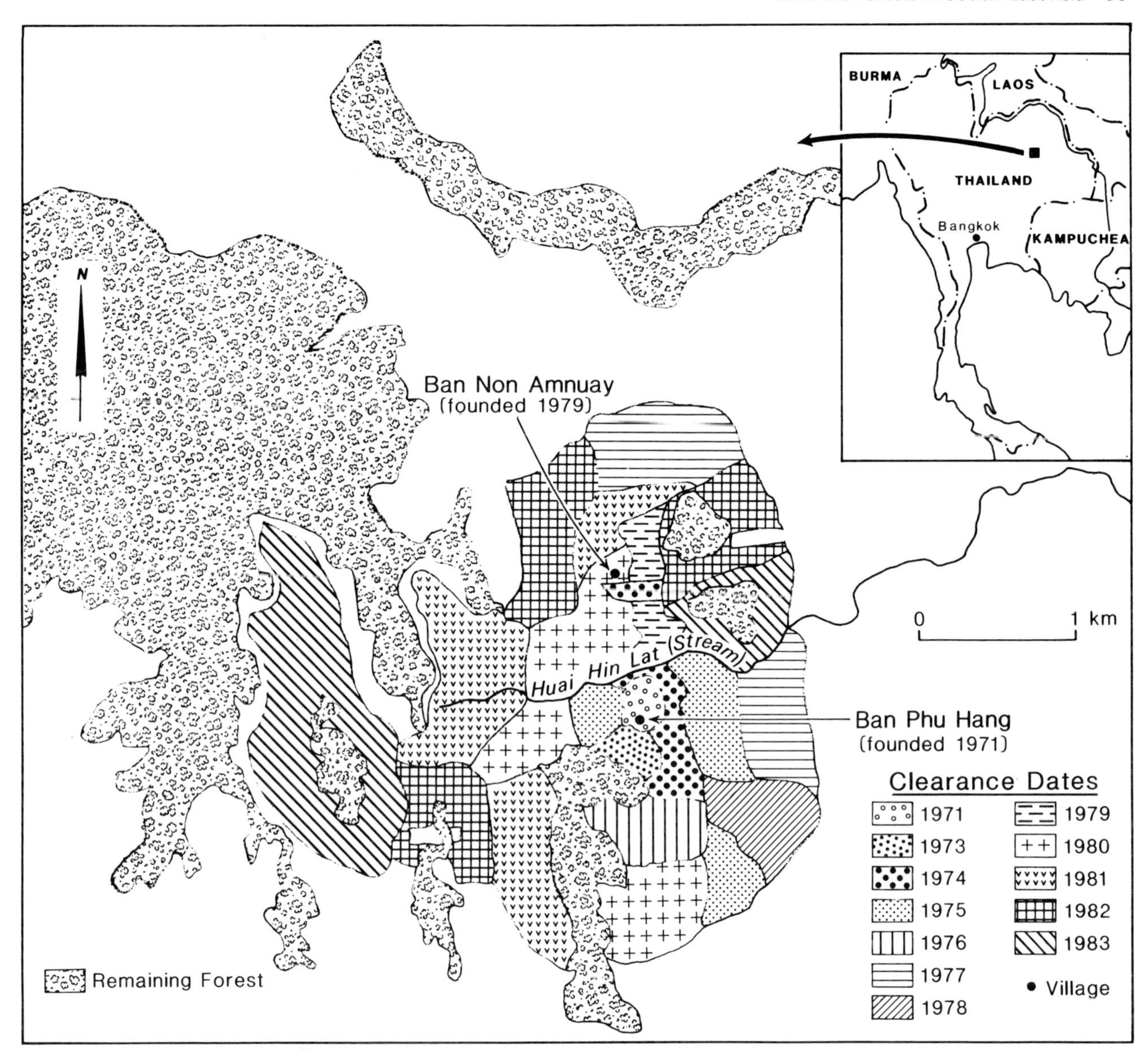

FIG. 3. The history of forest clearance in the Dong Mun Reserved Forest, District Nong Kung Si, Kalasin Province, North East Thailand. (*After* Sukaesinee Subhadhira *et al*., 1987; Fig. 70 and text.)

particularly sad reminder of this fact, especially when we recall that the reduction in its population has been largely brought about by desperate villagers and guerillas seeking food, and by the animal blowing itself up on mines cruelly laid in the forest.

The reduction of savanna forest over large areas to small 'island' remnants in a 'sea' of uncontrolled agricultural extensification and squatter exploitation inevitably reduces the ecological viability of the remaining patches, particularly with respect to the fauna, which above all depends on the maintenance of reasonable areas of savanna forest within a general forest matrix. Most of the main mammal species, for example, require a range of habitats for their survival and, for many, the ideal mix is a mosaic of tropical semi-evergreen rain forest, monsoon forest and savanna forest, broken by streams, and well-provided with shade and salt licks in addition to grazing. Sadly, few such areas remain, either to conserve or to manage. In Thailand, one surviving 'paradise' in this respect is the 'Huai Kha Khaeng Wildlife Sanctuary-Thung Yai Wildlife Sanctuary–Kroeng Kavia Non-Hunting Area' complex in Kanchanaburi and Uthai thani provinces. The loss of this would be a disaster on a world-scale, equivalent to the loss of a parallel World Heritage site, such as Kakadu National Park in the 'Top End' of the Northern Territory, Australia (Braithwaite & Werner, 1987), and the recent decision of the thirty-nine member inquiry team, chaired by Deputy Prime Minister Thienchai Srisamphan, to postpone for the time being the proposed development of the Nam Choan Dam in Thung Yai was a great relief to many (Paisal Sricharatchanya, 1988). The area, which also links with forests in bordering Burma, has some of the finest stands of savanna forest, with a complete range of fauna, including all the large browsers and grazers. The complex as a whole is clearly big enough

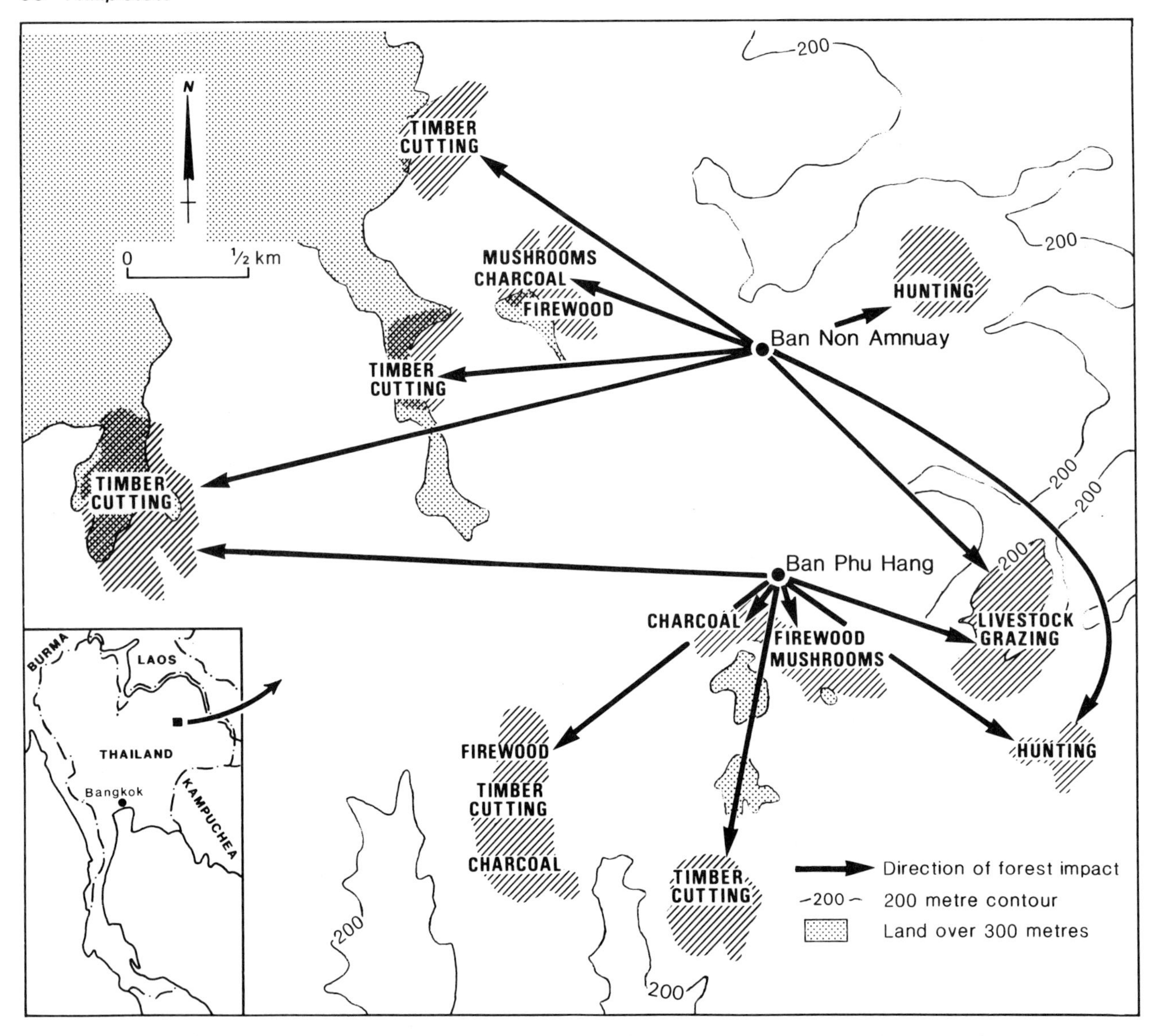

FIG. 4. Pressure on forest resources in the Dong Mun Reserved Forest, District Nong Kung Si, Kalasin Province, North-East Thailand. (*After* Sukaesinee Subhadhira *et al.*, 1987; Fig. 71 and text.)

to remain ecologically stable if it is now fully protected and managed. The proposed Nam Choan Dam would breach the essential integrity of the reserve and open up the area to the processes already outlined above for the Korat Plateau. In contrast, the many remaining small patches of savanna forest which dot the landscape and occupy the higher ground throughout Thailand are usually under too much stress, and are unlikely to survive in most areas, except as depauperate stands kept for a few specimen trees or as burial forests. As always, size is an important variable in assessing stability and resistance to stress.

Stress caused by specific ecological factors may also severely undermine the long-term stability of savanna forest ecosystems. It has already been emphasized that the key phanerophytes and much of the ground flora are clearly adapted to *moderate* levels of dry season fire, where regular annual or biennial burns feed on limited fuels; but it is also undoubtedly true that more intense and extensive fires caused by an excessive build-up of leaf litter and ground flora, especially the grasses and pygmy bamboos, may prove very destructive in even the best-adapted formations. For example, fire protection over 4 or 5 years will lead to the development of nearly pure stands of either *Themeda triandra* or *Arundinaria pusilla* A. Chev. & A. Camus, with a mean height of over 1 m and a range from 0.5 to 4.0 m, and a ground cover value of around 95%. In certain areas, such conditions may well improve the tree stocking in savanna forest for the short term, with a high level of seedling survival, but they will eventually trigger off the succession to a moister formation in regions where the savanna forest is invasive, although this is less likely in core areas. What *is* likely, however, is the ignition of a severe and highly damaging extreme ground cover burn, with temperatures in the region of 900°C at 0.5–1.0 m above the ground, and which moves at speeds of more than 3.0 cms^{-1}, even in near windless conditions (Stott, 1986, 1988a, b). Such conflagrations, which may have fireline intensities (I) of over 2000 kW m^{-1}, can lead to 'spotting', with the fire leaping across even wide fire breaks, and to the 'torching' of individual trees, particularly where the high grasses are clumped around the specimen involved. These fires will greatly stress the environment, often leading to the death of some tree taxa, and initiating accelerated erosion on steeper slopes. They may also cause the death of some of the larger fauna, and even of humans when trapped up-slope of the fire, and can lead to the destruction of property, where a wind change turns the flames from the forest to the wooden houses of a nearby village.

Yet it is the very policy of fire protection which will have led to this end result. Sadly, in mainland South-East Asia, the policing of forest fires will never prove totally effective, and even the best-protected stands will ignite at some time. The need to develop policies which work *with* the natural ecology of the savanna forests is therefore an urgent task. Studies are vitally required to establish the best cycle of prescribed burns in savanna forest: should they be annual or biennial?; early in the season or later?; early in the day or later?; and what is the maximum ground flora and litter depth that can be tolerated for prescribed burning? Again, we see a close parallel with the 'fire problem' in the eucalypt savanna woodlands of Northern Australia, where patterns of aboriginal burning may have much to teach us, and where expert advice is much more readily available (e.g. the SHRUBKILL programme devised by John Ludwig, see this issue, p. 203). The relationship between fire and erosion in savanna forest is also an important question, with even moderate fires tending to cause some gullying on slopes over 10° (Chunkao, 1969; Stott, 1986). Similarly, many types of savanna forest are adapted to some degree of inundation during the monsoon season, particularly stands dominated by *Dipterocarpus tuberculatus* growing on flat ground where the soil has a laterite layer some 50 cm down, and, of course, some examples of open savannas are specifically maintained by flooding. However, yet again, excessive and continuous flooding may undermine the ecosystem, leading to the death of selected tree species. Savanna forest is clearly a 'moderate' system in which 'excess' should be avoided.

Throughout the savanna forests of mainland South-East Asia, the fauna, and in particular the large grazers and browsers, are under stress. As already pointed out, most remaining patches of savanna forest are either too isolated or too tiny to maintain any fauna of note. Many are near 'deserts' with virtually no mammals, except the odd rodent, eastern mole (*Talpa micrura* Hodgson, 1841), and siamese hare, and no birds of note. The full fauna only persists in the few remaining large tracts which are still part of an ecological continuum from tropical semi-evergreen rain forest, through monsoon forest, to savanna forest. The absence of the larger fauna does not appear to disrupt the ecosystem as such, since fire, the 'red steer', acts as a natural replacement for grazing, but it is sad that the herds of deer and wild cattle which once roamed in and out of these forests are no longer a part of their general ecology. At a time when the herding of such animals for commercial considerations is increasingly seen as a viable option, this is especially to be regretted. The need to maintain the savanna forest 'gene pool' is thus an important consideration in planning future conservation strategies. Over-hunting has also caused many of the fauna to adapt their general behaviour, turning day-time animals into crepuscular feeders, and crepuscular animals into nocturnal foragers. Illegal hunting remains a severe stress throughout the region.

Sukwong (1974) found that most savanna forest tree species are resistant to pests and diseases. Lepidoptera larvae, however, feed on the leaves of *Dipterocarpus tuberculatus*, *Shorea obtusa*, *S. siamensis* and *Terminalia* spp., while a range of coleopterous insects attack the bark, sapwood and sometimes the heartwood of the main leaf-shedding dipterocarps. Mistletoes (Loranthaceae) attack somewhere in the region of 10% of all trees, but especially *Shorea siamensis*.

THE FUTURE OF THE SAVANNA FORESTS OF SOUTH-EAST ASIA

Unfortunately, savanna forest is not widely perceived as important among the general spectrum of ecosystems in mainland South-East Asia. Foresters tend to be blind to anything which does not have teak as a key taxon, and, in general, they have little understanding of the role of fire in the formation. Governments think of savanna forest, if they think of it at all, as a somewhat scruffy secondary formation of little beauty and value when compared with the great tropical semi-evergreen rain forests, which at least have obvious tourist potential. The management and conservation of savanna forests is therefore at a low ebb, although we do not know the precise situation in countries such as Burma, Laos and Kampuchea.

This is a regrettable situation, for it is a relatively easy formation to maintain and manage, and a rich one when part of a mosaic of forest communities. Moreover, it has great economic potential, with a range of tree species often rivalling teak and rosewoods in quality, dipterocarps which coppice effectively to produce small timbers, grazing for deer and wild cattle, and a large range of local forest products. The formation is also aesthetically portrayed in both ancient and modern literature throughout the region (e.g. the *Reamker*, the Cambodian version of the *Ramayana*; see the translation of Jacob, 1986), and holds a significant place in South-East Asian Buddhism, *Shorea siamensis* replacing the Indian *S. robusta* as the tree under which the Buddha was born, a narrative change brought about by a distinctive biogeographical fact. Savanna forest even has a significant place in the forests of Himavanta, as described in the great classic of Thai Buddhism, the *Traibhumikatha* or *Three worlds according to King Ruang* which was written by Phya Lithai at Sri Satchanalai in the fourteenth century A.D. (Reynolds & Reynolds, 1982). Perhaps the real future of savanna forest lies in the development of a Buddhist ethic of conservation and management in South-East Asia, the first signs of which are now apparent (see, for example, Chatsumarn Kabilsingh, 1987). The recent decision of the Thai Government to delay the building of the Nam Choan Dam may be such a turning point, a first step in the integration of the once uncivilized 'forest realm' into the civilized space of the Buddhist state (Stott, in press).

Practically, however, in order to maintain the place of the savanna forests in the ecology of the region, the following programme will be necessary:

Conserve rigorously the remaining large tracts of savanna forest which form part of major forest complexes, as in the Huai Kha Khaeng-Thung Yai-Kroeng Kavia sanctuaries. We must discover the 'Kakadu National Parks' of South-East Asia, and protect them as our heritage.

Carry out research into the prescribed use of fire in the formation and develop a suitable education programme to deal with this problem.

Maintain smaller patches as a resource for local products, as an element in agroforestry and social forestry programmes, and as a source of good timbers and fuel wood needing careful management.

In this way, a remnant of these unique forest savannas may be conserved.

REFERENCES

Allen, J.C. & Barnes, D.F. (1985) The causes of deforestation in developing countries. *Annal. Ass. Amer. Geogr.* **75**, (1), 163–184.

Barrington, A.H.M. (1931) Forest soil and vegetation in the Hlaing Forest Circle, Burma. (*Burma Forest Bulletin*, No. 25, *Ecology Series*, No. 1). Govt. printing and Stationery, Rangoon.

Biwater (1987) *Kingdom of Thailand, Green E-Sarn: investigation of a water resource development programme for North East Thailand.* (Final Report). Biwater House, Dorking, U.K.

Blasco, F. (1983) The transition from open forest to savanna in continental Southeast Asia *Tropical savannas* (ed. by F. Bourlière), pp. 167–181. Elsevier, Amsterdam.

Boulbet, J. (1982) *Évolution des paysages végétaux en Thaïlande du nord-est.* (Publ. de l'École Française d'Extrême-Orient, CXXXVI.) École Française d'Extrême-Orient, Paris.

Braithwaite, R.W. & Estbergs, J.A. (1985) Fire patterns and woody vegetation trends in the Alligator Rivers region of Northern Australia. *Ecology and management of the World's savannas* (ed. by J. C. Tothill and J. J. Mott), pp. 359–364. Australian Academy of Science, Canberra.

Braithwaite, R.W. & Werner, P.A. (1987) The biological value of Kakadu National Park. *Search*, **18**, (6), 296–301.

Champion, H.G. (1936) A preliminary survey of the forest types of India and Burma. *Ind. For. Rec.* (N.S.), **1**, 1–286.

Chandler, C., Cheney, P., Thomas, P., Trabaud, L. & Williams, D. (1983) *Fire in forestry*, Vols. I and II. John Wiley Interscience, Chichester, U.K.

Chunkao, K. (1969) The determination of aggregate stability by waterdrop impact in relation to sediment yields from erosion plots at Mae-Huad Forest, Lampang. (*Forest Research Bulletin* **4**), Kasetsart University, Bangkok.

Cole, M.M. (1986) *The savannas: biogeography and geobotany.* Academic Press, London.

Davis, J.H. (1960) *The forests of Burma.* University of Mandalay (with University of Florida), Mandalay.

Flenley, J.R. (1979a) *The equatorial rain forest: a geological history.* Butterworth, London.

Flenley, J.R. (1979b) Late Quaternary vegetation history of the equatorial mountains. *Prog. Phys. Geogr.* **3**, 488–509.

Flenley, J.R. (1982) The evidence for ecological change in the tropics. In: Whitmore T.C., Flenley, J.R. & Harris, D.R., The tropics as the norm in biogeography? *The Geographical Journal*, **148**, (1), 8–21.

Jacob, J.M. (1986) *Reamker (Ramakerti): the Cambodian version of the Ramayana.* (Oriental Translation Fund, New Series, **45**.) The Royal Asiatic Society, London.

Kabilsingh, C. (1987) How Buddhism can help protect Nature. *Tree of Life: Buddhism and protection of Nature*, pp. 7–16. Geneva.

Kaith, D.C. (1936) Manipur forests. *Indian Forester*, **62**, (9), 361–369; **62,** (10), 409–414.

Khemnark, C. (1979) Natural regeneration of the deciduous forests of Thailand. (*Tropical Agriculture Research Series*, **12**, 31–43.) Tropical Agriculture Research Center, Ministry of Agriculture, Forestry and Fisheries, Yatabe, Tsukuba, Ibaraki, Japan.

Kurz, S. (1876) *Preliminary report on the forest and other vegetation of Pegu.* Office of the Superintendent of Government Printing, Calcutta.

Kurz, S. (1877) *Forest flora of British Burma*, Vols I and II. Office of the Superintendent of Government Printing, Calcutta.

Kutintara, U. (1975) Structure of dry dipterocarp forest. Ph.D. thesis, Colorado State University, Fort Collins, Colorado.

Ludwig, J. (1988) *Shrubkill, Version 1.0: Users Guide.* CSIRO, Deniliguin, New South Wales. (With computer programme using PC-DOS 3.3.)

Martin, M.A. (1971) *Introduction à l'ethnobotanique du Cambodge.* Centre National de la Recherche Scientifique, Paris.

Mizukoshi, M. (1971) Regional divisions of monsoon Asia by Köppen's classification of climate. *Water balance of Monsoon Asia* (ed. by M. M. Yoshino), pp. 259–273. University of Tokyo Press.

Neal, D.G. (1967) *Statistical description of the forests of Thailand.* Military Research and Development Center, Bangkok.

Ogawa, H., Yoda, K. & Kira, T. (1961) A preliminary survey on the vegetation of Thailand. *Nature and Life in South East Asia*, **1**, 21–157.

Puri, G.S. (1960) *Indian forest ecology.* Oxford Book and Stationery Co., New Delhi.

Reynolds, F.E. & Reynolds, M.B. (1982) *Three worlds according to King Ruang: a Thai Buddhist cosmology.* (Berkeley Buddhist Studies Series, **4**.) University of California (Asian Humanities Press/Motilal Banarsidass), Berkeley, California.

Richards, P.W. (1952) *The tropical rain forest.* Cambridge University Press.

Rollet, B. (1953) Note sur les forêts claires de Sud de l'Indochine. *Bois et forêts des Tropiques*, **31**, 3–13.

Sabhasri, S., Khemnark, C., Aksornkoae, S. & Ratisoonthorn, P. (1968) *Primary production in dry-evergreen forest at Sakaerat, Amphoe Pak Thong Chai, Changwat Nakhon Ratchasima. I. Estimation of biomass and distribution amongst various organs.* (Report 27/2.) ASRCT, Bangkok.

Sangtongpraow, S. & Dhamanonda, P. (1973) Ecological study of stands in dry dipterocarp forest of Thailand. Bangkok. Unpublished Research Note reported to Kasetsart University [in Thai: English summary].

Schimper, A.F.W. (1903) *Plant geography upon a physiological basis* (Trans. W. R. Fisher from German edn. of 1898). Clarendon Press, Oxford.

Schmid, M. (1962) Contribution à la connaissance de la végétation du Vietnam: le massif sud-annamitique et les régions limotrophes. Doctoral thesis, University of Paris.

Schmid, M. (1974) *Végétation du Vietnam. Le massif sud-annamitique et les regions limotrophes.* (Memoires ORSTOM, 74.) ORSTOM, Paris.

Sen Gupta, J.N. (1939) *Dipterocarpus (Gurzan)* forests in India and their regeneration. *Indian For. Rec.* (NS), **3**, 61–164.

Smitinand, T. (1968) Identification keys to the genera and species of the Dipterocarpaceae in Thailand. *Nat. Hist. Bull. Siam Soc.*, **19**, 57–83.

Sricharatchanya, P. (1988) Thailand: Politics of Power. *Far Eastern Econ. Rev.* **31**, March 1988, 24.

Stamp, L.D. (1924) Notes on the vegetation of Burma. *Geogr. J.* **64**, 231–237.

Stamp, L.D. (1925) *The vegetation of Burma from an ecological standpoint*. Calcutta.

Stott, P. (1976) Recent trends in the classification and mapping of dry deciduous dipterocarp forest in Thailand. In: Ashton, P. & Ashton, M. (eds), *The classification and mapping of Southeast Asian ecosystems*, pp. 22–56. (Trans. IV. Aberdeen-Hull Symposium on Malesian Ecology). University of Hull, Department of Geography, Hull, U.K.

Stott, P. (1984) The savanna forests of mainland South East Asia: an ecological survey. *Prog. Phys. Geogr.* **8**, (3), 315–335.

Stott, P. (1986) The spatial pattern of dry season fires in the savanna forests of Thailand. *J. Biogeogr.* **13**, 105–113.

Stott, P. (1988a) The forest as Phoenix: towards a biogeography of fire in mainland South East Asia. *Geogr. J.* **154**, (3), 337–350.

Stott, P. (1988b) Savanna forest and seasonal fire in South East Asia. *Plants Today*, **1**, 196–200.

Stott, P. (1989) Lessons from an ancient land: Kakadu National Park, Northern Territory, Australia. *Plants Today*, **2**, 121–125.

Stott, P. (in press) *Müang* and *pàa*: elite views of Nature in a changing Thailand. (Paper presented to the *Thai Studies Symposium on local and élite cultural perceptions*, May 1988.) School of Oriental and African Studies, London.

Subhadhira, S., Apichatvullop, Y., Kunarat, P. & Hafner, J.A. (1987) *Case studies of human–forest interactions in Northeast Thailand*. (KU/KKU/Ford Foundation, 850–0391, Final Report 2.) North-east Thailand Upland Social Forestry Project, Khon Kaen.

Sukwong, S. (1974) Deciduous forest ecosystems in Thailand. (Paper presented to the UNESCO Symposium held on *Deciduous Forest Ecosystems*, November 1974). UNESCO, Sakaerat, Nakhon Ratchasima, Thailand.

Sukwong, S., Dhamanitayakul, P. & Pongumphai, S. (1975) Phenology and seasonal growth of dry dipterocarp forest tree species. *The Kasetsart Journal*, **9**, 105–113.

Taylor, J.A. & Dunlop, C.R. (1984) Plant communities of the wet–dry tropics of Australia: the Alligator Rivers region. *Proc. Ecol. Soc. Austr.* **13**, 83–127.

Thailand Development Research Institute (1987) *Thailand: Natural Resources profile*. Thailand Development Research Institute, Bangkok.

Vidal, J. (1956) La végétation du Laos. I. Le milieu (conditions écologiques). *Travaux du Laboratoire Forestier de Toulouse*, **5**, (1), 1–120. (Also available as a book, 1972. Éditions Vithagna, Vientiane.)

Vidal, J. (1960) La végétation du Laos. II. Groupements végétaux et flore. *Travaux Laboratoire Forestier de Toulouse*, **5**, (1), 121–5.

Wacharakitti, S. & Intrachandra, P. (1970) *Study on the coppicing power and growth of some valuable tree species in dry dipterocarp forest*. (Paper presented to the 3rd Forestry Conference at Chiang Mai, August 1970). Chiang Mai.

4/Savannas of north-east India

P. S. YADAVA *Department of Life Sciences, Manipur University, Imphal-795003, India*

Abstract. A survey of the savanna grasslands of north-east India is presented, including savanna types, life-forms, phenology, biomass and productivity, nutrients, fauna, and successional changes. The role of burning and grazing is stressed, and the management implications of ecological studies are considered.

Key words. Savanna grasslands, vegetation types, fire, grazing, north-east India.

INTRODUCTION

The Indian savannas have originated from woodland ecosystems through deforestation, abandoned cultivation and burning (Singh Hanxi Yang & Sajise, 1985; Misra, 1983; Gadgil & Meher Homji, 1985). These savannas are maintained at a sub-climax stage by repeated grazing and burning. This has led to the formation of mosaic types of savanna communities, depending upon the age and mode of origin, and the intensity of biotic disturbance. Since these communities represent different seral stages, they tend to differ in their species composition, productive potential, and nutrient cycling.

North-eastern India presents a wide variety of savanna ecosystems, depending upon various origins from a number of forest types, ranging from tropical rain forest, through sub-tropical and humid mountain forests to temperate forests (Puri, 1960; Champion & Seth, 1968).

North-east Indian savannas are spread over the plains, valleys and hilly regions of the north-eastern states of India, located between 22° 0′ N to 29° 5′ N latitude and 87° 7′ E to 97° 3′ E longitude. For this paper, north-east India comprises Arunachal Pradesh, Assam, Meghalaya, Manipur, Mizoram, Nagaland and Tripura States.

GENERAL GEOGRAPHY

North-eastern India comprises the eastern part of the Himalayan range, intercepted by plains, valleys and hilly terrains representing recent land (Pleistocene) of alluvial sediments created especially by the Brahmaputra river system running across the rising mountain ranges of the Himalayas, Pre-Cambrian (metamorphosed and crystalline rocks) and other sedimentary rocks more than 500 million years old. The Manipur hills, along with the Naga and Mizo hills, consist of mainly tertiary strata, and came into existence as a result of the Tertiary folding of sedimentary strata in the shallow Tethys Sea (some 40–90 million years ago).

The climate of the region is significantly influenced by the topography, and in particular of the great complex of mountains flanking the Himalayas. The climate is typically monsoonal. The monsoon rains start in the first week of June and continue late into October or even into November. The region also experiences rain showers in the winter season. The rainfall is not only heavy, but is also spread over a longer period of the year. A marked contrast exists between the dry season, which generally lasts from November to February, and the wet season from April to October. March is a transitional month between the cool dry and the wet periods. The rainfall is high in this region, and Cherapunji, which receives the highest rainfall in the world, is located here. Climatological data for selected stations are set out in Table 1.

The difference in mean maximum temperature varies from 7°C to 12°C in the various stations, showing that temperature fluctuations are very low, and the mean maximum temperature does not exceed 32°C. Annual rainfall ranges from 1000 to 4000 mm.

Soils are chiefly derived from metamorphic rocks, schists, quartzites and gneisses. The soils are acidic in nature, containing high percentages of organic matter and nitrogen, and are generally poor in phosphorus and calcium. Alluvial, red, lateritic mountain and hill soil types are the main soils found in north-eastern India (Raychaudhary, 1966).

Of the total land area of north-east India, forest comprises 54%, land not available for cultivation 21%, the net crop area being 16% and that under pasture and grazing 0.93%. However, 13.44% of the total land is occupied under shifting cultivation. The total population of livestock is 12.83 million. Out of this, cattle comprise 64.5%, goats 11.47%, buffaloes 6.68% and other, 17.35%. Cattle are used primarily for milk production and for farm work.

TABLE 1. Climatological data for selected stations in northeast India.

	Altitude	Temperature (0°C)		Rainfall	Wet	Dry
	(m)	Mean min.	Mean max	(mm)	month	month
Burnihat	100	12–24	25–32	1550	6	6
Imphal	785	4–24	22–29	1184	7	5
Shillong	1600	6–16	16–24	1800	7	5
Ukhrul	1800	3–15	16–28	1484	7	5

SAVANNA TYPES AND THEIR DISTRIBUTION

Bor (1940) classified the savannas of Assam into three habitat types:

(i) Tropical: In this habitat, grasses, generally found in forest or along forest margins, include *Pseudostachyium polymorphum*,* *Babusa pallida*, *Melocanna bambusoides*, *Lophatherum gracile* and *Centotheca leppecea*. *Panicum humidrom*, *Saccolepis interrupta*, *Hymenachne assamica*, *Elusine indica*, *Vassia aispiolata* and *Leersia hexandra* are the more common species found outside the forest on moist habitats. Typical savannas of the uplands consist almost entirely of *Imperata cylindrica*, but associated with this species are *Saccharum spontaneum*, *S. narenga*, *Apluda aristata*, *Cymbopogon pendulus*, *Arundinella* sp. and *Ophiurus megaphyllus*.

(ii) Subtropical habitats: In forest and along forest margins, species of *Dendrocalamus* and *Bambusa* abound, but the majority of bamboos belong to the genus *Arundinaria*. Herbaceous forest species includes *Oplismenus compositus*, *Cyrtococcum radicaus*, *Polygon monspelliensis* *Isachne albens*, *Ichnanthus vicinus* and *Panicum khasianum*. In moist habitats outside the forest, the most common species are *Alopecurus myosuroides*, several species of *Panicum*, *Eragrostis*, *Paspalidium*, *Echinochloa* and *Isachne*, *Deyeuxia elalio*, *Arundinella khasiana*, *Saccolespsis indica* and *S. myosuroides*, species of *Anthraxon*, *Dimeria fuscescens* and *Coelachne pulchella*.

In xerophytic habitats, *Imperata cylindrica* is the main species over large areas. Common associates are several species of *Themeda*, *Erianthus* and *Eulalia*, *Andropogon ascinoides*, *Sorghym nitidum*, *Saccharum neranga*, *Cymbopogon khasianus*, *Thyrsia* and *Zea*.

(iii) Temperate habitats: The common grasses are *Denthonia cachemyriana*, *Bromus asper*, *Calamagrostis emodis*, *Deyeuxia scabrescens*, *D. magarum*, *Capillipedium pteropechys*, *Tripogan filiformis* and several species of *Arundinaria*.

Certain tree species, such as *Careya arborea*, *Callicarpa arborea* and *Schima wallichii*, are scattered among the seven dominant grasses, namely *Saccharum narenga*, *Sorghum nitidum*, *Themeda triandra*, *Andropogon ascinodis*, *Erianthus longisetosus*, *Imperata*, *cylindrica* and *Ophiurum megaphyllus* in the Shillong plateau (Bor, 1942). Chief associates of the grasses in the community are *Atylosia elongata*, *Crotalaria humifusa*, *Desmodium sambuense*, *Eriosema chinensis*, *Flemingia latifolia*, *Indigofera astropurpurea*, *Lespedeza stenocarpa*, *Priotrapis cytrisoides*, *Alteris khasiana*, *Crinum amoenum* and *Hedychum spicatum*, while *Pheonix acaulis* is very common.

Rowntree (1954) reported two grassland and savanna types in Assam, namely:

(a) *Imperata–Saccharum–Themeda*, which is found on higher well-drained land throughout Assam, and which is dominated by *Apluda mutica*, *Arundinella bengalensis*, *Crotalaria striata*, *Eupatorium odoratum*, *Imperata cylindrica*, *Leea* sp., *Narenga porphyrocoma* and *Themeda arundinacea*.

(b) *Alphinia–Phragmites–Saccharum*, a community occurring on recent alluvium in the flood plains of rivers all over Assam. The dominant species are *Alpinia*, *Erianthus ravennae*, *Phragmites karka*, *Saccharum procerum*, *S. spontaneum*, *Albizzia procera* and *Bombax malabaricum*.

Champion & Seth (1968) have reported the following savannas and grasslands in Assam:

(i) Moist *sal* savannas: These are open *sal* forest with tall grasses in the east of the Brahmaputra valley with *sal* itself (*Shorea robusta*) and other tree species like *Lagestroemia parviflora*, *Wrightia tomentosa* and *Salmalia malabarica*. Among prominent grasses are *Themeda arundinacea*, *Imperata cylindrica*, *Cymbopogon nardus* and *Desmostachya bipinnata*.

(ii) Low alluvial savanna woodlands: These are most extensively found in the Brahmaputra valley and other parts of north-east India. The common tree species are *Salmalia malabarica* and *Albizzia procera* with *Themeda*, *Erianthus* and *Saccharum* as the characteristic genera of grasses.

(iii) Eastern wet alluvial grasslands: These treeless grasslands occur on the small alluvial sites which are being cut off by main rivers. They are flooded during the monsoon season but the stiff soil dries out completely during hot weather. This alternation appears to be too severe for tree growth and presumably grassland formation. In most cases, fire runs through the grass annually. *Pharagmites karka* and *Saccharum procerum* are generally the main grass species found in this grassland type where they form a very dense and high grass stratum.

(iv) *Syzygium* parkland or high savannas: These high savannas are distributed on heavy alluvium adjoining the *Sal* forests of the Brahmaputra valley. They tend to comprise very open stands of low branching trees about 3 m high, usually waterlogged during the rains; the ground cover dominants are *Imperata cylindrica*, *Saccharum spontaneum*, *Ophiuros* sp. and *Vetiveria* sp. These stands are burnt annually. The main shrubs and trees are *Glochidion assamicum*, *Leea edgievorthii*, *Syzygium ceresoideum* and *Emblica officinalis*.

* Nomenclatural authorities for plant species follow Hooker (1875–97).

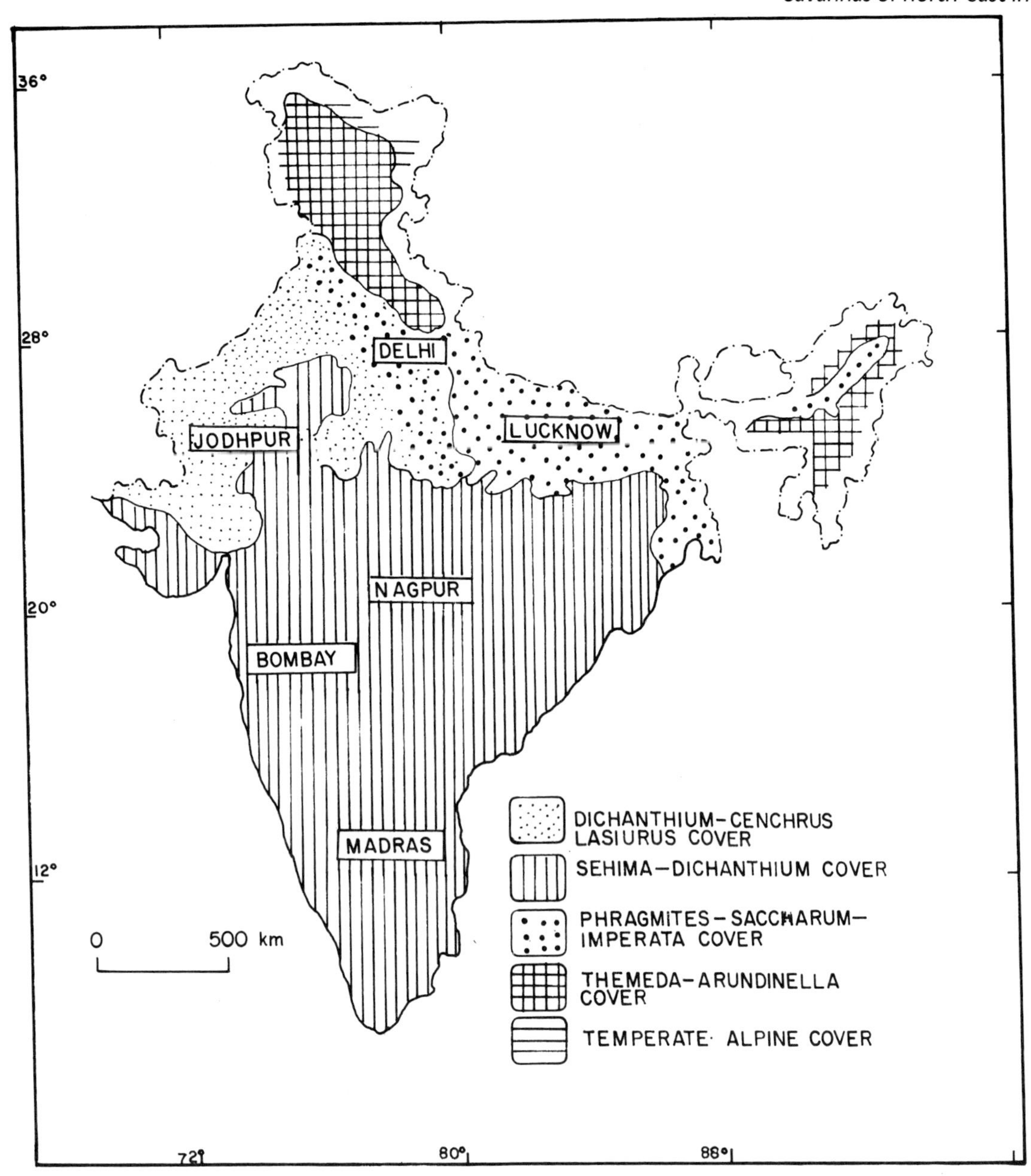

FIG. 1. The grass cover of India (after Dabadghao & Shankarnarayan, 1973).

This last type appears to be an intermediate stage between wetter and well-drained savannas in which tall trees such as *Salmalia* and *Albizzia* replace the low *Syzygium*.

(v) Assam sub-tropical hill savanna woodlands: Repeated shifting cultivation has altered the original forest and has led to the formation of open grasslands with scattered pines or broad-leaved forests.

(vi) Assam sub-tropical pine savanna: Pine occurs as scattered trees over grass, obviously as a result of biotic disturbances. *Agrostis* spp. and *Brachypodium sylvaticum* are the main grasses.

Out of the five broad savanna types recognized by Dabadghao & Shankarnaryan (1973) three savanna types, namely *Phragmites–Saccharum–Imperata*, *Themeda–Arundinella* and Temperate-Alpine, are found in the north-eastern region (Fig. 1).

The *Phragmites–Saccharum–Imperata* type covers the plains of the Ganga and Brahmaputra valley and extends into the plains of the Punjab covering all the states of the north-eastern region. The topography is level, low lying, ill-drained, with a high water table. Rainfall is up to 4000 mm per annum. The soil reaction varies from very acidic to mildly alkaline, with a pH range of 4.5–7.5. The cover consists of nineteen principal grass species, and fifty-six other herbaceous species, including sixteen legumes.

The common grasses are: *Bothriochloa intermedia*, *Chrysopogon aciculatus*, *Desmotachya bipinnata*, *Imperata cylindrica*, *Paspalum conjugatum*, *Phragmites karka*, *Saccharum arundinaceum*, *S. bengalensis*, *S. spontaneum*, *Sporobolus indicus* and *Vetiveria zizanoides*. Main shrubs are: *Clerodendron* sp., *Dendrocalamus hamiltonii*, *Lantana camara*, *Leea indica*, *Vitex negundo*, *Garcinia cowa* and

Miliusa sp. The main trees are: *Palaguium polyanthum, Diospyrus topoisa, Terminalia chebula, Tectona grandis, Eugenia* sp. and *Kayea assamica.*

The *Themeda–Arundinella* type covers the entire northern and north-eastern mountain area of the north-eastern states, including West Bengal, Uttar Pradesh, Punjab, Himachal Pradesh and Jammu, and Kashmir. It occurs in the altitudinal range between 350 and 2100 m above sea-level. The rainfall ranges from 1000 to 12,500 mm at Cherapunji in north-east India. There are sixteen principal grasses, including: *Arundinella bengalensis, A. nepalensis, Bothriochloa intermedia, Chrysopogon fulvus, Cymbopogon jwarancusa, Eragrostiella nardoides, Eragrostis nutans, Apuda mutica, Eulaliopsis binata, Heteropogon contortus, Ischaemum barbatum, Sporobolus indicus* and *Themeda anatheria.* The main shrubs are *Daphne cannabina, Cyclea* sp., *Crataeva nurvala, Rubus racemosus, Viburnum corieceum.* Key trees are: *Ficus nemoralis, Pinus khasya, Phoebe* sp., *Quercus* sp., *Schima wallichii, Castanopsis tribuloides* and *Alnus nepalensis.*

The Temperate Alpine type extends from the high hills of the northern mountain belt, comprising Jammu and Kashmir, Himachal Pradesh, Punjab, Uttar Pradesh, West Bengal, Assam, Meghalaya, Arunachal and Manipur. It occurs above 1500 m in north-east India and above 2100 m in north-west India. The occurrence of snow during winter is quite a common feature. There are fourteen principal grasses and sixty other herbaceous species, including six legumes. The common grasses are: *Agropyron canaliculan, Agrostis canina, A. filipes, A. munroana, A. myrianthus, Andropogon tristis, Calamagrostis epigejos, Chrysopogon gryllus, Danthonia jacquemontii, Phleum alpinum, Poa pratensis* and *Helictotrichon asperum.* The main shrubs include: *Rubus revens, Strobilanthes* sp. and *Berberis aristata.* The dominant trees are: *Rhododendron arboreum, Quercus* sp., *Betula utilis, Pyrus aucuparia Larix griffittiana, Picea spinulosa* and *Lyonia ovalifolia.*

The *Phragmites–Saccharum–Imperata* and the *Themeda–Arundinella* types are distinctly sub-tropical and the temperate alpine cover is distinct from these types (White, 1968). Recently Yadava & Kakati (1985) and Yadava & Singh (1988) reported *Imperata–Bothriochloa* and *Phragmites–Saccharum–Zizania* grassland communities in the plain and wetlands of Manipur respectively.

In the middle altitudes below 2000 m, the grasses are: *Anthoxa clarkei, Arundinella mutica, Catillipedium assimile, Eulalia palma, Holcus lanatus, Microstetgium cilliatum, Saccharum rufipilum, Sporobolus indicus* and *Themeda villosa.* Shrubs and trees are: *Viburnum corieceum, Luculis pineiana, Rubus racemosa, Quercus* spp., *Rhododendron arboreum* and *Alnus nepalensis.*

In the higher altitudes above 2000 m there is the *Mischanthus–Arundinella* community, with scattered trees of *Rhododendron griffithii* and *R. arboreum.* Besides *Mischanthus nepalensis* and *Arundinella mutica,* other associates are *A. triachata, A. nepalensis, Anemon revularis, Bambusa vulgaris, Cnicus involucratus* and *Lilium macklinae.*

TABLE 2. Percentage of perennial and annual species in certain dry and humid savanna sites.

Savannas	Perennial	Annual	Source
Dry savanna			
Pilani	31.0	69.0	Gill (1975)
Kurukshetra	37.5	62.5	Singh & Yadava (1974)
Humid savanna			
Imphal	55.0	45.0	Kakati (1985)

LIFE FORMS AND THE PHENOLOGY OF SAVANNA SPECIES

Life forms

The flora is generally dominated by annual species as compared to the perennial species in savannas experiencing strong seasonality in climate and a high intensity of grazing and burning. The number of perennial species was less in dry savannas as compared with humid savannas, and the situation is the reverse in the case of annual species (Table 2). Thus the number of perennial species increases with the increase in the rainfall at savanna sites.

Phenology

(a) *Grasses.* The sprouting and germination of seeds in the majority of the grass species starts during the pre-monsoon period, and continues with vigorous growth to complete their flowering and fruiting within the rainy season or in the post-monsoon period (Singh & Yadava, 1974; Yadava & Kakati, 1985).

Kakati (1985) has divided the species into five groups on the basis of their flowering stage in grassland and savanna at Imphal: (i) those which flower only in the moist summer season, e.g. *Imperata cylindrica*; (ii) flowering starts in summer and continues in the rainy season, e.g. *Paspalum orbiculare, Eragrostis atrovirens* and *Echinochloa colonum*; (iii) flowers only in the rainy season, e.g. *Desmodium bipinnata, Axonopus compressus, Sporobolus diander, S. indicus* and *Ischaemum rugosum*; (iv) flowering starts in the rainy season and continues up to the winter season, e.g. *Bothriochloa intermedia, Fimbristylis dichotoma* and *Mimosa pudica, Digitaria longiflora* and *Setaria glauca*; (v) those which flower only in the winter season, e.g. *Hemarthria compressa.*

(b) *Woody components.* The phenology of the woody species varies with the coexistence of evergreen and deciduous woody plants in the savannas. In the north-eastern region, trees drop their leaves in great quantity during the long dry period of December–March (cold and warm dry period) as compared with other parts of India, where leaf drop occurs in the warm and dry month of April (Boojh & Ramakrishnan, 1981; Shukla & Ramakrishnan, 1982; Singh & Singh, 1987; Yadava & Singh, 1988). Leaf-life for broad-leaf species is on average longer in north-eastern India (20 months) than in the central Himalayan trees (13 months). In north-eastern India, most of the forest tree

TABLE 3. Above ground biomass (g/m^2) of various species in different months.

Species	June	July	Aug.	Sept.	Oct.	Nov.	Dec.	Jan.	Feb.	March	April	May	June
Bothriochloa intermedia	123.0	213.2	328.9	343.2	183.0	190.4	192.4	124.8	145.4	218.2	234.0	246.5	283.8
Digitaria longiflora	4.5	0.3	1.7	1.2	1.7	0.2	—	—	0.9	1.5	1.0	—	—
Desmodium bipinnata	2.4	7.3	0.3	1.3	2.8	—	1.7	—	1.3	1.2	2.4	2.8	1.4
Eragrostis atrovirens	9.3	27.9	19.8	8.1	—	0.6	8.3	11.6	1.2	9.4	18.3	1.2	—
Fimbristylis dichromata	29.3	28.8	16.7	10.8	8.0	2.6	2.3	0.2	0.1	1.5	0.4	0.1	0.2
Hemarthria compressa	6.5	4.7	2.7	4.5	20.9	9.6	3.1	3.6	3.8	1.6	0.4	2.1	1.3
Imperata cylindrica	229.3	221.7	195.9	286.6	242.1	197.3	178.4	125.2	168.9	244.5	321.5	231.0	340.6
Kylligya triceps	2.0	1.0	1.3	0.3	1.3	0.5	0.1	0.8	0.03	2.5	1.3	1.1	2.7
Leersia hexandra	4.0	21.2	11.0	19.5	20.9	2.5	3.4	6.7	0.1	1.6	2.0	2.2	6.2
Paspalum arbiculare	0.5	9.1	2.4	30.3	19.0	11.4	5.3	19.4	3.5	3.6	3.9	7.1	14.7
Sporobolus indicus	—	—	3.0	—	1.8	—	—	—	3.6	14.8	12.5	6.4	1.7
Other grasses	—	—	2.2	—	9.6	—	—	—	—	0.7	1.3	2.8	2.3
Other forbs	—	0.5	0.6	0.1	0.9	0.6	3.3	—	—	28.1	52.0	39.9	26.9

species exhibit flowering during the warm and dry period with another small peak being observed in the autumn.

PLANT BIOMASS AND NET PRIMARY PRODUCTION

Work on plant biomass and net primary production have been reviewed by Yadava & Singh (1977), Singh & Joshi (1979) and Singh *et al.* (1985) for the tropical grasslands of India, but, sadly, little work has been done on plant biomass and net primary production in the savannas of north-eastern India (Yadava & Kakati, 1984, 1985). Therefore the data generated by Yadava & Kakati (1984, 1985), Yadava (1987) and Yadava & Singh (1988) have been used for discussion in this paper.

(a) Biomass

The majority of above-ground biomass is contributed by grasses, followed by sedges and forbs (Table 3). Out of twelve plant species studied, there were eight grasses, two sedges, and only one forb and shrub. The majority of the total live plant biomass was contributed by two dominant species, i.e. *Imperata cylindrica* and *Bothriochloa intermedia*. The percentage contribution of *Imperata cylindrica* varied from 36% (April) to 55.8% (June) whereas *B. intermedia* ranged from 35.7% (October) to 56% (August). The maximum percentage contribution in both the species was recorded during the flowering stage.

Above-ground biomass ranged from 292.3 g m^{-2} (January) to a maximum of 706.6 g m^{-2} in September. Standing shoots varied from 532.5 g m^{-2} (June) to 1158 g m^{-2} (January). Litter ranged from 167.0 g m^{-2} (November) to 231.8 g m^{-2} (June). Low amounts of litter during the rainy season indicate the disappearance of the old litter. The values of standing shoots are comparatively higher than of live shoots and litter in this savanna (Fig. 2).

Below-ground biomass exhibited two peaks in November and April, coinciding with the dry cool winter and early moist and warm summer. The maximum value of the below-ground biomass was recorded in April (1614 g m^{-2}), and the minimum in July (1000.6 g m^{-2}). Below-ground biomass values are comparatively higher than those of other Indian grass savannas (Yadava & Kakati, 1984). The amount of biomass in different components is of the order below-ground > standing shoots > live shoots > litter.

TABLE 4. Net above ground and below ground, and total net primary production in different seasons (g m^{-2}).

Season	Rainfall (mm)	ANP	BNP	TNP	Rate of production (g m^{-2} day^{-1})
Rainy	766	913.4	447.8	1361.1	11.2
Winter	153	36.4	412.2	448.6	3.0
Summer	404	619.2	36.2	655.5	7.1
Annual	1313	1569.0	896.2	2465.2	6.7

(b) Net primary production

There is clearly a seasonal influence on the growth pattern of major species, which is reflected by wide variation in the seasonal values of net primary production. The rate of above-ground net production was highest in the rainy season, and lowest in the winter season, while below-ground net production was also highest in the rainy season and lowest in the summer season (Table 4).

The maximum total net production was in August (656.7 g m^{-2}) and lowest in November (55.8 g m^{-2}). The rainy season contributed a maximum of 1361.1 g m^{-2} to the total net primary production of 2465.2 g m^{-2} with a minimum in winter (448.6 g m^{-2}).

System transfer functions. System transfer functions were calculated on the basis of annual production and its transfer in different components for one savanna from each of the dry and moist types and the data are presented in Table 5.

The accumulation of standing dead litter and its transfer to the litter compartment is faster in dry savanna than in moist savanna. There is complete disappearance of roots in dry savanna.

The major portion of the total net production is channelled underground in dry savanna whereas the reverse occurs in moist savanna. The net accumulation is, however, high in the case of moist savanna (41% of TNP) and very low in dry savanna (12% of TNP). The surplus

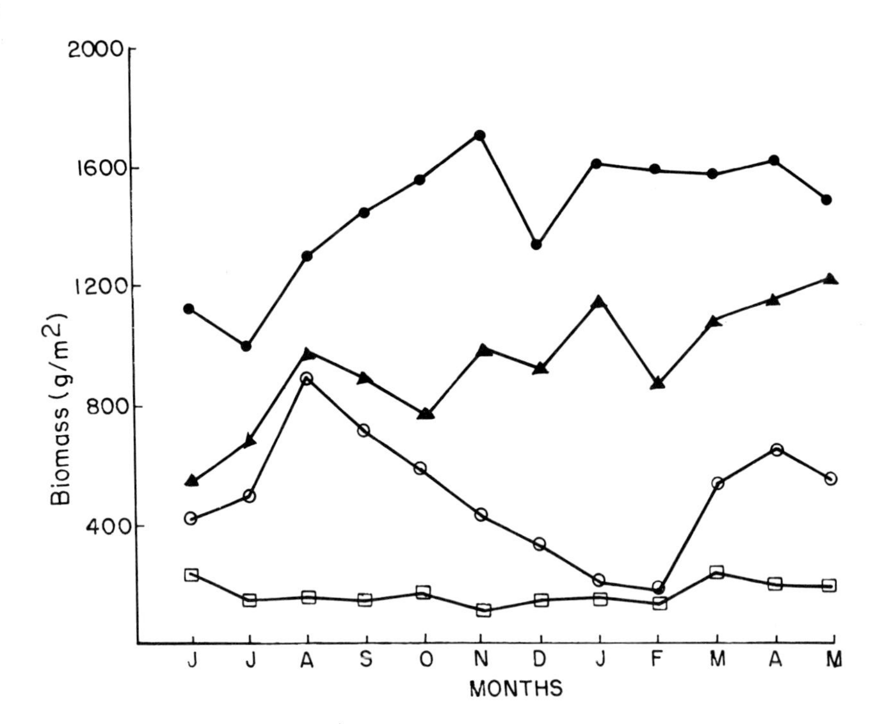

FIG. 2. Changes in the above-ground live biomass (○), standing dead (▲), litter (□) and below-ground biomass (●) in different months through the year.

TABLE 5. System transfer functions of dry matter for selected savannas.

Compartment	Pilani, dry savanna	Imphal, moist savanna
TNP–ANP	0.35	0.64
TNP–BNP	0.65	0.36
ANP–SD	1.00	0.81
SD–L	0.89	0.45
ANP–L	0.80	0.36
L–LD	0.81	1.19
BNP-RD	1.00	0.87
TNP-TD	0.88	0.59

TNP=Total net primary production; ANP=Above ground net primary production; BNP=Below ground net primary production; SD=Standing dead shoot production; L=Litter production; LD=Litter disappearance; RD=Root disappearance TD=Total disappearance.

amount of production could be utilized for grazing or burning in the savannas to maintain these savannas at a given level of succession, or it may lead to succession to woodlands.

Turnover rate of roots. The maximum amount of root biomass was replaced during the rainy season (31%), the lowest in the summer season (12%). The lower value in summer results from vigorous sprouting and upward translocation of reserve food material from underground parts. Annually 44% of roots were replaced.

NUTRIENT CYCLING

Nutrient cycling in climax woodland and savanna is being taken into consideration for comparison in our studies (Kakati, 1985; Yadava, 1987). The aerial perennial biomass of a 30-year-old stand of *Quercus* forest amounts to 1869.5 kg m^{-2} and includes 37.45 g of N, P, K, Na.

In *Quercus–Rhododendron* forest the amount of nutrients was 29.0 g m^{-2} for an aerial biomass of 2150.8 kg m^{-2}.

A comparison of the nutrient economy of three communities was carried out of the basis of the calculation of the amount included in a thousand kg of perennial biomass and the amount required (= amount taken up from soil) to build up this same weight of biomass (Table 6).

Quercus and *Quercus–Rhododendron* woodlands immobilize nearly the same amount of four cations; there are, however, some differences between these elements. The *Bothriochloa–Imperata* community immobilizes more nutrients in comparison to the woodland communities. There were significant differences between woodland communities and the grass savanna community in the uptake of nutrients, the highest value being recorded for *Quercus* woodland, the minimum in the grass savanna community.

The two woodland communities differed in the amount of nutrient uptake, yet required almost the same amount of nutrients for the production of biomass. But in the grass community, the amount of uptake is low, but requires 30% more nutrients to build up biomass in comparison with the woodland communities.

On calculation of immobilization in the perennial biomass as a percent of the uptake from soil, it showed that *Quercus* stands use more nitrogen and potassium, but low amounts of sodium. The phosphorus requirement, however, is almost the same. Grassland communities use high proportions of nitrogen and phosphorus.

The data for the standing quantities and cycling of nutrients in both forest and grassland savanna are given in Table 7. This shows that 40% of total nitrogen present in the

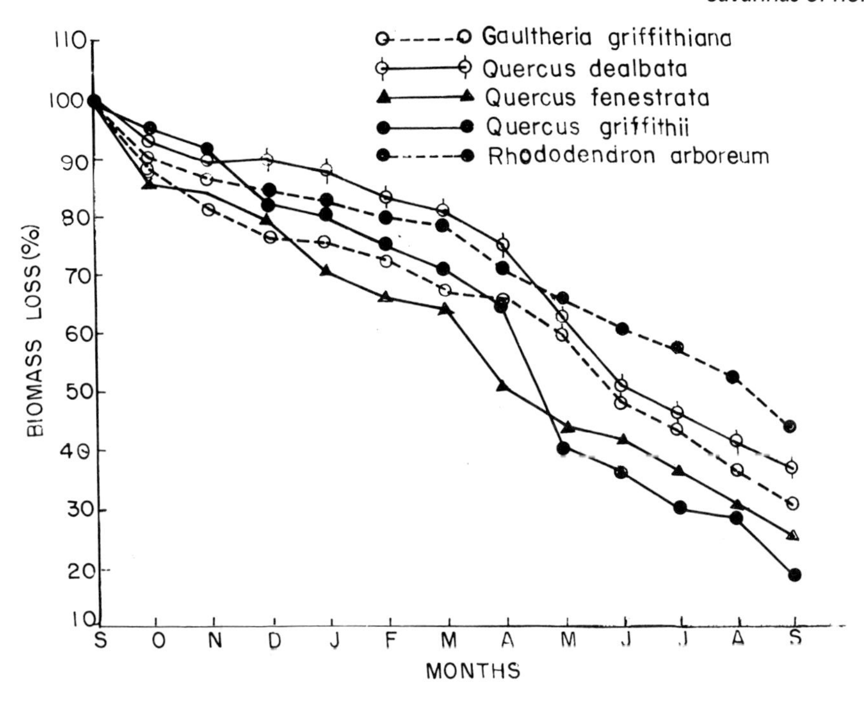

FIG. 3. Percentage of biomass remaining after different period of decomposition.

TABLE 6. Nutrients required (kg) to build a thousand kilograms of above-ground biomass in woodland and grass savannas.

	Fixation/immobilization					Absorption uptake				
Vegetation type	N	P	K	Na	Total	N	P	K	Na	Total
Woodland										
Quercus	7.5	0.6	2.2	0.2	10.5	15.3	1.0	3.8	0.6	20.70
Quercus–Rhododendron	8.8	0.3	2.0	0.2	11.3	13.1	0.4	2.7	0.2	14.4
Savanna										
Bothriochloa–Imperata	12.8	2.1	—	—	14.9	2.7	0.3	—	—	3.0

TABLE 7. Standing state and cycling of nutrients in sample forests and savannas in north-east India.

	Standing state (kg ha^{-1})			Cycling (kg ha^{-1} yr^{-1})		
Savanna type	Plant material	Soil	Total	Uptake	Release	Retention
Nitrogen						
Forest	2084.9	3168.0	5252.9	277.6	31.9	245.7
Savanna	447.7	6313.8	6761.5	446.8	303.7	143.1
Phosphorus						
Forest	66.0	86.4	152.4	8.6	1.5	7.1
Savanna	56.3	145.7	202.0	44.4	20.2	16.2

forest savanna resided in plant material, whereas in grass savanna it is only 6.63%. For phosphorus, 43% of total phosphorus in the forest savanna is retained in the plant system, with 28% in the grassland savanna. Thus the uptake of N and P is comparatively more in grass savanna that in forest savanna, and correspondingly release of nutrients is also greater in grass savanna than in forest savanna. The uptake of N and P is comparatively higher than the values reported by Singh, Singh & Yadava (1979) and Misra (1983).

Thus the conversion of forest land into savanna lands has a marked impact on the pattern of nutrient cycling and the nutrient budget. Therefore special attention is needed to study the long-term effects on these in developing sensible management practices for the savannas of the region.

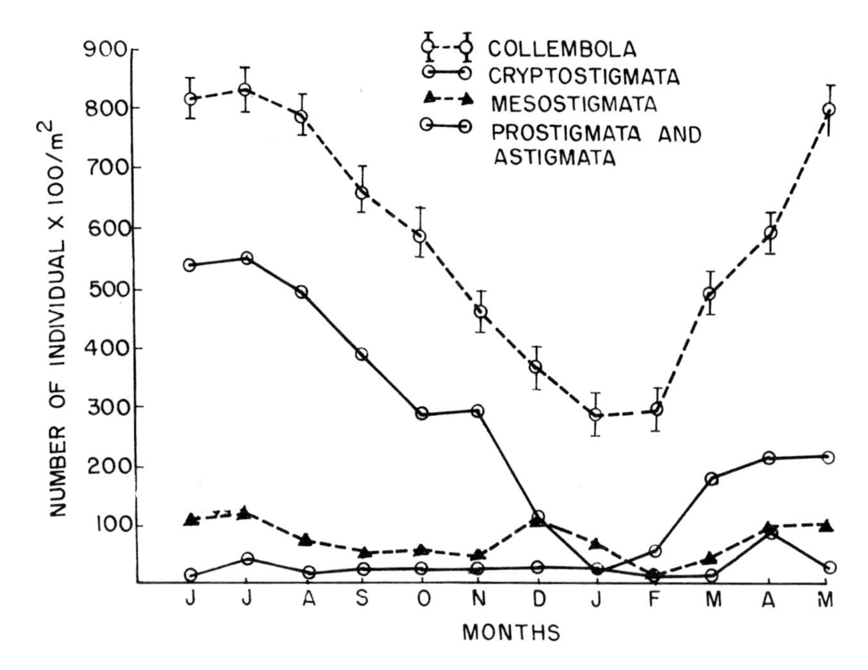

FIG. 4. Monthly variation in the population density of micro-arthropods in a forest stand.

TABLE 8. Protein value of grasses and fodder trees in the savannas of north-easst India (Kakati, 1985; Borthakur *et al.*, 1979).

Species	Crude protein (%)
Grasses	
Arundinella bengalensis	10.6
Axonopus compressus	14.5
Bothriochloa intermedia	7.8
Brachiaria sp.	7.0
Chrysopogon compressa	14.4
C. gryllus	7.0
Digitaria sp.	14.3
Hemarthria compressa	15.4
Leersia hexandra	11.9
Imperata cylindrica	10.0
Paspalum orbiculare	10.3
Sporobolus indicus	8.3
Thysanolaena maxima	10.21
Tree leaves	
Bauhinia malabarica	17.1
Bauhinia purpurea	24.0
Callicarpa arborea	20.6
Dalbergia sissoo	20.7
Macaranga thouars	15.1
Orbechia cumite	20.2
Veronia cinera	13.8
Vitex peduncularis	13.5

NUTRITIVE VALUE

Crude protein values of the important grass and tree fodder species growing in the north-eastern savannas are given in Table 8. These values are compartively higher than those reported for the semi-arid and dry sub-humid savannas (Yadava & Singh, 1977). The percentage of crude protein varies species to species, and also in different phenological stages, exhibiting a high value during the flowering and mature stages.

DECOMPOSITION

Recently Laishram & Yadava (1988) have studied the effects of initial lignin and nitrogen in the decomposition of the leaves of certain forest species, i.e. *Gaultheria griffithiana*, *Quercus griffithii*, *Quercus dealbata*, *Quercus fenestrata* and *Rhododendron arboreum*. The maximum weight loss was observed for *Q. griffithii* (82.49%), a broad-leaf deciduous species, the lowest for *R. arboreum* (41.15%), an evergreen species (Fig. 4). It was found that the rate of decomposition of the leaf litter was highly influenced by the initial lignin and nitrogen content of the litter.

SECONDARY PRODUCERS

The population density of soil arthropods was higher in the original forests than in the shifting cultivation (*Jhum*) sites in Meghalaya State. The population density of soil micro-arthropods was recorded to be at its maximum in May, coinciding with the high temperatures and pre-monsoon showers, whereas the minimum was in the January and February cool dry period. The soil fauna was reduced to half of that of the undistributed forest soil (Darlong & Alferd, 1982). In Manipur State, the maximum population of soil microarthropods was observed in July, the minimum in February (Fig. 3) (Thingbaijam, Yadava & Elangbam, 1986). Distributed sites slowly became colonized to the densities found in the adjacent forests. Collembolla and Acarina are the dominent orders in these forest and disturbed sites, and are influenced above all by soil moisture, temperature, and ground floor litter.

Laishram & Yadava (1988) have studied the seasonal changes in the soil respiration rates in the forests. The maximum value for soil respiration was obtained during the summer season (598 $mgCO_2m^{-2}h^{-1}$), the minimum in the winter season (424 $mgCO_2m^{-2}h^{-1}$). Thus, the moist summer period is more cogenial for the growth of micro-organisms in the soil. The value of soil respiration was lower in savanna (194 $mgCO_2m^{-2}h^{-1}$) than in the original forest (Kakati & Yadava, 1984).

VERTEBRATES

The north-eastern region is very rich in vertebrates, when compared with other parts of India. The large herbivores, such as rhinoceroses (1000), elephants (5000), wild buffaloes, brow antler deer, barking deer, sambhar and spotted deer are among the most prominent vertebrates in the region (Prater, 1968).

Bothriochloa–Imperata grass savanna may support 3.5 cattle ha^{-1} (Kakati, 1985). Biomass data on verebrates is unfortunately lacking, and it is not possible to speculate on their role in the short- and long-term functioning of the savannas in the region.

SUCCESSIONAL CHANGES UNDER BIOTIC DISTURBANCES

Successional patterns for various savanna cover types are suggested below (after Dabadghao & Shankarnarayan, 1973):

(i) Succession in the *Phragmites–Saccharum–Imperata* types is as follows:

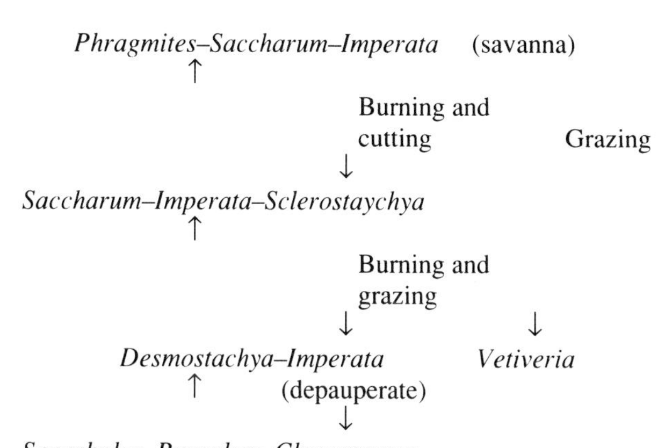

The *Phragmites–Saccharum–Imperata* type consists of tall, coarse, grasses, generally growing in swampy and wet places. On cutting and burning the grass is primarily used for thatching purposes. In disturbed conditions, the swampy areas invariably show the dominance of *Phragmites karka*, even when subjected to grazing to some extent. *Phragmites karka* is the first to go in drier habitats due to cutting and burning; it is replaced by *Saccharum*, *Imperata* and *Sclerostachya*. The introduction of grazing at this stage, along with burning, will favour the appearance of *Vetiveria zizanoides*. Continuous burning induces *Imperata cylindrica* to assume a depauperate form in humid situations. In north-eastern India, further grazing and burning induces yet further degredation to *Sporobolus*, with *S. indicus*, *Paspalum*, with *P. congugatum* and *P. orbiculare*, and *Chrysopogon*, with *C. aciculatus*. This indicates a change-over to mesophytes from hydrophytes.

Since this savanna cover favours tall, coarse and unpalatable species, and only on deterioration to a *Desmostachya* and *Imperata* (depauperate stage), these species can be utilized for grazing. Thus management practices should be designed to induce the change in habitat from hydrophytic to mesophytic. Burning, followed by heavy grazing, would induce the early palatable growth in these otherwise coarse species. Recently Deb-Roy (1986) has also recommended limited burning practice in the management of these grass species in the Manas Wild Life Sanctuary in Assam, which is the home of many endemic species, such as the pygmy hog, rhinocerus, and wild buffaloes, etc.

(ii) Succession in the *Themeda–Arundinella* type is as follows:

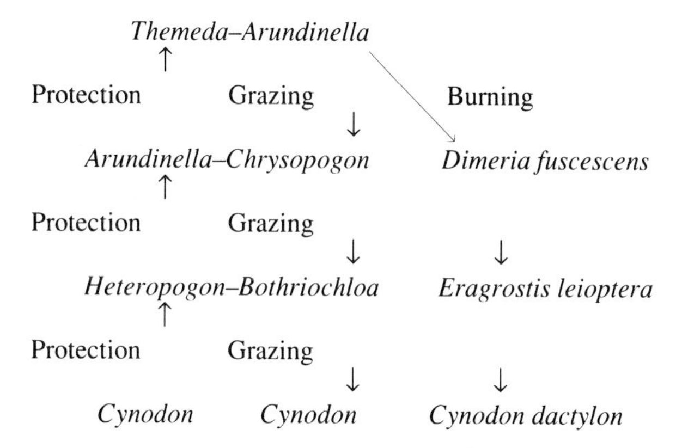

Themeda anathera, being the most desirable species, obviously deserves the highest consideration in the management of this type. Light grazing during the monsoon season and moderate grazing during the post-monsoon period does not appear to create any harmful effects.

Studies in the successional trends in the temperate alpine type of the north-eastern region is lacking, and further study is needed.

CONCLUSION

Most of the savannas of north-eastern India have been derived following the destruction of the forests through shifting cultivation and they then represent sub-climax communities induced by burning and grazing. Burning and grazing are the most important factors in both the moist and dry savannas. In general, the area experiences a high rainfall. The length of the growing season, along with species composition and the intensity of grazing and burning governs the net primary production. Mineralization is rapid in grass species, as compared to woody species. There are few studies on the functioning of the savanna ecosystems in north-eastern India. Therefore further reseach efforts are needed to fully understand the functioning of these savannas in relation to both stresses and disturbances.

REFERENCES

Boojh, R. & Ramakrishnan, P.S. (1981) Phenology of trees in a sub-tropical evergreen mountane in north-east India. *Geo. Eco. Top.* **5**, 189–209.

Bor, L.N. (1940) A list of the grasses of Assam. *Ind. For. Res. (Bot. N.S.)*, **1**, 47–102.

Bor, L.N. (1942) The relict vegetation of Shillong Plateau–Assam. *Ind. For. Res.* **3**, 152–195.

Borthakur, B.N., Prasad, R.N., Ghosh, S.P., Singh, A., Awasthi, R.P., Rai, R.N., Verma, A., Datta, H.H., Sachan, J.N. & Singh, M.D. (1979) *Proc. Agro-forestry Seminar, Indian Council of Agricultural Research, New Delhi*, pp. 109–131.

Champion, H.G. & Seth, S.K. (1968) *A revised survey of the forest types of India.* Govt. of India Publ., Delhi, 404pp.

Dabaghao, P.M. & Shankarnarayhan, K.A. (1973) *The grass cover of India.* ICAR, New Delhi, 713pp.

Darlong, V.T. & Alfered, J. (1982) Differences in arthropod population structure in soils of forests and Jhum sites of North-East India. *Pedobiologia*, **23**, 112–119.

Deb-Roy, S. (1986) Fire in wet grassland habitats of Assam. *Ind. For.* **112**, 191–197.

Gadgil, M. & Meher Homji, V.M. (1985) Land use and productive potential of Indian savanna. *Ecology and management of world savannas* (ed. by J. C. Tothill and J. J. Mott), pp. 107–113. Australian Academy of Science, Canberra.

Gill, G.S. (1975) Herbage dynamics and seasonality of primary productivity at Pilani, Rajasthan. Ph.D. thesis, Birla Institute of Technology and Science, Pilani, India.

Kakati, L.N. & Yadava, P.S. (1984) Soil respiration in the grassland eco-system at Imphal. *J. Curr. Bios.*

Hooker, J.D. (1875–97) *The flora of British India.* L. Breeve & Co., London.

Kakati, L.N. (1985) Structure and functioning of the grassland ecosystem at Imphal. Ph.D. thesis, Manipur University, Imphal, India.

Laishram, I.D. & Yadava, P.S. (1988) Lignin and nitrogen in the decomposition of leaf litter in a sub-tropical forest ecosystem at Shiroy hills in north-eastern India. *Plant and Soil*, **106**, 59–64.

Laishram, I.D. & Yadava, P.S. (1988) Soil and litter respiration in a mixed oak forest ecosystem at Shiroy hills, Manipur, in north-eastern India. *Plant and Soil.*

Misra, R. (1983) Indian savanna. *Tropical savanna* (ed. by F. Bourleere), pp. 151–166. Elsevier Scientific Publications, Amsterdam.

Prater, S.H. (1968) *The book of Indian animals.* Bombay Natural History Society.

Puri, G.S. (1960) *Indian Forest Ecology*, Vol. I. Oxford Book and Stationary Co., New Delhi.

Rowntree, J.B. (1954) An introduction to the vegetation of the Assam valley. *Ind. For. Res.* (N.S.), **9**, 1–87.

Raychaudhari, S.P. (1966) *Land and soil.* National Book Trust, New Delhi.

Shukla, R.P. & Ramakrishnan, P.S. (1982). Phenology of trees in a sub-tropical humid forest in north-eastern India. *Vegetatio*, **49**, 103–109.

Singh, E.J. (1988) Phytosociology, primary productivity and nutrient cycling in the sub-tropical forest ecosystem at Shiroy hills, Manipur. Ph.D. thesis, Manipur University, Imphal, India.

Singh, J.S., Hanxi Yang & Sajise, P.F. (1985) Structure and functional aspects of Indian and south east Asia savanna ecosystems. *Ecology and management of world's savanna* (ed. by J. C. Tothill and J. J. Mott), pp. 34–55. Australian Academy of Sciences, Canberra.

Singh, J.S. & Joshi, M.C. (1979) Primary production. *Grassland ecosystems of the world. Analysis of grassland and their uses* (ed. by R. T. Coupland). International Biological Programme, Vol. 18, pp. 197–218. Cambridge University Press.

Singh, J.S., Singh, K.P. & Yadava, P.S. (1979) Ecosystem synthesis. In *Grassland ecosystems of the world. Analysis of grasslands and their use.* (ed. by R. T. Coupland). International Biological Programme, Vol. 18, pp. 231–239. Cambridge University Press.

Singh, J.S. & Yadava, P.S. (1974) Seasonal variation in composition plant biomass and net primary production of a tropical grassland at Kurukshetra, India. *Ecol. Monogr.* **44**, 351–376.

Singh, J.S. & Singh, S.P. (1987) Forest vegetation of the Himalaya. *Bot. Rev*, **53**, 80–191.

Thingbaijam, B.S., Yadava, P.S. & Elangbam, J.S. (1986) Population density of soil arthropods in the sub-tropical forest ecosystem at Shiroy hill, Manipur. *Proc. Symp. Pest. & Env. Poll.* pp. 278–288.

White, R.O. (1968) *Grassland of the monsoon.* Faber and Faber, London.

Yadava, P.S. (1987) Ecological studies on forest ecosystem of Manipur. Manipur University, Imphal: Final Technical Report (DOE-L-14013/15182), p. 145.

Yadava, P.S. & Kakati, L.N. (1984) Biomass dynamics and net primary productivity in grassland community at Imphal. *Int. J. Biometeorol.* **28**, 123–145 (supplement).

Yadava, P.S. & Kakati, L.N. (1985) Seasonal variation in herbage accumulation, net primary productivity and system transfer function in an Indian grassland. *Ecology and management of world's savanna* (ed. by J. C. Tothill and J. J. Mott), pp. 273–276. Australian Academy of Science, Canberra.

Yadava, P.S. & Singh, J.S. (1977) *Grassland vegetation: Its structure, function, utilization, and management.* Today and Tomorrow's Printers and Publishers, New Delhi, India.

Yadava, P.S. & Singh, E.J. (1988) Some aspects of ecology of forest in Shiroy, Manipur in north-eastern India. *Int. J. Ecol. Environ. Sci.* **14**, 103–113.

5/Tropical grasslands of Sri Lanka and India

M. A. PEMADASA *Department of Botany, University of Ruhuna, Matara, Sri Lanka*

Abstract. A general survey is made of the comparative ecology, structure, dynamics and productivity of grasslands of Asia with emphasis on those of Sri Lanka and India. Nearly 20% of the land-surface of Asia and more than 10% of that of Sri Lanka and India are under grasslands, which include both lowland and montane communities. They are the product of successional progression on abandoned cultivated land or forest clearings due to human activities and/or natural forces such as fire, landslide, alternate flooding and wild animals; most of them occupy a tension-zone of biotic and anthropogenic forces working on potential forest lands.

In Sri Lanka, the majority of grasslands occur in the dry zone with an annual rainfall of less than 1675 mm and a dry period of 2–5 months during May–September. They are both natural and man-made. The natural grasslands are edaphic climaxes where the soil conditions are unfavourable for the development of woody vegetation. The man-made grasslands are the result of clearing of forest for human settlement and shifting cultivation, but also of deliberate fires.

The Sri Lankan grasslands are in a constant state of flux owing to the seasonality of climate, mainly the rainfall, human interference, and invasion by gramineous weeds, particularly *Pennisetum polystachyon*. The water-stress accentuated by prolonged droughts, nutrient deficiencies, mainly that of nitrogen and phosphorus, aggravated by erosion and biotic interference are the major forces which prevent the development of woody vegetation so that the succession culminates and is maintained at the grassland-stage.

Key words. Grasslands, ecological dynamics, productivity, biotic pressure, *Pennisetum polystachyon*, South Asia.

INTRODUCTION

Nearly 20% of the landscape of the earth is covered by grass-dominated vegetation, which includes communities with a prominent tree component (savannas) and those without trees (steppe). Savannas occur in tropical and subtropical climatic regions, while the other grasslands are more widespread.

Richards (1976) concluded that most tropical grasslands are the deflected successions following continued destruction of woody vegetation by burning and shifting cultivation. However, in Sri Lanka at least, some grasslands may be regarded as edaphic climaxes.

It is conservatively estimated that nearly 25% of the Asian landscape was under grasslands in the past; much of this is now being converted to agricultural land.

In tropical Asia, grasslands are ecologically indispensable for they are the most efficient and effective natural biotic force which can rehabilitate disturbed and degraded land, protect and conserve exposed soils, and convert a hostile physical environment into one which is more hospitable for the development of woody vegetation if left undisturbed.

In this paper a brief general survey is made of the structure, dynamics and productivity of the tropical grasslands of Asia, with major emphasis on those of Sri Lanka and India; other Asian grasslands are given only a mention.

GRASSLANDS OF SRI LANKA

The grasslands of Sri Lanka can be categorized as montane (the so-called patanas), savanna and lowland grasslands. The patanas cover about 650 km^2 of the south-central highlands. On the eastern parts of the central mountains, the patanas merge into savannas around 300–500 m altitude. The encircling lowlands support a wide variety of grass-dominated vegetation.

Patana grasslands

Patanas occupy edaphically, climatically, topographically and altitudinally diverse hillsides above 500 m elevation. They have been the subject of repeated studies (e.g. Pearson, 1899; de Rosayro, 1946; Holmes, 1951; Pemadasa, 1982).

The general opinion is that most patanas are the result of deflected succession following the destruction, for anthrophogenic requirements, of an early woody vegetation. In addition to these disclimaxes (Pearson, 1899; Holmes,

1951), there are others considered to be edaphic climaxes where the edaphic conditions have prevented the development of woody vegetation (de Rosayro, 1946).

The earlier classifications of patanas are only of historical interest (Pemadasa, 1982). The most reasoned characterization (Mueller-Dombois & Perera, 1971) has recently been confirmed by Pemadasa & Mueller-Dombois (1979, 1981) using the multivariate analyses of ordination and classification. The resulting five patana types being:

Humid-zone dry patana (mainly in the Western Basin)
Summer-dry-zone dry patanas (in the Uva Basin)
Intermediate patanas (on the eastern escarpment at 1500–2000 m)
Lower wet patana (at 2000–2330 m)
Upper wet patana (above 2330 m)

Environmental heterogeneity. The humid-zone dry patanas receive an annual rainfall of over 2100 mm from both the SW and NE monsoons with a single dry spell in February. In the summer-dry-zone dry patanas rainfall (1450–1750 mm) is primarily convectional-cyclonic (October–November) and only secondarily NE-monsoonal, with a short dry spell in February and a longer more severe one in June–September. Though similar in rainfall, the upper wet patana is up to 2°C cooler than the lower one. The wet and humid-zone dry patanas differ altitudinally and hence in temperature relations. The intermediate patanas are somewhat intermediate between the wet and dry patanas in altitude, temperature and rainfall (up to 2000 mm); also they lack the February dry spell.

Both the wet and dry patanas have red-yellow podzolic soils, but with and without a prominant A_1 horizon respectively. The humic acids in the blackish top soils are higher in the upper than the lower wet patanas. The dry patana soils are eroded, denuded, truncated with exposed rocks and poor in nutrients. The hard pan-like quartz-gravel subsoil layer is more marked in the summer-dry zone than the humid-zone dry patanas. The patana soils are generally poor in nutrients.

Floristic heterogeneity. The tall grass-covers of the humid-zone dry patana are dominated by the dense-standing bunch-grass *Cymbopogon nardus* (L.) Rendle at lower altitudes (500–700 m) and the tussock-forming palatable grass *Themeda tremula* (Nees & Stend.) Hack. above 900 m. The matrix-filling subsidiary species include the grasses *Dimeria gracilis* Nees & Stend, *Eulalia trispicata* (Schult.) Henrard and *Imperata cylindrica* C. E. Hubb, the herbs *Desmodium heterocarpum* (L.) DC., *D. triflorum* (L.) DC., *D. triquetrum* (L.) DC. and *Vernonia* spp., and semi-woody shrubs *Psuedium guajava* L. and *Wickstroemia indica* (L.) C. A. Mey.

The heavily-grazed closely-appressed short-turf of the summer-dry-zone patana is dominated by creeping perennial grasses, such as *Alloteropsis cimicina* (L.) Staff., *Bracharia distachya*, *Chrysopogon aciculatus* (Retz.) Trin., *Digitaria longifolia* (Retz) Beauv. and *Eragrostiella secunda* (Nees) NLB, and the sedge *Fimbristylis nigrobrunnea* Thw. Both *Cymbopogon nardus* (L.) Rendle and *Themeda tremula* occur locally in places exposed to less severe grazing.

The two wet patanas are dominated by *Arundinella villosa* Arn. *ex* Nees and *Chrysopogon zeylanicus* (Stend) Thw., *Ischaemum indicum* (Houth) Mert. and *Justicia procumbuns* L., are absent and *Dicanthium polyptycum* (Stend) A. Camus. and *Tripogon bromoides* Roem. & Schult., are more abundant in the upper wet patanas than in the lower one.

The cattle-grazed, stubble-tussock, short-turf (up to 30 cm) of the lower wet patana is co-dominated by *Arundinella villosa* and *Chrysopogon zeylanicus*, while in the upper wet patana the fast-growing ungrazed latter species forms more prominent tussocks than those of the former.

The intermediate patanas are co-dominated by *Arundinella viollosa*, *Chrysopogon zeylanicus*, *Cymbopogon nardus* and *Themeda tremula*.

Careya arborea Roxb. and *Knoxia platicarpa* Arn. occur sporadically in all the patanas.

Savannas

Most grasslands occurring north and east of the Uva Basin with an annual rainfall of 1450–2000 mm are regarded as savannas because of the prominence of the fire resistent trees *Careya arborea*, *Phyllanthus emblica* L., *Terminalia belerica* Gaertn.) Roxb. and *T. chebula* Roxb., which supports the view that they are the result of the fire-destruction of forests (de Rosayro, 1950).

The upland savannas (300–500 m altitude) are mainly of tall grasses (up to 1.5 m) like *Cymbopogon polyneuros* (Stend.) Staff., *Themeda tremula* and *T. triandra* Forsk., with occasional clumps of *Cymbopogon nardus*.

The lowland savannas, occurring below 300 m, in the eastern plains are locally called *talawa* and *damana*; they are primarily of *Imperata cylindrica*, with species of *Aristida* sp., *Panicum* sp. and *Cymbopogon* spp. occurring locally. The scattered trees include *Pterocarpus marsupeum* Roxb. and *Syzygium cumini* (L.) Skeels.

Szechowyez (1961) believes that the lowland savannas are maintained naturally as grasslands because of the frequent failures of the monsoons and prolonged droughts plus the prevalence of shallow soils with a low moisture storage capacity, which are unfavourable for the development of forest.

Lowland grasslands

The lowland (<300 m) surrounding the central highlands and covering nearly four-fifths of the island have a variety of grass-dominated communities differing floristically and physiognomically owing to the diversity of climate. They are the main grazing lands and are therefore more correctly regarded as pastures. On the basis of climatic differences they are characterized as wet, intermediate, dry and arid pastures.

Wet pastures. Occurring extensively, mostly in coconut cultivations of the western, south-western and southern lowlands, the wet pastures receive an annual rainfall of over 2300 mm; even the driest month February receives nearly 100 mm of rain. They are biotic climaxes maintained as grass-cover by management. The short turf (up to 20 mm) consists mainly of the carpet grasses *Axonopus compressus* (Sw.) Beauv., *A. affinis* Chase and love-grass *Chrysopogon aciculatus*, with *Desmodium heterophyllum* (L.) D., *D. triflorum*, *Asytacia gangetica* (L.) T. Anders and *A. variabilis* (Nees.) Trim. as common associates.

The continued anthropogenic interference by way of management helps the temporary establishment of short-lived composite weeds such as *Ageratum conyzoides* (L.), *Eleutheranthera ruderalis* (Sw.) Schultz-Bip. and *Vernonia cinerea* (L.) Lees. If left unmanaged, *Eupatorium odoratum* (L.) readily colonizes virtually destroying the grass-cover.

Intermediate pastures. Widespread in the north-western lowlands, which are mostly under coconut cultivation, the intermediate pastures are the result of forest-clearing for planting coconut. They remain grasslands because of management and cattle-grazing, and are, accordingly, biotic climaxes. The rainfall is around 2000 mm with a distinct dry spell in February. They share grasses occurring in both the wet and dry pastures, such as *Aristida setacea* Retz., *Axonopus compressus*, *Chrysopogon aciculatus*, *Cyrtococcum trigonum* (Retz.) A. Camus and *Digitaria* sp. Continued interference helps temporary establishment of composite weeds occurring in the wet pastures.

Dry pastures. The dry pastures cover nearly 7200 km^2 of the northern, north-central and north-eastern lowlands with a mean annual rainfall of around 1400 mm due to NE monsoons; the SW monsoon during June–September is dry.

A great variety of grass-dominated communities can be recognized in edaphically distinct habitats (Senaratna, 1956). Alluvial habitats near rivers, banks of tanks and forest margins support populations of *Imperata cylindrica* long-neglected land has the vigorously growing dense-tufted *Aristida setacea*, recently opened land with favourable moist soils carries a luxuriant turf of *Chloris inflata*, while *Cyrtococcum trigonum* grows on the floors of open forest. Subordinate grasses occurring in varying amounts in most of these pastures include *Cynodon dactylon* (L.) Pers., *Echinochloa colonum* (L.) Link., *Paspalum* sp. and *Zoysia matrella* (L.) Merr.

These pastures are believed to be disclimaxes resulting from the abandonment of cultivated land during ancient times and frequent human interference has prevented the establishment of forest. They are the major grazing land for cattle.

Contrasting sharply with these disclimaxes are edaphic climaxes called *villus* occupying perennially moist, humus-rich depressions, dominated by succulent creeping grasses like *Iseilema laxum* Hack., with *Paspalidium flavidum* (Retz.) A. Camus. and *Fimbrystilis* spp. The edaphic features prevent the development of forest.

Arid pastures. The extensive sandy and muddy flats of the north-western and south-eastern coasts in Hambantota, Yala, Puttalum and Mannar support a mosaic of short grass-cover dominated by *Cynodon dactylon*, *Eragrostis tenella*, *Zoysia matrella* or *Chloris barbata* Sw., with scattered trees of *Acacia eburnea* (L.f.) Willd., *A. planifrons* Wright and Arn. and *Salvadora persica* L. These habitats receive the lowest annual rainfall (<1000 mm) with prolonged droughts during May–August; mean temperature is around 27°C, but may reach 35°C. They are most likely to be climatically-determined climaxes resulting from hostile environments which are favourable for woody vegetation. They are the main-grazing lands of Sri Lanka.

Biotic pressures

The most predominant biotic pressures are those of man and domestic animals, mainly cattle, buffaloes and goats. Continued destruction of grass-cover for agriculture, burning, removing herbage for animal fodder and grazing has accentuated erosion, aggravating soil impoverishment and facilitating the establishment of weeds.

Most of the montane grasslands, savannas and some lowland grasslands are being continuously and increasingly opened for temporary human settlement and agriculture. Subsequent to their abandonment, fast growing weeds such as *Pennisetum polystachyon* Schum., *Eupatorium odoratum* and *Tithonia diversifolia* (Hemst) A. Goay colonize, so disturbing the natural regeneration of grasslands.

Amarasinghe & Pemadasa (1982) provided evidence of human interference by way of burning and cutting in assisting *Pennisetum polystachyon* to supress natural grasses, such as *Cymbopogon nardus* and *Themeda tremula* in humid-zone dry patana. In fact, some hillsides are now covered with almost pure populations of *Pennisetum polystachyon*.

The patanas on the central highlands are repeatedly being put under vegetable crops; the abandoned land is quickly colonized by *Eupatorium riparium* L.

The annual burning of upland savannas to facilitate the collection of fruits of *Terminalia belerica*, *T. chebula* and *Phyllanthus emblica*, which are used extensively in local medicines, helps maintain grass-cover and prevents formation of forest. More recently, however, the burning of savannas has resulted in rapid invasion by *Pennisetum polystachyon*.

Lowland grasslands, if left abandoned, are quickly invaded by *Eupatorium odoratum*. Its rapid and aggressive development into a closed canopy is extremely harmful to the persistence of grass-cover which is ultimately replaced by a useless monoculture of the weed.

Productivity

Of the factors governing the growth and productivity of Sri Lankan grasslands, rainfall, soil fertility and anthropogenic and other biotic pressures are of major importance. The humid-zone dry patanas grow more luxuriantly with a higher standing crop than do the summer-dry-zone dry patanas; this difference is mostly related to the differences in species composition. In general, the standing short-turf crop of the wet patanas is much less than the taller turf of dry patanas.

Amarasinghe & Pemadasa (1983) estimated the annual

productivity of humid-zone dry patana to be 68,000–111,000 kg ha^{-1}, which is much less than the values of around 3,810,000 kg ha^{-1} recorded by Ambasht, Maurya & Singh (1972) for some Indian grasslands.

The low productivity of montane and coastal grasslands has been shown to be due chiefly to the limitations of nitrogen and phosphorus (Pemadasa, 1981, 1983), for the addition of these two nutrients improve the performance of turves under glasshouse conditions. Appadurai (1969) reported improvement of biomass production in response to the addition of fertilizer to the established swards of *Bracharia brizantha* in hillside habitats. Amarasinghe & Pemadasa (1982) found that soil moisture had considerable influence on the growth of humid-zone dry patana grasses. Amarasinghe & Pemadasa (1983) also reported a reduction of standing crop as a result of repeated cutting and burning. Apparently, continued and frequent human interference is extremely harmful to the persistence of patanas.

Conservation or exploitation?

Nearly 12,000 km^2 of a total land area of 65,600 km^2 of the island is under grass-cover. With explosive expansion of demographic pressure, the tiny island is burdened with the opposing priorities of conservation of vegetation and exploitation of biotic resources. At the moment grasslands are being cleared for temporary agriculture (not even shifting cultivation) and settlement is so haphazard and unsystematic that their future prospects are extremely bleak. Carelessly planned development projects and general negligence and indifference of the laymen too are a growing threat to their existence. Accordingly, the most urgent compromise must be to protect the remaining patches of grassland while systematically exploiting the already cleared and disturbed areas for socioeconomic development – otherwise we will destroy both our ecology and economy at one and the same time.

INDIAN GRASSLANDS

In India, the majority of grasslands are regarded as savannas which constitute landscape systems of grass-cover with scattered islands of trees and shrubs occupying an ecologically intermediate position between forest and steppe. In spite of the prevalence of a climate which, except in the alpine zones of the Himalayas, favours tropical forest, about 72 million hectares of India are under grasslands (Misra, 1978). In northern India the annual rainfall decreases from 3000 mm in the east to less than 300 mm in the west, with the duration of drought increasing from 3 to 9 months (Misra, 1981).

Historically, the entire Indian subcontinent should have been under the cover of woody vegetation, but today only about 20% of the land area is regarded as forest land, which shows the extent of forest clearing. According to Misra (1981), more than 50% of the so-called forest land is, in fact, under savannas or more generally grasslands. These grasslands are believed to remain stable owing to the human and biotic pressures operating through cattle-grazing, fire, and the removal of forest products.

Luxuriantly-growing grasslands cover extensive areas of the Vindhyan Hills of the Chandraprabha Sanctuary, Varanasi. Of an annual rainfall of around 1050 mm, nearly 90% occurs during June–September with a mean air temperature of about 26°C (Srivastava, 1976). Thus, the climate is rather favourable for the development and persistence of grasslands.

According to Ambasht & Singh (1981), the intensified urbanization and agricultural development during the last two to three decades have interfered considerably with the natural woody vegetation and this has resulted in a successional progression towards the establishment of Vindhyan grasslands. Despite the apparent homogeneity of climatic and edaphic conditions, the grasslands differ markedly in their physiognomy, structure, floristic composition, species-richness, species-dominance, and standing-crop, largely because of differences in micro-climate, soil depth, degree of erosion, and biotic interference.

Physiognomy

The tall perennial grasses, such as *Dichanthium* spp. and *Bothriochloa pertusa* (L.) A. Camus., grow as short tussocks bearing a large number of tillers as a result of grazing by cattle, goats, sheep and other ungulates. In response to severe and continued grazing, these perennials may behave as annuals. The grasses dry up and may even become parched during the 2–5-month period of drought, but they recover to grow vigorously during the monsoons.

The intense grazing causes the normally large trees of *Anogeissus pendula* (Roxb. ex DC.) Wall. to grow as small bushes with radiating branches orginating from a short shoot. *Butae monosperma* (Cam.) Janb. and *Diospyros melanoxylon* Roxb. are characterized by stunted shoots with a fire-tolerant bark. The thorny species of *Acacia*, *Prosopis* and *Zizyphus* grow as low bushes because of browsing by camels. The growth of all these tree species is further restricted by the low rainfall and prolonged droughts leading to near arid conditions.

Floristic-composition and structure

The luxuriantly-growing dry grasslands are characterized by the abundance of such grasses as *Aristida adscensionis* L., *Bothriochloa pertusa*, *Chrysopogon fulvus* Trin., *Dicanthium annulatum* (L.) A. Camus, *Desmostachya* sp., *Heteropogon contortus* (L.) Beauv, *Digitaria bifasciculata* (Retz.) Pers., *Imperata cylindrica*, *Themeda triandra* and *Vetiveria zizanoides* (L.) Nash. The plants which can tolerate fire and heavy grazing include shrubs of *Adhatoda*, *Calotropis*, *Dodonea*, *Holarrhena*, thorny bushes of *Carrisia*, *Euphorbia*, *Mimosa*, *Randia* and *Zizyphus*, palms such as *Phoenix* and bamboos. Among the common tree species *Anogeissus latifolia* (Roxb. ex DC.) Wall., *Diospyros melanoxylon* (Roxb.), *Lannea* sp. and *Pterocarpus marsupium* are the most prominent.

The intensity of grazing is known to influence the species-dominance of grass-covers (Ambasht & Misra, 1981). The protected grasslands are dominated by *Desmostachya* and *Heteropogon contortus*, forming a closed community with bushy shrubs of *Zizyphus jujuba* (L.) Mid., while those exposed to moderate grazing are co-dominated by *Bothriochloa pertusa* and *Aristida adscensionis*. The grass-covers exposed to heavy grazing are rather open with *Desmostachya* sp. and *Digitaria* sp. occurring as the subsidiary species among the luxuriantly-growing leguminous weed *Cassia tora* (L.). Mesic grasslands are dominated by *Vetiverea zizanoides*, while *Aristida cyanantha* L. is abundant on undulating land.

Productivity

The estimated standing-crop yield of Indian forests ranges from 53,850 to 101,600 kg ha^{-1} (Misra, 1981). While that of grasslands is less than 101,600 kg ha^{-1}. In contrast, however, the annual primary productivity of both the forests and grasslands is around 14,200 kg ha^{-1}. Clearly, the grasslands are as productive as forests despite the fact that trees enjoy prolonged photosynthesis for about 9 months while the grass leaves are active only during 3–4 rainy months from July onwards. According to Misra & Singh (1978), the photosynthetic efficiency of grasslands is very high because most of the grasses are C_4 species. Moreover, the grasses have a larger green area because the ears are photosynthetically as efficient as leaves.

Ambasht *et al.* (1972) estimated the annual productivity of some grasslands in the dry zone of India to be around 38,000 kg ha^{-1}, a yield much higher than that reported by Misra (1981). According to Ambasht & Singh (1981), the annual productivity of ungrazed Vindhyan grasslands is around 42,675 kg ha^{-1} compared with about 15,250 kg ha^{-1} for moderately grazed grass-covers. Obviously, the productivity differs greatly between grasslands. This should be expected, because of the differences in the environmental conditions. Grassland productivity is known to be controlled by annual rainfall and its monthly distribution, soil fertility, floristic composition, and grazing pressure (Murphy, 1975).

The Indian grasslands have been found to be much more productive than crops such as rice, bajra, barley, gram and mustard cultivated in lands that were once under grass-cover (Misra, 1981). Apparently, the maintenance, management and improvement of grasslands may be more profitable than cropping.

Biotic pressure

There is no other natural vegetation which experiences such acute biotic pressures as grassland. In fact, most grasslands originate and remain without developing into woody vegetation because of the continued biotic interference. Of the biotic forces influencing the dynamics of Indian grasslands, humans, cattle, buffaloes, goats, sheep, camels, donkeys and ponies are of major importance. Obviously, almost every species must be predated by one predator or the other, so that these grasslands are under continued considerable biotic pressure.

According to Pandeya & Misra (1978), in 1971, India had a population of 260 million cattle grazing on 77 million hectares of grassland and 600 million people in a geographical area of 326 million hectares. Mistra (1981) reports that the arid zone of India has 10.23 million people and 16.44 million livestock and that this represents a very unfavourable ratio of land–man–livestock. Thus, the biotic pressure on Indian grasslands appears to be rather high. Yadav & Singh (1977) presumed that up to 70% of the annual herbage is removed by cattle. Wild animals are believed to account for about 10% of the herbage removal.

In India, the human density is higher in the humid than in the dry and arid regions, while the reverse is true of cattle density; thus, the influence of cattle is more acute in the grasslands of the dry and arid than in the wet zones.

Humans interfere directly and indirectly with the dynamics of grasslands. The herdsmen bring their cattle to feed on grass-cover in the Vindhyan Hills so that they are subject to severe grazing during the active growing season. In addition, the villagers collect young leaves of *Diosphyros melanoxylon* to be used as wrappers for country cigarettes, called beedi. The grasses and timber are also used for building huts.

Clearly, the grasslands of India are under considerable biotic pressure caused by both man and his beasts, which seems to play a decisive role in their natural maintenance, so preventing the development of woody vegetation.

OTHER ASIAN GRASSLANDS

Ecologically invaluable grass-dominant communities are found in, for example, Bangladesh, Borneo, Burma, Java, Pakistan, Sumatra and southern China.

In Burma there are savannas occurring in clay habitats receiving up to 1000 mm of rain annually. The *Andropogon* predominant communities are characterizied by scattered trees of *Tectonia, Terminalia* and *Acacia*.

In the Indo-Malayan region, the re-establishment of woody vegetation is prevented by repeated burning, intensive grazing and/or intermittent cultivation, which enable the persistence of grasslands as biotic climaxes. The open grasslands, principally of *Imperata cylindrica*, are known locally as *alang alang, cogon, kunai* or *lalang*. In the Malay penninsula, the *Imperata* grasslands are converted to *Neprolepsis*-dominated vegetation, if left undisturbed. Early this century, nearly 40% of the landscape of the Philippines was under grass-cover (Brown, 1919, cited by Richards, 1976), dominated by *Imperata* and/or *Saccharum spontaneum* L., which appeared subsequent to the destruction of dipterocarp forest. Much of the grassland has subsequently been cropped and cultivated. Grassland and cropland revert to forest if not interfered with by biotic forces, especially fires. The successive fires favour grasses, while, in between fires, trees tend to spread. According to Whitmore (1975), because of the dry season, the semi-evergreen and secondary forests are more easily destroyed by fires and replaced by grasslands.

In Sumatra, the frequency of fires is believed to affect the vegetational dynamics. Fires at intervals of 5–20 years help maintain pine forests, while more frequent fires favour the establishment of grasslands.

Both in Bangladesh and Pakistan, there are extensive grasslands dominated by *Chrysopogon aucheri* Retz. and

Cymbopogon jwaranusa (L.) Rendle., with annual grasses such as *Sporobolus marginatus* (Retz.) Beauv., *Cenchrus biflorus* L., *C. pennisetiformis* L. and *Aristida adscensionis*. Especially in Pakistan, most disturbed hill-sides are sheltered by dwarf grasslands.

In Borneo there are extensive grass-mats floating in lakes; these are believed to be fire climaxes resulting from the destruction of fresh-water swamp forests.

Accordingly, it is reasonable to conclude that the tropical grasslands of Asia are maintained as they are because of continued biotic pressures, deficiencies in major nutrients, the seasonality of the rainfall, and suboptimal moisture conditions, which prevent the development of woody vegetation.

REFERENCES

Amarasinghe, A. & Pemadasa, M.A. (1982) The ecology of a montane grassland in Sri Lanka. II. The pattern of four major species. *J. Ecol.* **70**, 17–23.

Amarasinghe, A. & Pemadasa, M.A. (1983) The ecology of a montane grassland in Sri Lanka. VII. Biomass production. *Ceylon J. Sci.* **16**, 15–21.

Ambasht, R.S., Maurya, A.N. & Singh, U.N. (1972) Primary production in certain protected grasslands of Varanasi, India. *Tropical ecology with an emphasis on organic production* (ed. by P. M. Golley and F. B. Golley), pp. 43–50. Athens.

Ambasht, R.S. & Misra, K.N. (1981) Conservation studies of a hilly grassland. *Tropical ecology and development* (ed. by J. I. Furtado), pp. 133–139. Kuala Lumpur.

Ambasht, R.S. & Singh, A.K. (1981) Productivity studies of grasslands. *Tropical ecology and development* (ed. by J. I. Furtado), pp. 155–159. Kuala Lumpur.

Appadurai, R.R. (1969) *Grassland farming in Ceylon*. Gunasena, Colombo, Sri Lanka.

De Rosayro, R.A. (1946) The montane grasslands (patanas) of Ceylon. 2,3,4. *Tropical Agriculturist*, **102**, 4–16, 81–94, 139–148.

De Rosayro, R.A. (1950) Ecological conceptions and vegetational types with special reference to Ceylon. *Tropical Agriculturist*, **151**, 108–121.

Holmes, J. (1951) The grass, fern and savannah lands of Ceylon, their nature and ecological significance. Imperial Forestry Institute Paper No. 28.

Misra, R. (1978) Productivity and structure of Indian savannas. *Proceedings of Institute of Ecology, Jerusalam* (Abstracts).

Misra, R. (1981) Forest savanna transition. *Tropical ecology and development* (ed. by J. I. Furtado), pp. 141–154. Kuala Lumpur.

Misra, G. & Singh, K.P. (1978) Some aspects of physiological ecology of C_3 and C_4 grasses. *Climpse of ecology* (ed. by J. S. Singh and B. Gopal), pp. 201–206. Jaipur.

Mueller-Dombois, D. & Perera, M. (1971) Ecological differentiation and soil fungal distributions in the montane grasslands of Ceylon. *Ceylon J. Sci.* **9**, 1–41.

Murphy, P.G. (1975) Net primary productivity in tropical terrestrial ecosystems. *Primary productivity of the biosphere* (ed. by H. Leith and R. H. Wittaker), pp. 217–231. Springer, New York.

Pandeya, S.C. & Mistra, R. (1978) Structure and function of grazing land and (savanna) ecosystems of India. *Proceedings of Institute of Biology, Jerusalam* (Abstracts).

Pearson, H.H.W. (1899) The botany of the Ceylon patanas. *J. Linn. Soc. Lond.* **85**, 430–463.

Pemadasa, M.A. (1981) The mineral nutrition of the vegetation of a montane grassland in Sri Lanka. *J. Ecol.* **69**, 125–134.

Pemadasa, M.A. (1982) Effects of added nutrients on the vegetation of two coastal grasslands in the dry-zone of Sri Lanka. *J. Ecol.* **71**, 725–734.

Pemadasa, M.A. (1983) Grasslands. *Ecology and biogeography of Sri Lanka* (ed. by C. H. Fernando), pp. 99–131. The Netherlands.

Pemadasa, M.A. & Mueller-Dombois, D. (1979) An ordination study of montane grasslands of Sri Lanka. *J. Ecol.* **67**, 1009–1022.

Pemadasa, M.A. & Mueller-Dombois, D. (1981) An association-analysis of montane grasslands of Sri Lanka. *Austr. J. Ecol.* **6**, 111–121.

Richards. P.W. (1976) *The tropical rainforest*. Cambridge University Press.

Senaratne, S.D.J.E. (1956) Regional survey of the grasslands of Ceylon. *Proceedings of Kandy Symposium, UNESCO*, pp. 175–180. Paris.

Srivastava, M. (1976) Working plan of the Varanasi Forest Division, pp. 222.

Szechowyez, R.W. (1961) The savannah forest of the Gal Oya catchment, Ceylon. *Ceylon Forester*, **5**, 17–22.

Whitmore, T.C. (1975) *Tropical rain forests of the Far East*. Clarendon Press, Oxford.

Yadev, P.S. & Singh, J.S. (1977) Grassland vegetation. *Progress in ecology* (ed. by R. Misra, K. P. Singh and J. S. Singh), p. 182. New Delhi.

Section II
Ecological Determinants of Savannas: Abiotic and Biotic

There is an intrinsic heuristic value in understanding just what combinations of environmental factors produce, change and maintain particular savanna types. Further, such understanding is necessary for wise management of this important biome. General models, coupled with comparative data and insights gained in studies of individual savannas, will assist in studies and management of other savannas. Of the four main determinants of savannas – water, nutrients, herbivory, and fire – the tendency has been to concentrate on the factors most obvious or frequent in a particular savanna, making general models and generalizations difficult. Nevertheless, all authors of the papers in this section have extended studies, concepts and/or hypotheses about determinants of savannas, either by discussing interactions among factors and/or comparing to a wide range of other savanna types.

Medina and Silva deal with a wide range of savanna types (from 15% to 80% tree cover) in northern South America, and attempt to understand the coexistence of trees and grasses. Since nutrients are poor in the entire region, they attribute savanna structure mainly to plant responses to available water and fire, and to their interactions with plant phenology, architecture and life history. In contrasting sites, the southern African savannas, generally considered dry but nutrient-rich, Scholes deals explicitly with the origin and maintenance of nutrient-rich patches, and proffers that biotic feedbacks maintain them, thus affecting overall savanna structure and function. An interesting juxtaposition is the performance of a plant which evolved (most likely) in a dry, nutrient-rich savanna now growing in a relatively wet, nutrient-poor savanna; this is afforded in Bilbao and Medina's study of nitrogen-use efficiency of an African grass, an aggressive invader in South American savanna pastures.

Holt & Coventry discuss factors which are important in decomposition processes of organic matter in the vast nutrient-deficient soils of Australia – soil microorganisms, termites and fire. That termites are important decomposers is echoed by Andersen & Lonsdale.

Andersen & Lonsdale point out that the major grazing animals in Australia are probably grasshoppers. Further, there does not seem to be any 'ecological compensation' for the lack of large native grazing ungulates in Australia, as the overall abundance and composition of herbaceous insects in Australian savannas appear similar to those of other savannas. Freeland points out that the introduced ungulates in northern Australia have carrying capacities greater than would be predicted, and greater than the species achieve in their native habitats, most likely due to release from natural predators and pathogens.

Although there is no dedicated chapter on fire, it has been included as a determinant in many papers. For example, Holt & Coventry discuss fire as an agent in nutrient cycling, and Medina & Silva consider fire a determinant of savanna structure through mediation of tree/grass interactions. In Section III dealing entirely with tree/grass ratios, papers by Blackmore *et al.*, Menaut *et al.*, and Adámoli *et al.* consider the role of fire in the balance of plant types. Braithwaite, McKeon *et al.* and Stott in Section I, discuss the role of fire in the origin and maintenance of savannas. Also in Section I, McKeon *et al.*, and in Section IV, Winter, Burrows *et al.* and Ludwig, all deal with the use of prescribed burns in pastoral management.

P.A.W.

6/Savannas of northern South America: a steady state regulated by water–fire interactions on a background of low nutrient availability

ERNESTO MEDINA and JUAN F. SILVA *Centro de Ecología, Instituto Venezolano de Investigaciones Científicas, Aptdo. 21827, Caracas 1020-A, and CIELAT, Universidad de los Andes, Mérida, Venezuela*

Abstract. Savannas in northern South America (the Orinoco Llanos) are found on a variety of highly leached substrates, from tertiary sediments to alluvial soils, with markedly seasonal rainfall, 800–2500 mm/year. Physiognomic types range from tree-less grasslands to woodland type communities. Dominant tree species are evergreen and sclerophyllous being favoured by low soil fertility and fire against deciduous, mesophyllous trees. Water appears not to be a limiting factor for established trees since leaf flushing and flowering take place in the dry season.

Tree/grass ratios increase with soil water availability during the dry season. Areas with high water table, or in which a large fraction of the previous rainfall is accessible to tree roots, have higher tree densities than savannas with soils of low water retention capacity and/or deep water tables. Tree recruitment is dependent on their capability to withstand grass-root competition during early growth stages, and to reach deeper soil layers to guarantee water availability during dry periods. Therefore tree seedling establishment appears associated with flush germination during sequences of humid years. Duration of the season with plant available moisture in any year determines the productivity of the herbaceous layer, particularly the perennial grasses. The establishment of perennial grass seedlings depends on the length of PAM and is closely related to the phenological pattern of the species.

The duration of PAM regulates specific and phenological diversity of the grass layer. An extension of the dry season may impair populations of early growers, whereas an early end of the rainy season affects negatively late species. These effects are both direct, and are mediated by species competitive interactions. Longer PAM periods determine higher probability for successful growth and reproduction of annual species.

Regular occurrence of fire determines low diversity of the tree layer and may affect its productivity if it occurs after leaf flushing and the initiation of flowering. Fire also appears to maintain vigour of the herbaceous layer, its exclusion leading to deleterious changes. Fire frequency is also associated with length of PAM. The interplay of dry, fire-prone years, and wet, fireless years probably determines short-term changes in the composition of the herbaceous layer. Grass productivity is enhanced by fire during the middle of the rainy season, when there is still some water left in the upper soil layers, and the rainy season begins before water reserves are exhausted.

Key words. Fertility, productivity, water relations, phenology, architecture, Llanos, South America.

INTRODUCTION

The Orinoco Savannas (Llanos) are located in a large geosyncline in northern South America, limited by the Guiana shield to the south, the Andean Cordillera to the west and the coastal Caribbean Cordillera to the north. This region from the Guaviare river in Columbia to the eastern coast of Venezuela comprises around 500,000 km^2 dominated by savannas (Beard, 1953; Blydenstein, 1967; Sarmiento, 1983a). Areas of savannas are also found in the Guayana region of Venezuela.

As typical savannas, the Llanos have a characteristic physiognomy combining an open tree layer and a continuous herbaceous layer. The latter is largely dominated by the graminoid growth form (grasses and sedges). Trees are of low stature, with rather narrow and tortuously branched stems. The relative density of trees varies widely from the tree-less savanna grassland to the savanna woodland with a 15–80% tree cover (Blydenstein, 1962; Hueck, 1971; Sarmiento, 1984) (Fig. 1). The herbaceous layer varies also considerably in density, in extreme cases with less than 50% cover, leaving large areas of bare ground. It is known that phenology, biomass allocation and production patterns of these savanna components are fundamentally different, strongly suggesting divergent strategies in the exploitation of environmental resources. Graminoid species, with an intensive root system, exploit the upper soil layers and strictly follow a growth cycle associated with seasonality of

FIG. 1. Savannas in the Llanos. (a) Dense tree-savanna on clay-sandy, acid soils. The trees are *Curatella americana* (centre) and *Bowdichia virgilioides*, the tallest tree being *c*. 5 m. The grass layer is dominated by *Thrasya petrosa*, *Trachypogon plumosus* and *Axonopus purpusii*. 50 km south of El Tinaco, Edo. Cojedes, on the road to El Baul. Beginning of dry season, November 1989. (b) Same site and time. In the centre, *Curatella americana* (barely visible senescing leaves), left *Byrsonima crassifolia* (new foliage), and *Bowdichia virgilioides* (new foliage). Note the white inflorescences of *Thrasya petrosa* in the grass layer. (c) Same location as (d), detail. Dwarf *Curatella americana* and bare soil patches between grass tufts (grass layer *c*. 60 cm high). (d) *Trachypogon*-savanna, dominated by *Trachypogon plumosus*, with scattered dwarf trees of *Curatella americana* and *Byrsonima crassifolia*. General view, in the eastern savanna of Venezuela, near Chaguaramas, *c*. 150 km south of Maturin. On sandy, deep soils, with very low levels of organic matter. January 1990.

rainfall, while trees, with extensive, less efficient root systems, are able to exploit both water and nutrients from deeper soil layers. As a result they show a growth cycle dissociated from the availability of rain water (Walter, 1973; Walker & Noy-Meir, 1982; Medina, 1982b; Sarmiento & Monasterio, 1983; Sarmiento, Goldstein & Meinzer, 1985).

Physiognomy, phenology, and patterns of resource utilization of savanna vegetation result from complex environmental–biological interactions. These interactions may be understood from the conceptual model put forward by RSSD program (Frost *et al*., 1986) which consider plant available moisture (PAM) and nutrients (AN) as basic determining factors. Performance of savannas within a given area of this plane (the PAM-AN plane) is modulated by the recurrence of fire and the frequency and intensity of herbivory.

Regional studies have emphasized the relationships between savanna physiognomy and the operating geomorphological and edaphic processes (FAO, 1965; Goosen, 1971; Cochrane & Sánchez, 1981; Sarmiento, Monasterio & Silva, 1971; Sarmiento, 1983a). In some large areas the homogeneity of geomorphological and pedogenetic processes result in a particular savanna physiognomy, such as in the tree-less savannas of the eastern plateaus of the Orinoco Llanos and in the alluvial plains of the Apure region. In other cases, an intricate pattern of land forms results in a mosaic of savanna physiognomies, such as in the piedmont savannas of western Venezuela (Sarmiento *et al.*, 1971; Silva, Monasterio & Sarmiento, 1971). Landscape characteristics influence development of savanna vegetation by affecting basic nutrient availability (geological origin and degree of nutrient leaching) and water availability through variations in drainage and soil water retention capacity. However, the interplay of these variables has been explicitly established only on a general basis. Savanna soils appear to regulate primary production potential because of their low natural fertility, but also their texture and depth frequently affect water availability and duration of the growing period. Soil water retention capacity may exacerbate drought stress during period of low rainfall (in sandy soils) or increase the extent and intensity of flooding in heavy soils during the peak of the rainfall season.

The main aspects of phenology, water relations and nutrient economy, in connection with productive processes in neotropical savannas have been reviewed recently (Cole, 1986; Medina, 1982a, b; 1987; Sarmiento, 1983a, 1984; Tothill, 1985). Therefore, in this paper we will concentrate on the problem of coexistence, and variations in densities, of trees and grasses. Our point of view is that the Orinoco savanna soils have a widespread low fertility status, therefore patterns of variation in ecosystem structure and productivity are more correlated with the duration of plant available moisture and the frequency and timing of fire. In addition, the phenological and architectural characteristics of savannas plants play a significant role in determining responses to environmental stresses.

ENVIRONMENTAL FRAMEWORK AND ECOSYSTEM PROCESSES

Water availability and savanna physiognomy

There are few studies on the geographic variation of annual rainfall in the savannas area of northern South America (Burgos, 1967; Walter & Medina, 1971; Sánchez & García, 1969). Total rainfall varies widely, and ranges from 800 to 2500 mm being concentrated in a wet season of 5–8 months duration (Monasterio, 1970; Susach, 1984). Mean annual rainfall is frequently correlated with the length of the rainy season and the number of rainy days, as in several other areas in the tropics (Medina, 1986).

The range of savanna types found within the Orinoco Llanos (seasonal, semi-seasonal, and hyperseasonal; Sarmiento & Monasterio, 1975) and their structural and floristic variations appear to be regulated by a combination of soil properties and rainfall patterns. As pointed out above, trees and herbs have different rooting patterns. Therefore, in what follows we will consider top soil characteristics to refer to the grass component responses, while deeper soil layers will be taken into account for the woody components.

Texture, structure, depth of profile, relative topography, and slope control drainage and water storage capacity of soils under a strongly seasonal rainfall distribution (Silva & Sarmiento, 1976). Variations in total annual rainfall have different effects depending on soil properties. Higher annual rainfall favours the growth of plants adapted to flooding but is deleterious to seasonal savanna species. In savannas with good water storage capacity and moderate to good drainage, increases in annual rainfall lengthens period when PAM is sufficient for growth. This effect is less pronounced in soils with lower water storage capacity and good to excessive drainage, where the length of PAM is shorter. The relationship between water availability and grass production may be demonstrated experimentally extending the PAM period. Natural *Trachypogon* grasslands irrigated during the dry season are able to maintain a green biomass similar to that of the rainy season, indicating that there is no intrinsic seasonal reduction in grass growth capability (San José & Medina, 1976).

Established trees can use underground water during the dry season; therefore, they are not affected by the length of the period when PAM in the upper soil layers is sufficient for growth. We assume that annual rainfall, depth of permeable substrate and drainage regulate tree density in these savannas. Deep permeable substrates result in a water table beyond the reach of tree roots, and therefore in a grassland savanna physiognomy. At the other extreme, a superficial impervious layer (f.i. lithoplinthic hardpans) does not allow tree root development, and again the result is a tree-less grassland. In high rainfall areas with poorly drained soils (seasonal floods), or in low rainfall areas with very permeable soils (longer dry seasons) we also find grasslands. In moderately drained soils increased rainfall leads to higher tree densities (woodlands).

Trees develop new leaves and flower between the middle and the end of the dry season (Foldats & Rutkis, 1975; Monasterio & Sarmiento, 1976). During this season new leaves show moderate photosynthetic rates and pronounced water losses through transpiration (Vareschi, 1960; Foldats & Rutkis, 1975; Sarmiento *et al.*, 1985). This behaviour indicates that roots of savanna trees have free access to water from moderately deep soil layers; however, stomatal conductance is reduced by the lower air water vapour partial pressure during the dry season. Reduction of transpiration due to lower stomatal conductance seems to be compensated by the increased leaf–air water pressure deficit during the dry season (Medina, 1982a; Sarmiento *et al.*, 1985). The development of large leaves under the high radiation, high temperature environment characteristic of the Orinoco savannas is also a clear indication that the energy balance of the leaves is guaranteed by sufficient water availability allowing evaporative leaf cooling. However, it seems that leaves produced during the dry season have smaller areas than those which grow after the onset of rains (Montes & Medina, 1977).

There is virtually no information on the physio-ecological characteristics of tree seedlings, in regards to their growth rates and resistance to water stress. However, tree seedling establishment appears to be heavily constrained by grass root competition and risks of drought and fire. Survival of tree seedlings generated during a given rainy season depend on the water availability in the top soil the domain of the herbaceous layer roots. Probability of seedling establishment depends on their capability to reach moist soil layers beyond the grass root zone, and on the build up of underground energy reserves which allow regrowth of aerial biomass after fire or drought. We expect this probability to be directly correlated with the length of the rainy season. Moderately drained soils with medium to high water tables and long rainy seasons result in higher recruitment and therefore higher tree densities. Towards shorter average PAM lengths, recruitment would depend on the occurrence of flush germination taking place during wetter years of lower mortality. During average or drier years, only well established trees would persist.

From these considerations it is concluded that the evaluation of the water availability in the Orinoco savannas has to take into account both climatic and edaphic parameters. Ideally soil water availability should be measured directly at different soil depths, to include the whole rooting zone. This type of measurement has been conducted by San José & Medina (1975) and Sarmiento & Vera (1977) for savannas in Central and Western Venezuela, respectively. These estimations require frequent periodic measurements of soil conductivity (or neutron radiation absorption), which have to be calibrated against gravimetric determinations of soil water content. This may be the reason why this type of data is so scanty in the savanna literature. Climatological approaches have been frequently utilized to measure PAM in several tropical regions with different degrees of success (Lal, 1987). Lack of agreement with vegetation performance derive from the fact that climatological measurements of PAM frequently do not take into account the soil properties in a given area. An approach to solve this problem would be to consider the limit between dry and humid seasons, calculated by any of the different methods available in the literature (Lal, 1987), as representative for soils of moderate drainage. Differences in texture can be taken into account allowing a given percentage of variation around the average climatological figure. We have done that with a particularly simple moisture index based on average rainfall and temperature data proposed by Bailey (1979). A number of savanna meteorological stations covering the range of rainfall in the Orinocos Llanos show variations of average periods when PAM is sufficient for growth from 150 to 229 days, but average extremes span from 128 to 251 days (Table 1). There is a reasonable degree of agreement of established PAM length with meteorological data and those estimated through direct and frequent measurements of soil water content, considering that the mean meteorological index has been calculated on the basis of average rainfall and temperature (Table 2).

TABLE 1. Average duration of the growing season (PAM length in days) in Orinoco savannas with contrasting annual rainfall. Calculated on the basis of the monthly Bailey index* allowing 20% variation in rainfall/evaporation ratio due to soil texture and water retention capacity.

Station	Annual rainfall	Beginning	End	Duration	Average extremes
Puerto Ayacucho	2144	27 April	11 Dec.	229	214–251
Caicara	1526	8 May	2 Nov.	178	167–193
Barinas	1400	22 April	3 Dec.	225	211–240
Calabozo	1239	13 May	19 Nov.	190	177–207
Valle de la Pascua	1025	25 May	22 Sept.	150	128–173

* Bailey index $si = 0.018P/1.045T$; the value for rainfall/evaporation balance is 6.37 on a year basis and 0.53 on a monthly basis.

TABLE 2. Comparison of PAM length based on direct soil water content measurements and on the Bailey moisture index calculated with long-term average rainfall and temperature data (Bailey, 1979).

Station	Estimated PAM duration (days)		
	Soil water content	Monthly Bailey index	
		Average	Extremes
Garza*	270		
Barinas*	239	225	211–240
Bononoito*	225		
Calabozo†	222	190	177–207

*Soil depth 100 cm (Sarmiento & Vera, 1977).
†Soil depth 70 cm (San José & Medina, 1975).

Nutrient availability and productivity of savannas in the Llanos

Tropical savanna soils in northern South America can be considered homogenously dystrophic because of their geologic origin and the leaching effect of relatively high rainfall (Blydenstein, 1967; Sarmiento, 1984; Cole, 1986). Therefore, it is expected that variations in actual organic matter production are probably less related to the fertility status of the soils than to other factors such as water availability and fire regime. Evaluation of soil fertility status has been a matter of debate because of disagreements on how to compare soils with different proportion of fertility factors such as organic matter, nitrogen, and phosphorous content, sum of bases, pH and aluminum mobility (Medina, 1988). Often fertility status of savanna soils has been evaluated on the basis of analyses of the upper soil layers, which are obviously of more interest for agricultural purposes, but may bear little relation with the development of natural vegetation. It seems that a useful index for the estimation of the fertility status of savanna soils is the sum of bases, which allow the separation of distrophic, mesotrophic and eutrophic soils (Huntley, 1982; Sarmiento, 1990). This criterion is acceptable provided that an agreement is reached on the methodology for its measurement, particularly the cation extraction procedures and the determination of the cation exchange capacity. On the other hand, care should be taken to sample soil profiles beyond the grass root zone, because upper soil layers are strongly affected by the process of organic matter production of the grass cover. This last statement may be visualized using the detailed soil analyses performed by Susach (1984) in his study of grassland productivity in Orinoco savannas in the southern

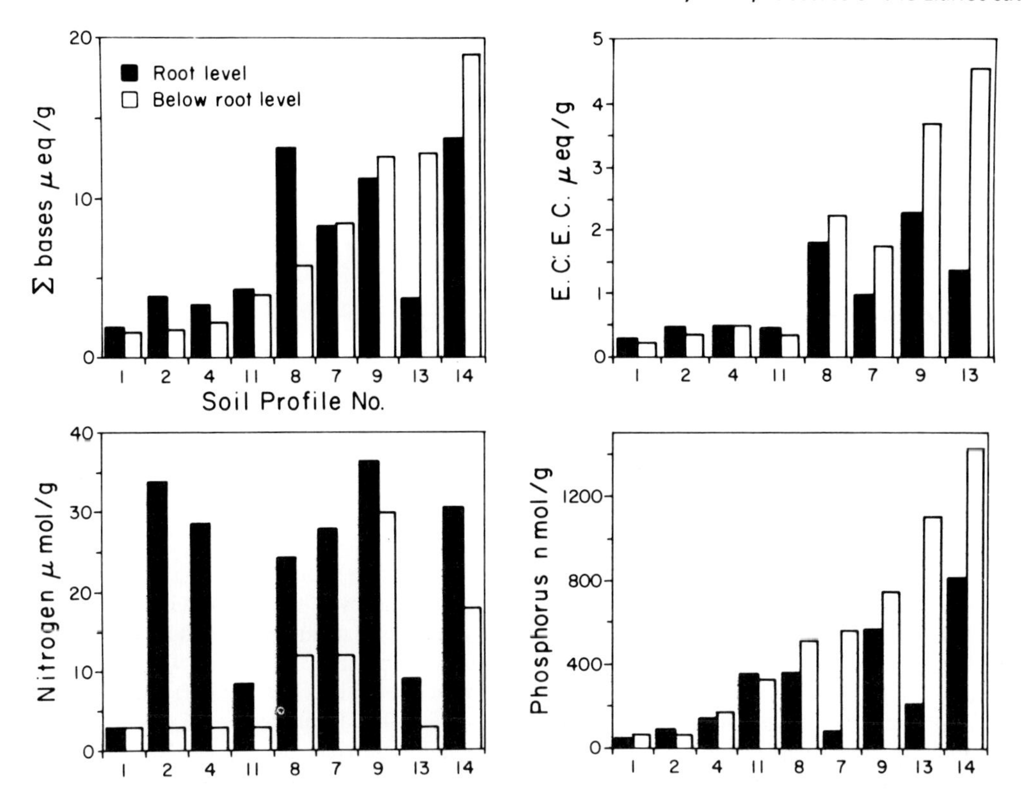

FIG. 2. Fertility factors in savanna soils. Analyses differentiate among soil properties at the root level and below the root level in each savanna (original data from Susach, 1984). Soil profiles are ordered according to increasing Σ bases in the soil layer below the root level. Soil profile number indicate soil pits with the following classification: Profile 1: Typic quartzipsamment. *Trachypogon plumosus* (H & B) Nees.+trees; root depth 45 cm; Profile 2: Undefined. *T. plumosus* (H & B) Nees., tree-less; root depth 30 cm; Profile 4: Aquic quartzipsamment. *T. plumosus* (H & B) Nees., tree-less; root depth 70 cm; Profile 7: Aeric tropaqualf. *T. plumosus* (H & B) Nees.+ trees; root depth 40 cm; Profile 8: Oxic paleustult. *T. vestitus* Anders.+ trees; root depth 18 cm; Profile 9: Typic plinthaqualf. *T. plumosus* (H & B) Nees.+ *T. vestitus*, tree-less; root depth 30 cm; Profile 11: Oxic rhodustalf. *T. plumosus* (H & B) Nees.+ trees; root depth 110 cm; Profile 13: Lithoplinthic ustorthent. *T. vestitus* Anders.+ *Byrsonima verbascifolia* (L.) H.B.K.; root depth 45 cm; Profile 14: Typic tropaqualf. *Axonopus anceps* (Mez.) Hitchc., tree-less; hyperseasonal; root depth 35 cm.

Guárico State in Venezuela (Fig. 2). It is clear that soil characteristics within the grass root layer are strongly affected by plant activity, thereby obscuring the true nutrient status of the soils considered. Notice that soil N content is always higher within the root zone than below, while these relationships are not so clear with extractable P. Availability of P is the other nutrient appearing to be limiting natural savanna productive potential. Interestingly, extractable P in this soil sample set is significantly correlated with the sum of bases, but the coefficient of determination is far stronger for the soil layer immediately below the root level (r^2=0.606 at the root level, and r^2=0.957 below the root level). These relationships should be further investigated to develop a robust soil fertility index for natural savannas.

In Brazilian savannas variations in tree densities have been associated with the nutritional capacity of the soils, particularly P deficiency and associated Al mobility (Goodland & Pollard, 1973; Lópes & Cox, 1977). Open grasslands (campos limpos) are supposed to grow on soils with higher Al/Ca ratios than those of dense woodlands (Cerradão) or the semi-deciduous forest found in the Brazilian Cerrado areas. Doubts on these soil–vegetation correlations have been raised because soil samplings were made within those soil layers directly influenced by the vegetation (Medina, 1982b; Montgomery & Askew, 1983). Nevertheless, high Al mobility in many South American soils in contrast to Australian and African soils (Sánchez & Isbell, 1978) may play an important role in selecting plant species which resist Al toxicity either through exclusion at the root level or detoxifying it after uptake (Medina, 1982b; Haridasan, 1987). Nutrient availability certainly regulates grassland aboveground biomass production, and overall P deficiency has been clearly demonstrated for both native and introduced pastures in the Orinoco Llanos (Medina, Mendoza & Montes, 1978; San José & García-Miragaya, 1981). Nitrogen availability seems to be limited both by the accumulated pool in the soil and its rate of mineralization, and its deficiency may be aggravated by the recurrence of fire (Medina, 1982a, 1987).

There are not enough productivity data of savannas in the Llanos to establish if variability of production values obtained in different sites can be accounted for by differences in soil nutrient availability. Rather, variations in aboveground biomass accumulation have been often associated with interannual variations in the length of PAM. In addition, lack of reliable data on underground biomass production prevents any definitive conclusion at

TABLE 3. Soil fertility (Σ bases below root level), duration of PAM and maximum aboveground biomass accumulation in regularly burnt seasonal and semi-seasonal savannas.

Savanna site	Σ bases (μeq/g)	Estimated PAM duration (days)	Aboveground production (g/m^{-2} year)	
Boconoito (ultic haplustalf)	1.5	211	534	(1)
Puerto Ayacucho (typic ustipsamment)	3.6	214	393	(3)
Calabozo		190	198–635	(4)
Barinas (oxic paleustalf)	6.3	228	590	(1)
Cabruta (1) (quartzipsamment)	6.5	167	178–288	(2)
Garza (oxic paleustalf)	10.7	225	604	(1)
Cabruta (2) (typic tropaqualf)	19.0	167	688	(2)

(1) Sarmiento & Vera, 1979; Sarmiento, 1984; (2) Susach, 1984; (3) Guinand & Sánchez, 1979, with soil data from Blancaneux, Hernández & Aranjo, 1977; San José & Medina, 1975, 1976; Medina *et al.*, 1978.

this time. Some examples obtained in different savannas in Venezuela (Table 3) indicate that dense swards are produced in relatively richer soils with short lengths of PAM (Cabruta 2) as well as in poorer soils with longer lengths of PAM (Barinas). In a given savanna a wide range of values of aboveground biomass production have been measured in different years (Calabozo). The sum of bases for the savannas included, however, show that all sites are extremely low in nutrient availability (Montgomery & Askew, 1983).

Fire effects

Regular occurrence of fire selects those tree species that resist burning of aboveground biomass. Areas protected against fire for more than 20 years show a significant increase in tree density, both of fire-resistant common savanna species, and fire-susceptible species from the surrounding semi-deciduous forest (San José & Fariñas, 1983). The number of tree species and stem density appears to be increasing quite rapidly in the last 6 years of protection (Fariñas & San José, 1987) (Table 4).

These data suggest that the predominance of evergreen trees, characteristic for neotropical savannas, is related to their fire resistance. Deciduous trees, with phenological cycles associated with the distribution of rainfall, appear to be more fire sensitive, and are therefore excluded from the set of woody species inhabiting regularly burned savannas. This observation is relevant because the predominance of sclerophyllous, evergreen trees has been considered a consequence of selection in nutrient poor environments (Montes & Medina, 1977).

TABLE 4. Number of stems/ha in a protected plot in the Biological Station at Calabozo, Edo. Guárico (Fariñas & San José, 1987).

Species	1962	1969	1977	1983
Savanna trees (evergreen)	92	174	270	1010
Forest species (evergreen)	1	3	11	32
Forest species (deciduous)	0	77	229	1319
	93	254	510	2361

Fire and water are not independent factors because probability of fire decreases with increasing lengths of the rainy season. These interactions may influence tree density as discussed above.

Fire is also an important factor of mortality in some grass species, specially those with short rhizomes close to the surface (Silva & Castro, 1988). However, the exclusion of fire leads to deleterious effects on the perennial grass layer. In the Biological Station at Calabozo, within areas protected from fire and grazing for more than 20 years, perennial grasses are dying back, leaving large patches of empty ground covered by dead plant material. Studies on the demography of perennial grasses in burned and protected plots showed that lack of fire result in higher seedling mortality, a reduction in size and vigour in adults, and a trend toward local extinction (Silva *et al.*, unpublished). The pattern of change in grassland composition after fire protection is a short-term increase in diversity (after 8 years, San José & Fariñas, 1983) and a long-term reduction in the relative dominance of grasses. In these *Trachypogon*-grasslands the relative density of *Trachypogon plumosus* (H&B) Nees. was reduced from 57% at the beginning of the protection to about 14% after 20 years, while that of *Axonopus canescens* (Nees. & Trin.) Pilger increased from 16% to 44% during the first 16 years of protection, receding to 33% 6 years later (Fariñas & San José, 1987).

The case study of savanna protection against fire in Calabozo shows that the long-term accumulation of dead biomass leads to the death of the whole grass layer. The duration of the die back process may differ among species. The interplay of fire prone years, and wet, fireless years probably determines short-term changes in the composition of the herbaceous layer. Also, the patchy nature of savanna fires may result in a mosaic of savanna composition associated to slight changes in species dominance.

Burning of grasslands during the dry season results in a stimulation of the production of aerial shoots and changes in the production pattern of the sward. Accumulation of dead standing biomass and a reduction in the total amount of green foliage as the dry season approaches has been shown to be the normal pattern of biomass development in several savannas in the Llanos (San José & Medina, 1975; Bulla, Mirando & Pacheco, 1980; Nazoa & López-Hernández, 1981; Susach, 1984). Several authors have shown that burning results in an increase of the aerial biomass production (Blydenstein, 1962; San José & Medina, 1975; Medina *et al.*, 1978; Guinand & Sánchez, 1979), but frequently the contrary has been observed (Blydenstein, 1962; Susach, 1984). It should be noticed that the effect of burning on aerial biomass development is different from the regrowth induced by cutting (Medina *et al.*, 1978). The reason for these contradictory results may reside in timing of burning, which is associated with the soil water availability (Medina, 1982b), and community composition. Burning too early in the dry season stimulates grass growth leading to a depletion of soil water reserves, and eventually death of the new shoots produced, thereby decreasing underground reserves for the next growth period. Burning too late in the dry season delays the process of new shoot development, resulting again in a reduction in aerial biomass production during the following rainy season. However, these responses may be modified by the phenology of the particular species involved. The pattern of root biomass production appears to be less affected by burning (Guinand & Sánchez, 1979; Susach, 1984), although in some cases positive responses to burning during the middle of the dry season have been observed (San José & Medina, 1975). There are many technical difficulties for an accurate assessment of belowground productivity in grasslands. However, the pattern of variation of total belowground biomass (including roots and rhizomes) points to a clear reduction of belowground biomass when aerial biomass is actively developing, followed by an accumulation towards the beginning of the dry season (Susach, 1984) (Fig. 3). Seasonal changes of total underground biomass are difficult to interpret, since they are the result of mortality and decomposition of old roots, translocation of nutrients and organic compounds and the development of new roots.

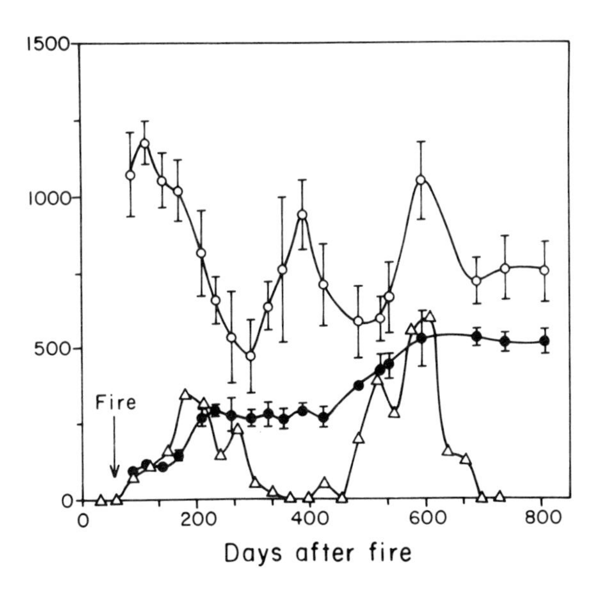

FIG. 3. Seasonal variations in above- and belowground biomass in savannas dominated by *Trachypogon plumosus* (H & B) Nees. Δ, Rainfall (mm); ○, belowground biomass (g/m²); ●, aboveground biomass (g/m²). (Data from Susach, 1984.)

POPULATION DYNAMICS AND STRUCTURAL FEATURES

Variations in physiognomy and plant cover are determined by physico-chemical factors. However, we know very little as how these factors operate upon savanna structure. To understand the processes and mechanisms involved we have to consider the role of some biological properties and interactions of savanna plants. Firstly, we will refer to phenology and its role in the effects induced by rainfall regime in savannas of different composition. Then, we will consider plant architecture, its relations to phenology and to interactions between species.

Phenology

Studies on the floristical and phenological diversity of Orinoco savannas have shown that despite similarities in seasonal growth, perennial grasses show important differences in growth form and phenology, both vegetative and reproductive (Sarmiento & Monasterio, 1983). Other differences are related to their demographic properties (Silva & Ataroff, 1985). Niche differentiation along these and other axes may be responsible for specific diversity. This biological and ecological diversity seems to mediate in the changes in plant cover and biomass induced by environmental factors. Sarmiento & Monasterio (1983) described the temporal division of the niche in the grass layer of a seasonal savanna on the basis of four phenological types: precocious species which flower at the transition between the dry and the wet seasons; early species, which flower during the first 2 months of the wet season; intermediate species, flowering during mid-season months; and late species, which flower during the last 2 months of the wet season. These phenological types correspond to reproductive and vegetative endogenous rhythms of aboveground biomass. The over imposed seasonality of water affects growth on a quantitative basis and regulates the ratio of living to dead biomass.

The dynamics of the proportion and productivity of the different components of the grass layer is modulated by the on- and offset of rains and the duration of the intervening period (PAM). Sarmiento (1983b) showed that the specific diversity is significantly higher in humid savannas (>1600 mm/year) than in the dry savannas (<1100 mm/year). Also, precocious, early, and late species showed higher relative frequencies in humid savannas, whereas intermediate species showed higher relative frequencies in dry savannas.

Perennial grasses with different phenologies show very different seasonal regrowth curves (Sarmiento & Monasterio, 1983; Silva, 1983). When precocious species grow under field conditions their growth is maximum during the first 2 months of the rainy season, when the other species are still very low. When they grow alone, without neighbour interference, the growth period is longer and production higher (Gallardo, 1983; Canales & Silva, 1987; Raventos & Silva, 1988). Silva (1987) discussed the possible effects of changes in the length of the rainy season on the population dynamics of grass populations differing in both their reproductive phenology and their reliance on

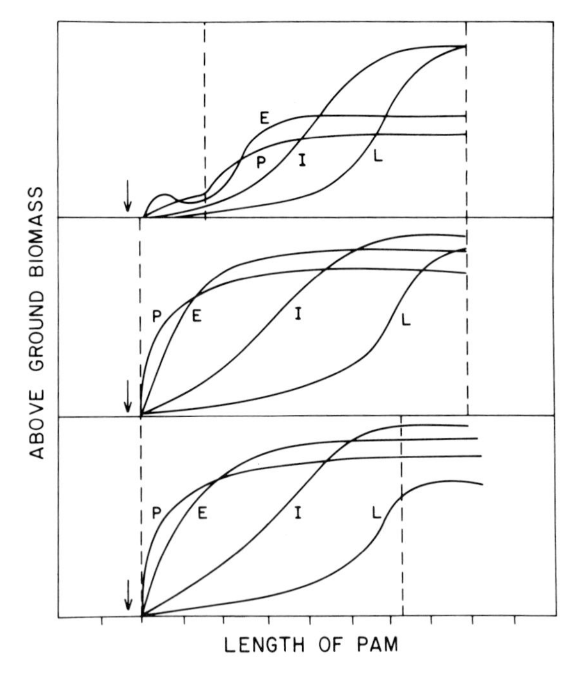

FIG. 4. Hypothetical effects of the changes in the length of PAM on the annual regrowth of aboveground biomass of different phenological types of perennial grasses (P=precocious; E=Early; I=Intermediate; L=Late). Arrows indicate the occurrence of fire. PAM length is indicated by the area between the broken lines. The middle set of curves represent a long rainy season (approx. 8 months). The upper set shows an extension of the dry season, while the lower set represents a reduction of the rainy season.

seed reproduction. A decrease in the length of PAM reduces the annual growth and reproduction of some phenological types without direct effects upon others. Beyond certain limits, this may not only reduce the population biomass but may result in local extinction. Other coexisting species may be benefited. In Fig. 4 we show a graphic hypothesis of the effects of changes of PAM upon the annual regrowth of the different phenological types of perennial grass species. A delay in the onset of rains, would seriously diminish the possibilities of precocious and early species to produce green foliage, since when the competitors are small water limits their growth potential and when water is available, competitors are growing fast. This affects not only their annual vegetative production but also their population recruitment since they flower very early in the wet season. By the same token, a shorter wet season (ending earlier) would affect late species since they would experience water stress during the peak of growth and reproduction. Intermediate species would not be affected by these changes and therefore they would take advantage of the decline of competitor populations (Sarmiento, 1983b).

Architecture

Plant architecture is another axis of differentiation of savanna species. It plays an important role in the interactions between plant and environment and between species, but few studies are available. Recent data (Silva, 1987, and unpublished) show that the savanna grass architecture is closely related to phenology as well as to competitive interactions between species.

A first level of relation between phenology and architecture is the pattern of underground biomass allocation. Since precocious and very early species flower and produce most of their foliage during the first 3 months of the rainy season (Fig. 4), they are expected to have underground reserves (both energy and nutrients) to sustain this rapid growth. That is, they depend on energy and nutrient accumulation from the previous growing season. The other species regrow more slowly, reaching their peaks of growth at different times later in the wet season. The longer the lag phase, the lesser would be the requirement for an underground reserve to sustain this growth because nutrients would be available from the soil and the cost of new foliage is covered by current photosynthesis. There seems to be a clear correlation between the ratio of below/aboveground biomass and flowering month as shown in Fig. 5.

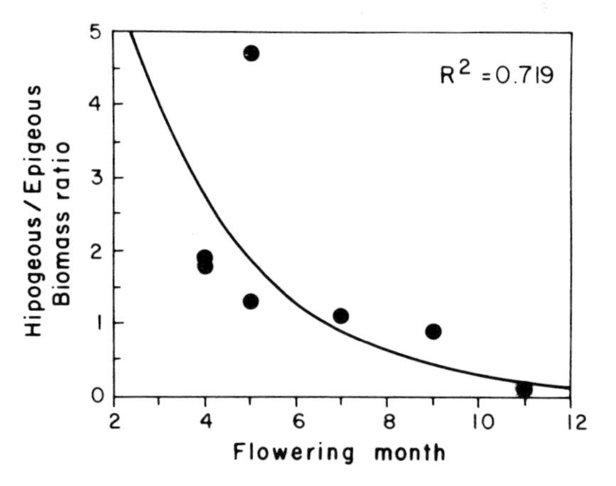

FIG. 5. Hipogeous/epigeous biomass ratios as a function of flowering time in seven perennial grass species. (Data from Sarmiento & Monasterio, 1983; Silva, unpublished.)

Aerial architecture of perennial grasses could be classified into two general types: basal and erect plants. In the basal types, the elongated culms bear small leaves and the foliage is produced from short, basal internodes. In the erect type, elongated culms bear long leaves, and as the culm grows upwards the mass of leaf surface is displaced upwards. Precocious and early grasses show a basal architecture. For the other species, the later they flower the more erect their habit and the higher their culms (Silva, 1987). Differences in final height between species are due to the number of internodes in the culms, which are predetermined several months earlier, when the apex differentiate into inflorescence. These differences result in various patterns of vertical distribution of foliage throughout the season. In Fig. 6 we show profiles of several perennial species by the end of the growing season, ordered from precocious to late species. Basal-precocious species occupy their aerial space very rapidly whereas this process takes more time in the erect species depending on their height.

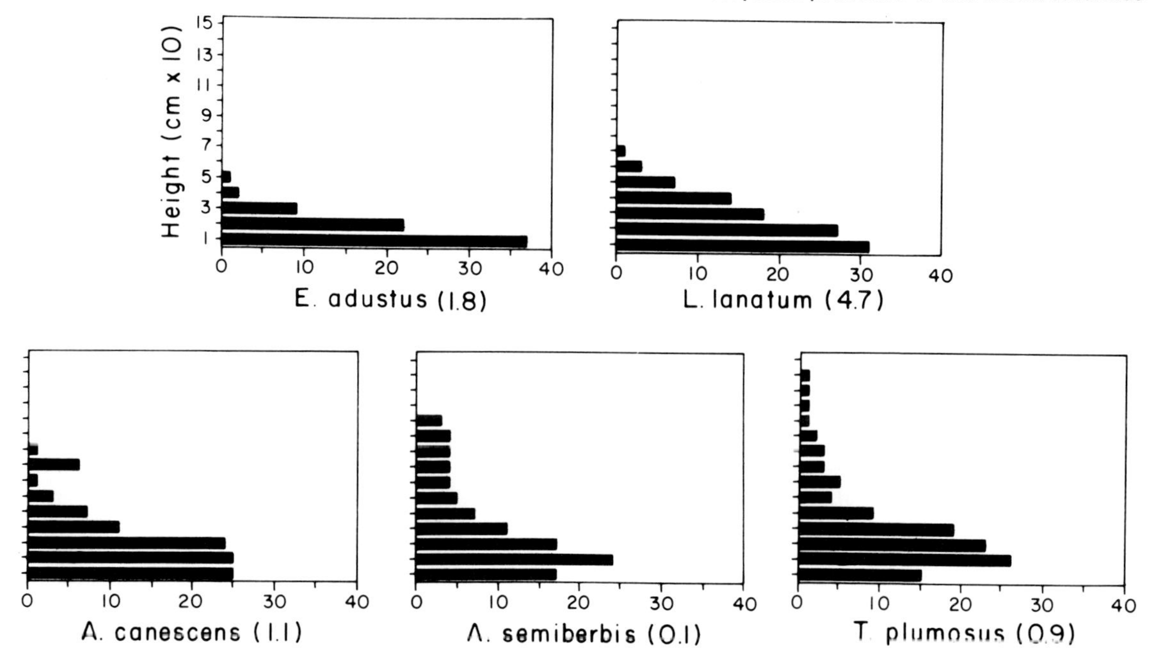

FIG. 6. Vertical distribution of aerial biomass as a percentage of total in five grass species with different flowering times: *Elyonurus adustus* (Trin.) Eckm., *Leptocoryphium lanatum* (H.B.K.) Nees., *Axonopus canescens* (Trin.) Pilger, *Andropogon semiberbis* (Nees.) Kunth, *Trachypogon plumosus* (H & B) Nees. Number in parentheses indicate hipogeous/epigeous biomass ratios. (Data from Raventos & Silva, 1988; Sarmiento, 1984.)

Relative height is important in the competitive interactions between species. In savanna grasses, taller species interfere in the growth of the lower ones more than vice versa, despite the fact that the former reach their peaks of growth later. Growing close to other species results in changes in the pattern of foliage distribution and a reduction of the season growth, in both precocious and late species. Basal species seem to have higher productive potential than taller (late growers) species but since interference from late growers is higher all types attain the same level of annual production of aboveground biomass when growing together under natural conditions (Raventos & Silva, 1988).

Annual grasses

Annual grasses constitute another component of the grass layer, being affected by changes in PAM. Common annual species belong to the genera *Andropogon, Aristida, Dictyomis, Eragrostis* and *Gymnopogon* (Ramia, 1974; Sarmiento & Monasterio, 1983). Annual grasses share several common architectural and ecological traits. They are erect, but not taller than a few decimetres, and largely restricted to the bare ground left between the clumps of perennial grasses, where they form high density patches. They germinate as soon as rains start and grow fast during the rainy season. In contrast to the behaviour of annual grasses described for African savannas, in the Llanos annual grasses are restricted to savannas with longer PAM. Studies on the population ecology and the physiology of these annual grasses are badly needed to understand the dynamics of perennial–annual strategies in neotropical savannas.

REFERENCES

Bailey, H.P. (1979) A simple moisture index based upon a primary law of evaporation. *Geograf. Ann.* **3–4**, 196–215.

Beard, J.S. (1953) The savanna vegetation of northern tropical America. *Ecol. Monogr.* **23**, 149–235.

Blancaneaux, P., Hernández, S. & Araujo, J. (1977) *Estudio edafológico preliminar-Sector Puerto Ayacucho, Territorio Federal Amazonas.* Serie Informes Científicos D.G.I.I.A./IC/01. Ministerio del Ambiente y de los Recursos Naturales Renovables. Caracas.

Blydenstein, J. (1962) La sabana de Trachypogon del alto Llano. *Bol. Soc. Ven. Cienc. Nat.* **23**, (102), 139–206.

Blydenstein, J. (1963) Cambios de la vegetación después de la protección a la quema. I. Aumento anual del material vegetal en varios sitios quemados y no quemados en la Estación Biológica. *Bol. Soc. Ven. Cienc. Nat.* **23**, (103), 233–238.

Blydenstein, J. (1967) Tropical savanna vegetation of the Llanos of Columbia. *Ecology*, **48**, 1–15.

Bulla, L., Miranda, R. & Pacheco, J. (1980) Producción, descomposición, flujo de materia orgánica y diversidad en una sabana de banco del Módulo Experimental de Mantecal (Edo. Apure, Venezuela). *Acta Cient. Venezolana*, **31**, 331–338.

Burgos, J.J. (1967) Regiones bioclimáticas para la ganadería de Venezuela. *Agronomía Tropical*, **15**, 139–167.

Canales, M.J. & Silva, J. (1987) Efecto de una quema sobre el crecimiento y demografía de vástagos en Sporobolus cubensis. *Acta Oecologica, Oecologia Generalis*, **8**, 301–401.

Cochrane, T.T. & Sánchez, L.F. (1981) Clima, paisajes y suelos de las sabanas tropicales de Sur América. *Interciencia*, **6**, 239–244.

Cole, M. (1986) *The savannas: biogeography and geobotany.* Academic Press, London.

FAO (1965) *Reconocimiento edafológico de los Llanos Orientales de Colombia*, Tomo II. FAO, Rome.

Fariñas, M. & San José, J.J. (1987) Efectos de la supresión del fuego y el pastoreo sobre la composición de una sabana de Trachypogon en los Llanos del Orinoco. *La capacidad bioproductiva de sabanas* (ed. by J. J. San José and R. Montes), pp. 513–545. Centro Internacional de Ecología Tropical, Unesco-IVIC, Caracas.

Foldats, E. & Rutkis, E. (1975) Ecological studies of chaparro (*Curatella americana* L.) and manteco (*Byrsonima crassifolia* HBK) in Venezuela. *J. Biogeogr.* **2**, 159–178.

Frost, P., Medina, E., Menaut, J.C., Solbrig, O., Swift, M. & Walker, B.H. (1986) *Responses of savannas to stress and disturbance*. International Union of Biological Sciences. Special Issue No. 10, Paris.

Gallardo, H. (1983) Patrones de crecimiento, alocación de recursos y energía en tres gramíneas de las sabanas estacionales: *Sporobolus cubensis, Trachypogon plumosus* e *Hyparrhenia rufa*. Magister Scientia thesis, Universidad los Andes.

Goodland, R. & Pollard, R. (1973) The Brazilian Cerrado vegetation: a fertility gradient. *Ecology*, **61**, 219–224.

Goosen, Ir. D. (1971) *Physiography and soils of the Llanos Orientales, Columbia*. International Institute for Aerial Survey and Earth Sciences (ITC), Enschede, The Netherlands.

Guinand, L. & Sánchez, P. (1979) *Productividad primaria, fenología y composición florística de un tipo de sabana situada en el Territorio Federal Amazonas*. Escuela de Biología, Fac. de Ciencias, UCV, Caracas.

Haridasan, M. (1987) Distribution and mineral nutrition of aluminium-accumulating species in different plant communities of the Cerrado region of Central Brazil. *La capacidad bioproductiva de sabanas* (ed. by J. J. San José and R. Montes), pp. 309–349. Centro Internacional de Ecología Tropical, Unesco-IVIC, Caracas.

Hueck, K. (1971) Verbreitung, Ökologie und wirtschaftliche Bedeutung der 'Chaparrales' in Venezuela. *Ber. Geobot. Inst. Rübel*, **32**, 192–203.

Huntley, B.J. (1982) Southern African Savannas. *Ecology of tropical savannas* (ed. by B. J. Huntley and B. H. Walker), pp. 101–119. Ecological Studies 42, Springer, Berlin.

Lal, R. (1987) *Tropical ecology and physical edaphology*. John Wiley & Sons. Chichester.

Lópes, A.S. & Cox, F.R. (1977) Cerrado vegetation in Brazil: an edaphic gradient. *Agron. J.* **69**, 828–831.

Medina, E. (1982a) Nitrogen balance in the Trachypogon grasslands of Central Venezuela. *Plant and Soil*, **67**, 305–314.

Medina, E. (1982b) Physiological ecology of neotropical savanna plants. *Ecology of tropical savannas* (ed. by B. J. Huntley and B. H. Walker), pp. 308–335. Ecological Studies 42, Springer, Berlin.

Medina, E. (1986) Forests, savannas and montane tropical environments. *Photosynthesis in contrasting environments* (ed. by N. R. Baker and S. P. Long), pp. 139–171. Elsevier Science Publ. B.V. (Biomedical Division).

Medina, E. (1987) Nutrients: requirements, conservation and cycles in the herbaceous layer. *Determinants of savannas* (ed. by B. H. Walker), pp. 39–65. IUBS Monographs Series No. 3, Chapter 3, IRL Press, Oxford.

Medina, E. (1988) The concept of nutrient availability. *Research procedure and experimental design for savanna ecology and management* (ed. by B. H. Walker and J.-C. Menaut), pp. 6–8. Responses of Savannas to Stress and Disturbance. Publication No. 1. IUBS and Unesco-MAB. CSIRO Printing Center, Melbourne, Australia.

Medina, E., Mendoza, A. & Montes, R. (1978) Nutrient balance and organic matter production in the Trachypogon savannas of Venezuela. *Tropical Agriculture (Trinidad)*, **55**, (3), 243–253.

Monasterio, M. (1970) Ecología de las Sabanas de América Tropical. II. Caracterización ecológica del clima en los llanos de Calabozo, Venezuela. *Rev. Geográfica* (ULA, Mérida), **9**, (21), 5–38.

Monasterio, M. & Sarmiento, G. (1976) Phenological strategies of plant species in the tropical savanna and the semi-deciduous forest of the Venezuelan Llanos. *J. Biogeogr.* **3**, 325–356.

Montes, R. & Medina, E. (1977) Seasonal changes in nutrient content of leaves of savanna trees with different ecological behaviour. *Geo-Eco-Trop.* **4**, 295–307.

Montgomery, R.F. & Askew, G.P. (1983) Soils of tropical savannas. *Tropical savannas* (ed. by F. Bourliere), pp. 63–78. Ecosystems of the World. Elsevier Publ. Co., Amsterdam.

Nazoa, S. & López-Hernández, D. (1981) Contenido nutricional en sabanas de *Trachypogon* sp. cercanas a Puerto Ayacucho, Venezuela. *Acta Biol. Venezuelica*, **11**, (1), 21–50.

Ramia, M. (1974) *Plantas de las Sabanas Llaneras*. Monte Avila, Caracas.

Raventos, J. & Silva, J. (1988) Architecture, seasonal growth, and interference in three grass species with different flowering phenologies in a tropical savanna. *Vegetatio* (in press).

San José, J.J. & Fariñas, M. (1983) Changes in tree density and species composition in a protected Trachypogon savanna in Venezuela. *Ecology*, **64**, 447–458.

San José, J.J. & García-Miragaya, J. (1981) Factores ecológicos operacionales en la producción de materia orgánica de las sabanas de *Trachypogon*. *Bol. Soc. Ven. Cienc. Nat.* **36**, (139), 347–374.

San José, J.J. & Medina, E. (1975) Effects of fire on organic matter production and water balance in a tropical savanna. *Tropical ecological systems* (ed. by F. B. Golley and E. Medina), pp. 151–164. Ecological Studies 11. Springer, Berlin.

San José, J.J. & Medina, E. (1976) Organic matter production in the *Trachypogon* savanna at Calabozo, Venezuela. *Tropical Ecology*, **17**, 113–124.

Sánchez, G. & García, J. (1969) Regiones mesoclimáticas en el Centro y Oriente de Venezuela. *Agronomía Tropical*, **18**, 429–440.

Sánchez, P.A. & Isbell, R.F. (1978) Comparación entre los suelos de los trópicos de América Latina y Australia. *Producción de pastos en suelos ácidos de los trópicos* (ed. by L. E. Tergas and P. A. Sánchez), pp. 29–58. CIAT, Colombia.

Sarmiento, G. (1983a) The savannas of Tropical America. *Tropical savannas* (ed. by F. Bourliere), pp. 245–288. Ecosystems of the World. Elsevier Publ. Co., Amsterdam.

Sarmiento, G. (1983b) Patterns of specific and phenological diversity in the grass community of the Venezuelan tropical savannas. *J. Biogeogr.* **10**, 373–391.

Sarmiento, G. (1984) *The ecology of neotropical savannas*. Harvard University Press, Cambridge, Mass.

Sarmiento, G. (1990) Ecología comparada de ecosistemas de sabanas de América del Sur. *Actas del ler. Simposio Regional del Programa RSSD de la IUBS* (ed. by G. Sarmiento). Mérida, Venezuela.

Sarmiento, G., Goldstein, G. & Meinzer, R. (1985) Adaptive strategies of woody species in neotropical savannas. *Biol. Rev.* **60**, 315–355.

Sarmiento, G. & Monasterio, M. (1975) A critical consideration of the environmental conditions associated with the occurrence of savanna ecosystems in Tropical America. *Tropical ecological systems* (ed. by F. B. Golley and E. Medina), pp. 223–250. Ecological Studies 11. Springer, Berlin.

Sarmiento, G. & Monasterio, M. (1983) Life forms and phenology. *Tropical savannas* (ed. by F. Bourliere), pp. 79–108. Ecosystems of the World. Elsevier Publ. Co., Amsterdam.

Sarmiento, G., Monasterio, M. & Silva, J. (1971) Reconocimiento ecológico de los Llanos Occidentales. I. Las unidades ecológicas regionales. *Acta Científica Venezolana*, **22**, 52–61.

Sarmiento, G. & Vera, M. (1977) La marcha anual del agua en elsuelo en sabanas y bosques tropicales en los Llanos de Venezuela. *Agronomia Tropical*, **27**, 629–649.

Sarmiento, G. & Vera, M. (1979) Composición, estructura, biomasa y producción primaria de diferentes sabanas en los Llanos occidentales de Venezuela. *Bol. Soc. Ven. Cienc. Nat.* **34**, (136), 5–41.

Silva, J. (1983) *Contrastes ecológicos entre gramíneas codominantes de una sabana tropical*. Trabajo de Ascenso, Facultad de Ciencias, ULA, Mérida.

Silva, J. (1987) Responses of savannas to stress and disturbance: species dynamics. *Determinants of tropical savannas* (ed. by B. H. Walker), pp. 141–156. IUBS Monographs Series No. 3. IRL Press, Oxford.

Silva, F. & Ataroff, M. (1985) Phenology, seed crop, and germination of coexisting grass species from a tropical savanna in Western Venezuela. *Acta Oecologica, Oecologia Plantarum*, **6**, 41–51.

Silva, J. & Castro, F. (1988) Fire, growth, and survivorship in *Andropogon semiberbis* (Mees.) Kunth. *J. Tropical Ecol.* (in press).

Silva, J., Monasterio, M. & Sarmiento, G. (1971) Reconocimiento ecológico de los Llanos Ocidentales II. El norte del Edo. Barinas. *Acta Cient. Venezolana*, **22**, 60–71.

Silva, J. & Sarmiento, G. (1976) Influencia de factores edáficos en la diferenciación de las sabanas. Análisis de componentes principales y su interpretación ecológica. *Acta Cient. Venezolana*, **27**, 141–147.

Susach, F. (1984) Caracterización ecológica de las sabanas de un sector de los Llanos Centrales bajos de Venezuela. Tesis Doctoral, Universidad Central de Venezuela, Facultad de Ciencias, Caracas.

Tothill, J.C. (1985) American savanna ecosystems. In: *Ecology and management of the world's savannas* (ed. by J. C. Tothill and J. J. Mott). Australian Academy of Science.

Vareschi, V. (1960) Observaciones sobre la transpiración de árboles llaneros durante la época de sequía. *Bol. Soc. Ven. Cienc. Nat.* **21**, 128–134.

Walker, B. & Noy-Meir, I. (1982) Aspects of stability and resilience of savanna ecosystems. *Ecology of tropical savannas* (ed. by B. J. Huntley and B. H. Walker), pp. 556–590. Ecological Studies 42. Springer, Berlin.

Walter, H. (1973) *Die Vegetation der Erde in öko-physiologischer Betrachtung*. Band I. *Die tropischen und subtropischen Zonen*. VEB Gustav Fischer Verlag, Jena.

Walker, H. & Medina, E. (1971) Caracterización climática de Venezuela sobre la base de climadiagramas de estaciones particulares. *Bol. Soc. Ven. Cienc. Nat.* **29**, 211–240.

7/The influence of soil fertility on the ecology of southern African dry savannas

R. J. SCHOLES *Resource Ecology Group, Department of Botany, University of the Witwatersrand, WITS 2050, Republic of South Africa*

Abstract. In dry savannas, soil fertility has a controlling influence on the slope of the relation between annual rainfall and annual aboveground herbaceous production. It also influences many other aspects of their structure and function; such as the species composition, morphology, forage chemistry and degree and type of herbivory. In dry savannas there is a wide range in soil fertility status, related at a regional scale to differences in parent material and the age of the land surface. At a landscape scale, catenal processes result in large fertility contrasts between ridgetop and bottomland positions. There are several other processes which lead to the creation and maintenance of nutrient-rich and nutrient-poor patches within the landscape, including patterns of herbivory, the activities of prehistoric man and nutrient cycling through trees and termite mounds. While the base-richness of the parent material is initially important in determining soil fertility, biological activities involving phosphorus, nitrogen and carbon appear to be important in the creation and maintenance of localized areas of enhanced soil fertility on base-poor substrates, and could provide a mechanism for the long-term stability of such nutrient hot-spots.

Key words. Soil fertility, nutrient status, dry savannas, southern Africa.

INTRODUCTION

This paper is concerned with the influence of soil fertility on the structure and function of southern African dry savannas. It will briefly review current knowledge on the subject, and will then examine some of the mechanisms which may be responsible for the observed patterns of nutrient distribution and usage.

It is conventional to divide African savannas into dry and wet forms, in the belief that significant ecological differences exist between the two types. The dividing criterion is necessarily arbitrary. A useful dividing limit is the annual rainfall at which the strong linear dependence of annual herbaceous production on annual rainfall in dry savannas begins to level off. This occurs between 700 and 900 mm MAP, with the lower limit occurring on sandy substrates and the higher on fine textured soils. This limit also broadly coincides with the transition from a grass layer dominated by malate-forming C4 variants (NADP-ME; principally the Andropogoneae) to malate formers (NAD-ME and PEP-CK; Chlorideae, Panicoideae) (Ellis, Vogel & Fuls, 1980; Johnson & Tothill, 1985). Dry savannas are therefore by this definition primarily water limited. The overwhelming obviousness of the rainfall dependence has tended to obscure the importance of edaphic factors in determining dry savanna structure and function.

Water supply of grasses growing in semi-arid regions can usually be treated as a series of discrete events, with abrupt beginnings and relatively well-defined ends (Fig. 1). This is because photosynthesis and transpiration in savanna grasses growing under conditions of high potential evapotranspiration and high radiant flux densities continue at near-maximum rates until the plant available water is depleted below a critical level; at which stage the rates of both decline rapidly (Scholes, 1987). There is some interspecific variability in the maximum rates and the water content at which stomatal closure and wilting occurs, but viewed at the scale of a growing season, the pattern is one of pulses of rapid growth alternating with dormancy. Water limitation (in the strict sense that a small increase in water supply will lead to a proportional increase in production) is only apparent for the tail end of the wet period. During the rest of the wet period the growth rate of the plant is principally controlled by soil nutrient status, either directly, or indirectly through the maximum intrinsic growth rate of the species found growing on that soil. In dry savannas there is

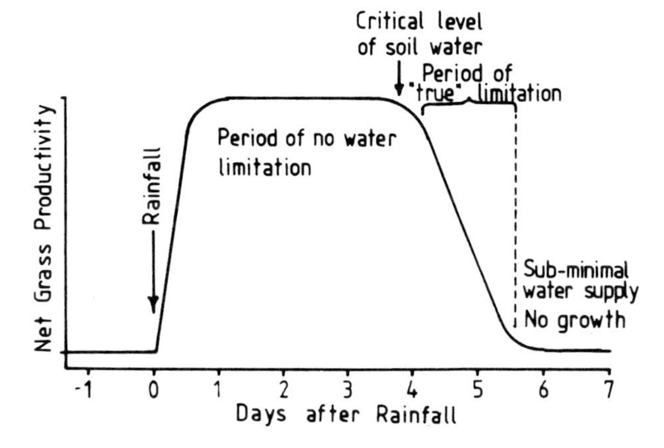

FIG. 1. An idealized illustration of the influence of declining plant available water on net grass productivity as the soil dries out following rainfall.

therefore a temporal alternation of water and nutrient limitation. This situation can be considered as water supply controlling the duration of grass production, but nutrient supply controlling the growth rate during productive periods. An alternate way of looking at the same phenomenon is to interpret the slope of the herbaceous production versus precipitation relation as an index of site fertility, whereas the mean annual precipitation less the x-intercept of the relation is an index of site water availability. The two together define the herbaceous productivity of the site (Fig. 2).

THE FERTILITY CONTINUUM

The general association of infertile soils (Ultisols and Oxisols in the Soil Taxonomy system; USDA, 1975) with the high rainfall areas of tropical Africa, and more fertile soils (Vertisols and Alfisols) with the more arid areas led to the concept of moist/dystrophic savannas on the one hand, and arid/eutrophic savannas on the other (Huntley, 1982). This distinction may have had its intellectual origin in the South African practice of differentiating between 'sweet' and 'sour' veld. The former has a forage protein content sufficient for cattle maintenance throughout the year, while the latter requires protein supplementation during the winter months (Booysen & Tainton, 1978). Sweetveld occurs in lowland, warm, arid areas on fertile soils, whereas sourveld occurs in highland, cool, wet infertile areas. This classification is applied to both grasslands and savannas, but is of limited usefulness when applied to savannas under non-cattle herbivory. Both sweetveld and sourveld support indigenous ungulates throughout the year. As a generalization, the eutrophic/dystrophic distinction appears to be valid and is reflected at many levels of ecosystem structure and function. When explicitly linked to the dry/moist distinction, however, it tends to hide the fact that within dry savannas there are extensive areas of infertile soils (such as the Kalahari sands), while within wet savannas there are extensive fertile soils (such as the volcanic soils of East Africa).

Confining the discussion to dry savannas, the infertile variants are found on the deep sand deposits of the Kalahari basin, on sands derived *in situ* from sandstones, and on the

TABLE 1. Characteristic features of dry savannas on fertile and infertile substrates.

Feature	Nutrient-poor	Nutrient-rich
Soils		
% Organic carbon	0.2–1.0	1.0–3.0
Mineralogy	Quartzitic or kaolinitic	Smectitic
Sum of bases	<20 cmol (+)kg^{-1}	>20 cmol (+)kg^{-1}
Parent material	Sands, sandstones, granites (upslope)	Basalts, dolerites, shales, granites (bottomlands)
Vegetation		
Tree taxonomy	Combretaceae and Ceasalpinoideae dominate	Mimusoideae dominate
Leaf type	Simple or compound	Compound
Leaf size	>15 mm	1–15 mm
Root:shoot ratio	High	Lower
Grass palatability	Low	High
Tree anti-herbivore strategy	Chemical (tannins, polyphenolics, etc.)	Structural (thorns)
Woody biomass	15–25 Mg ha^{-1}	5–15 Mg ha^{-1}
Herb layer water use efficiency	2–5 kg mm^{-1}	5–10 kg mm^{-1}
Litter layer	Conspicuous	Inconspicuous
Consumers		
Herbivory: mammal	Low	High
insect	Sporadic outbreaks	Continuously high
Soil fauna	High, termite dominated	Low, ant dominated

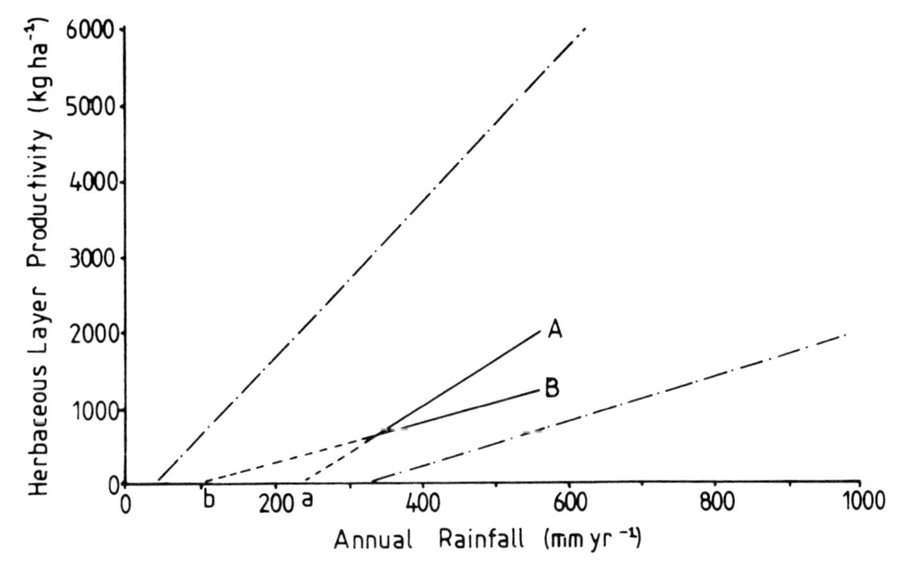

FIG. 2. The relationship between above-ground annual herbaceous production and annual rainfall. The broken lines indicate the upper and lower limits of the data for semi-arid regions reported by Rutherford (1980). The line marked A represents the relation obtained on a fertile, clayey soil, while B represents a sandy, infertile soil in the same climatic region (data from Scholes, 1987). The slope of the two relations can be used as an index of relative fertility and has values of 5 and 2.5 kg mm^{-1} for A and B respectively. The mean annual precipitation (500 mm in this case) less the x-intercept (a and b) provides an index of relative aridity: 300 mm and 400 mm for A and B respectively.

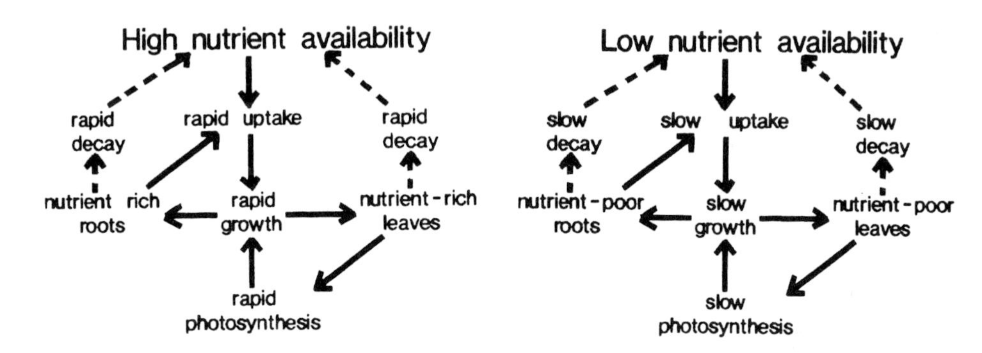

FIG. 3. A model proposed by Bryant & Chapin (1986) to account for differences in herbivore defence strategy on fertile and infertile substrates in the Arctic tundra. The broken arrows represent a modification of the model for savannas.

sandy upper catenal positions on soils derived from granites and gneisses. The fertile dry savannas occur on soils derived from basic igneous substrates (typically basalts, gabbros and dolerites), shales and mudstones, alluvium, and in the bottomlands of granite catenas. Table 1 contrasts the characteristic features of dry savannas on infertile and fertile substrates. It is offered as a working hypothesis, based mainly on observations in southern Africa.

Many dry African savannas can be unequivocally categorized as being nutrient-rich (eutrophic) or nutrient-poor (dystrophic) according to these criteria. This may be a coincidental consequence of the areal distribution of fertile and infertile soils; however, intuitively (and on the basis of soil test data) soil fertility would be expected to represent a continuum rather than two discrete classes. At this stage it is not clear whether these two variants are genuine alternatives, or merely represent the end-points of a continuum of fertility states. An intermediate state is recognized in South Africa as 'mixed bushveld', and some very widespread communities, such as *Colophospermum mopane* savanna, are not easy to classify.

A possible mechanism leading to distinct classes may be the presence of feedback loops and thresholds in the biological cycling of nutrients, tending to separate savannas into rapid-turnover or slow-turnover forms. A model proposed for Alaskan tundra vegetation by Bryant & Chapin (1986) illustrates the effects of such feedbacks, and may be equally applicable to savannas (Fig. 3).

The distinction between nutrient-rich and nutrient-poor may hold for wet as well as dry savannas, and this is the basis for the suggestion (Frost *et al.*, 1986) that savannas be classified according to the position they occupy in a water-supply versus nutrient-supply plane. In reality these axes tend not to be independent for several reasons. The first has already been mentioned: on an old and stable land surface such as the African shield, high rainfall areas tend to have weathered and infertile soils. The second is that in dry climates, soils with a high clay content are both more fertile and more xeric than sandy soils receiving the same rainfall, since more water is lost by runoff and evaporation from the heavier-textured soils.

NUTRIENT DISTRIBUTION WITHIN THE LANDSCAPE

Nutrient distribution is markedly patchy within the dry savanna landscape, and the consequences of this unevenness are particularly important in the generally infertile variants. Even in areas with extremely homogeneous parent materials, such as the Kalahari sands, geomorphological processes have resulted in the deposition of nutrient rich areas such as the Mubabe depression and the pans of the Central Kalahari. On granitic materials a classical semi-arid zone catena with nutrient-poor uplands and nutrient-rich bottomlands is widespread. Savanna fauna make extensive use of this patchiness in their feeding behaviour. Therefore to classify entire regions or landscapes as fertile or infertile on the basis of their geology may be useful at a generalization level (as for instance is done by Bell, 1982, and East, 1984), but within a landscape it tends to conceal the reality that both consist of mosaics to different proportions of fertile and infertile sites. It is suggested here that the existence of this mosaic and the interaction between sites of nutrient enhancement and nutrient depletion is fundamental feature of savanna structure and function.

It appears that the creation and long-term persistence of fertile patches within infertile savannas is a widespread phenomenon. Several processes other than the geomorphological examples mentioned above can lead to their establishment. Nutrient redistribution through the feeding movements of mammalian herbivores have been shown to result in nutrient concentrations (du Toit, 1988). The foraging and mound-building activities of termites result in an accumulation of bases and fines in the vicinity of the mound (Malaisse, 1978); a fact which is capitalized upon by small farmers. The broad lateral spread of tree roots, which can exceed the canopy spread by a factor of seven (Rutherford, 1982), results in an accumulation of nutrients in the sub-canopy habitat, which is frequently exploited by a different suite of grasses. The traces of pre-colonial man are everywhere evident; cattle penning, firewood collection and iron smelting activities have led to major nutrient buildups (Blackmore, Mentis & Scholes, 1990).

Nutrient-rich patches of human origin on sandstone in the Northern Transvaal (Fordyce, 1980) and on Kalahari sands in Botswana (Denbow, 1979) have been dated on the basis of archaeological evidence and ^{14}C analysis to be 700–1000 years old. The nutrient concentrations are surprisingly persistent, given the low ion exchange capacity of the soil. Despite the dry climate, the low water-holding capacity of the soil permits regular movement of water through the profile. In the absence of a reinforcing mechanism, it would be expected that the nutrient accumulations would be dissipated by leaching. At a smaller scale, it is obvious from aerial inspection of 20-year-old maize lands on sandy soils that the previous locations of termite heaps and trees have a persistent effect on crop growth. What mechanisms could account for the reinforcement of nutrient accumulations once they have been established, in the face of the dispersing effect of leaching, tree rooting or herbivory?

THE NUTRIENT DYNAMICS OF AN ENRICHED SITE

The following observations arise out of the intensive study of nutrient-enriched patches of anthropogenic origin at Nylsvley, South Africa. They illustrate the types of mechanisms found to be important there; other mechanisms may be involved elsewhere.

The nutrient flux rates that have been measured in the enriched patches are greater than those measured in the adjacent nutrient-poor areas (Scholes & Scholes, 1989). Conventional wisdom suggests that rapid turnover should accelerate rather than retard the nutrient decline. There are several mechanisms which may contribute to persistence:

Immobilization. The raised organic matter content of enriched sites (and the increased clay content typical of termite mounds) lead to a greater ion exchange capacity, which helps to retain nutrients. Some of the other key nutrients, nitrogen and phosphorus, are in organic or otherwise non-leachable forms, which would retard the rate of loss.

Nutrient import. Mammalian herbivores at Nylsvley have been observed to spend up to six-fold more time on the nutrient-rich patches than would be expected from a random distribution model. This is attributed to greater palatability of the grasses dominant in these patches. Productivity on the sites is only about twice that in the surrounding nutrient-poor areas. Therefore it is likely that defecation exceeds consumption on the sites, and the nutrient flux due to herbivory is inwards, not outwards. If this flux were sufficient to balance losses due to other factors, the nutrient accumulation would persist or increase.

Allelopathy. The distribution of woody plants on the sites, the apparent absence of roots of trees of the surrounding types (despite their known ability to root over great distances) and the distinctive smell of roots from the sites all suggest allelopathy may have a role in the defence of the nutrient accumulations from depletion by adjacent plants.

These suggested mechanisms raise a number of questions regarding the necessary conditions for the persistence of a nutrient accumulation. Must the nutrient accumulation exceed a certain concentration or absolute size for the mechanisms to be effective? Is the ratio of nutrient-enriched to nutrient-depleted areas important?

CONCLUSIONS

Within the dry savannas great differences in soil fertility exist. These differences have profound effects on the species composition, physiognomy, structure and functioning of the savanna. The origin of the fertility differences can be geological, geomorphological, anthropogenic, or as a result of biotic activity within the community. The scale of fertility variation can be from regional down to a few metres. The formation of localized nutrient enriched patches is suggested to be of great importance to the function of the savanna as a whole, particularly in generally nutrient-poor savannas. The processes involved in their maintenance are at this stage not demonstrated, but it is suggested that biotic feedbacks could be important in establishing the enriched patch as a new stable state. There is evidence that such patches can be created by man, and that they are extremely persistent once established.

REFERENCES

Bell, R.H.V. (1982) The effect of soil nutrient availability on community structure in African ecosystems. *Ecology of tropical savannas* (ed. by B. J. Huntley and B. H. Walker), pp. 193–216. Ecological Studies; Springer, New York.

Blackmore, A.C., Mentis, M.T. & Scholes, R.J. (1990) The origin and nature of nutrient-enriched patches within a nutrient-poor savanna. *J. Biogeogr.* (in press).

Booysen, P. de V. & Tainton, N. (1978) *Pasture management in South Africa.* Shooter & Schuter, Pietermaritzburg.

Bryant, J.P. & Chapin, F.S., III (1986) Browsing–woody plant interactions during Boreal forest plant succession. *Forest ecosystems in the Alascan Taiga* (ed. by K. van Cleve), pp. 213–225. Springer, New York.

Denbow, J.R. (1979) *Cenchrus ciliaris*: an ecological indicator of Iron Age middens using aerial photography in eastern Botswana. *S. Afr. J. Sci.* **75**, 405–408.

du Toit, J.T. (1988) Patterns of resource use within the browsing ruminant guild in the central Kruger National Park, Ph.D. thesis, University of the Witwatersrand, Johannesburg.

East, R. (1984) Rainfall, soil nutrient status and biomass of large African savanna mammals. *Afr. J. Ecol.* **22**, 245–270.

Ellis, R.P., Vogel, J.C. & Fuls, A. (1980) Photosynthetic pathways and the geographical distribution of grasses in South West Africa/Namibia. *S. Afr. J. Sci.* **76**, 209–213.

Frost, P.G.H., Menaut, J.C., Walker, B.H., Medina, E., Solbrig, O.T. & Swift, M. (1986) Responses of savannas to stress and disturbance. *Biology International*, Special Issue 10, 82pp.

Fordyce, B. (1980) The prehistory of Nylsvley. Progress report to the National Programme for Environmental Science. Typescript 12pp. CSIR, Pretoria.

Huntley, B.J. (1982) Southern African savannas. In: *Ecology of tropical savannas* (ed. by B. J. Huntley and B. H. Walker). Ecological Studies: Springer, New York.

Johnson, R.W. & Tothill, J.C. (1985) Definitions and broad geographic outline of savanna lands. In: *Ecology and management of the world's savannas* (ed. by J. C. Tothill and J. J. Mott). Australian Academy of Science, Canberra.

Malaisse, F. (1978) High termitaria. *Biogeography and ecology of southern Africa* (ed. by M. J. A. Werger), pp. 1279–1300. Junk, The Hague.

Rutherford, M.C. (1980) Annual plant production–precipitation relations in arid and semi-arid regions. *S. Afr. J. Sci.* **76**, 53–56.

Rutherford, M.C. (1982) Woody plant biomass distribution in *Burkea africana* savannas. *Ecology of tropical savannas* (ed. by B. J. Huntley and B. H. Walker), pp. 120–142. Ecological Studies; Springer, New York.

Scholes, M.C. & Scholes, R.J. (1989) Phosphorus mineralisation and immobilisation in savannas. In: *Proceedings of the Phosphorus Symposium, Pretoria, September 1988.*

Scholes, R.J. (1987) Response of three semi-arid savannas on contrasting soils to the removal of the woody component. Ph.D. thesis, University of the Witwatersrand, Johannesburg.

USDA (1975) *Soil taxonomy: A basic system of soil classification for making and interpreting soil surveys.* Agriculture Handbook 436, Soil Conservation Service, United States Department of Agriculture, Washington.

8/Nitrogen-use efficiency for growth in a cultivated African grass and a native South American pasture grass

BIBIANA BILBAO and ERNESTO MEDINA *Centro de Ecología y Ciencias Ambientales, Instituto Venezolano de Investigaciones Científicas, Aptdo. 21827, Caracas 1020-A, Venezuela*

Abstract. Frequency and density of introduced African grass species in disturbed neotropical ecosystems reveal their adaptability and competitive capacity compared to the native savanna grass species. A nutritional hypothesis has been advanced to explain the success of these species. Deforested areas provide periods of short duration of high nutrient availability, while in savannas protected against fire a marked increase of nutrient and organic matter content in the upper soil layers occurs. There is no direct evidence indicating that the production potential of introduced African grasses is lower than that of native grasses when they are grown in soils of low fertility. A study of a widely cultivated African grass species, *Andropogon gayanus* Kunth, and an occasionally cultivated native South American species, *Paspalum plicatulum* Michx., was undertaken to compare their growth rates and nitrogen use efficiency when grown in nutritionally poor savanna soils of Cojedes State in central Venezuela, and fertilized with intermediate levels of nitrogen + potassium, phosphorus + potassium, and a combination of nitrogen + phosphorus + potassium. Production potential of *A. gayanus* Kunth was larger and more limited by P than that of *Paspalum plicatulum*. The African grass is capable of extracting larger quantities of N from the unfertilized savanna soil than the native grass, and this capability was greatly enhanced by the addition of P + K. Both species did not respond to the fertilization with N and K alone. Differences in growth potential are based on greater nitrogen use efficiency and lower shoot/root ratios in *A. gayanus*.

Key words. Nitrogen nutrition, productivity, grasses, Llanos, South America.

INTRODUCTION

Frequency and density of introduced African grasses in disturbed ecosystems of tropical South America shows the adaptability and competitive capacity of these exotic species. The invasion of disturbed areas by exotic African grasses is so strong that Parsons (1972) described the process as an 'africanization' of the American tropics. This aggressive behaviour has been explained on the basis of resistance to fire and grazing, allelopathic effects, and reproductive mechanisms (Baker, 1978).

Half-a-dozen African species constitute the main set of invaders: *Hyparrhenia rufa* (Nees) Stapf, *Melinis minutiflora* Beauv., *Panicum maximum* Jacq., *Brachiaria mutica* (Forsk.) Stapf., *Pennisetum purpureum* Schumacher and *Digitaria decumbens* Stent. These species are frequently found invading deforested areas and along roadsides. Some of them have been also observed to invade grasslands protected against fire and grazing. That is the case of *Hyparrhenia rufa* in the Savanna Biological Station in the Central Venezuelan Llanos (Cruces, 1977) and of *Melinis minutiflora* in protected Cerrado areas near São Paulo, Brazil (Coutinho, 1982).

A nutritional hypothesis has been advanced to explain the success of these pastures. Disturbed forest areas are transiently nutrient rich, due to the increased mineralization of organic matter after disturbance. On the other hand, in protected savannas, elimination of fire results in an increased organic matter content of the upper soil layers leading to higher availability of nitrogen and probably other nutrients. Under these conditions the introduced grasses, with higher nutritional requirements, would be able to outcompete native grasses because of their larger potential for organic matter production. However, there is no direct evidence of the production potential of introduced African grasses in savanna soils of low fertility in tropical America, as compared with the performance of native grasses.

Within the Venezuelan RSSD project (Frost *et al.*, 1986), the nutritional eco-physiology of native and introduced grasses is being investigated to understand their competitive abilities under natural savanna conditions. Here we present a comparative study of the effect of fertilization on the nitrogen use efficiency for growth by two grass species which are becoming popular for establishing improved pastures in Venezuela: *Andropogon gayanus* Kunth. (African introduction) and *Paspalum plicatulum*

TABLE 1. Soil characteristics of the savanna site at Sembra, Edo. Cojedes, Venezuela. Samples were taken from 0 to 30 cm depth.

Sand (%)	Silt (%)	Clay (%)	pH		N (mg/g)	P (ppm)	Organic matter (%)
			H_2O	KCl			
21.7	30.5	47.8	4.4	3.7	1.47	2.3	1.62

Σ bases (μeq/g)	Al (μeq/g)	CEC (μeq/g)	Base saturation (%)	Al saturation (%)
4.5	27.6	35.0	13.2	78.5

Michx. (South American native). Two hypothesis were tested:

(1) Native species of non-flooded savannas in tropical America have a production potential adjusted to low nutrient supply during growth, mainly phosphorus. Therefore, expected increases in forage yield brought about by fertilization should be more dependent on nitrogen supply.

(2) Efficiency of nitrogen use by African introduced grasses is larger than by native species. When cultivated under similar soil nutrient supply, the latter should have a lower production potential than the introduced grasses.

STUDY AREA

The study site was located in a *Trachypogon* spp. dominated savanna (tropical seasonal savanna) in Cojedes State, Venezuela (68° 15′W, 9° 25′N). The site has a typical tropical climate with very small annual temperature variations, and a strongly seasonal distribution of rainfall with 80–90% falling between May and October (10 years temperature average is 26.9°C, 20 years average rainfall is 1322 mm). The experimental area is located on a flat alluvial plain (120 m a.s.l.) formed on pleistocenic deposits. The soil has been classified as an Haplustult (Soil Taxonomy, 1975). The soil is acid, has a clay texture, a very low sum bases, and a high percentage of Al saturation (Table 1).

MATERIALS AND METHODS

The species selected for comparison were: (a) *Andropogon gayanus* var. *bisquamulatus* Kunth, a perennial cespitous grass, native of Western Africa, commonly growing at elevations lower than 1000 m with rainfall ranging between 700 and 1500 mm. This species tolerates drought seasons from 2 to 9 months (Bogdan, 1977). (b) *Paspalum plicatulum* var. *longipilum* Michx., perennial cespitous grass, distributed from southern U.S.A. and the Antilles to northern Argentina. It is found up to 1800 m a.s.l. in areas with rainfall ranging from 750 to 2000 mm (Guzmán-Pérez, 1986). Generally it grows naturally in seasonal savannas but may tolerate short-term shallow flooding (Ramia, 1974).

Grass tussocks of each species were obtained from cultivated fields near the experimental area and were transplanted in twenty (10×10 m) plots and grown under the following conditions: (1) control; (2) nitrogen + potassium (NK): 70 kg N/ha + 30 kg K/ha; (3) phosphorus + potassium (PK): 102 kg P/ha + 30 kg K/ha; (4) nitrogen + phosphorus + potassium (NPK): N + P + K as before.

Measurement of organic matter production and calculation of growth indices were performed as indicated by Evans (1972). Four grass tussocks per treatment, and per species were harvested at each sampling time. The biomass was divided in: (a) green and dead leaves; (b) living and dead stems; (c) root and rhizomes; and (d) inflorescences. Efficiency of nitrogen use for growth (UNR) was calculated as:

Unit Nitrogen Rate (UNR): kg dry matter mol^{-1} N day^{-1}

$$\mathrm{UNR} = \frac{dW}{dT} * \frac{1}{[N]} = \frac{(W_2 - W_1)}{(T_2 - T_1)} * \frac{(\ln[N]_1 - \ln[N]_2)}{([N]_1 - [N]_2)}$$

Data were analysed with two-way analysis of variance, followed by the *a posteriori* test of Newman-Keuls to establish significance of the differences between treatments (Snedecor & Cochran, 1984). Analyses were carried out with SYSTAT 2.0–1984 version statistical program. Levels of significance are taken for $P = 0.05$ and 0.01.

RESULTS

Biomass accumulation

The treatments PK and NPK resulted in significant higher total biomass accumulation in both species ($P < 0.05$) (Fig. 1). However, in *A. gayanus* the increase was due almost entirely to the addition of P (relative growth increase [RGI] by PK = 2.2; by NPK = 2.5); while in *P. plicatulum* both N and P seemed to be limiting production potential (RGI by PK = 2.3 and NPK = 4.0). Those effects were detected also in total leaf biomass, but not in roots and rhizomes.

Leaf/stem and shoot root ratios

Leaf/stem ratios increased rapidly after planting. At the peak of flowering (160 days in *A. gayanus*; 112 days in *P. plicatulum*) these ratios were reduced significantly due to

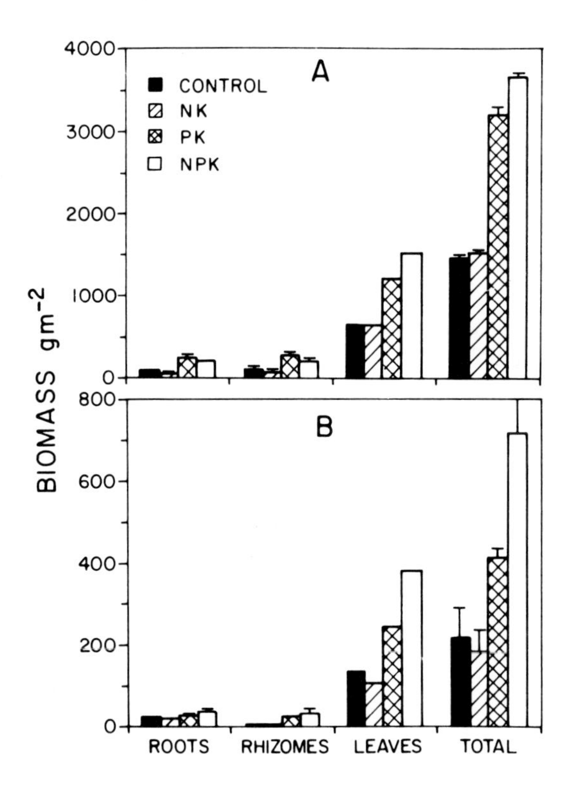

FIG. 1. Peak biomass accumulation separated in above- and below-ground organs of (A) *Andropogon gayanus* Kunth and (B) *Paspalum plicatulum* Michx. with different nutrient supplies.

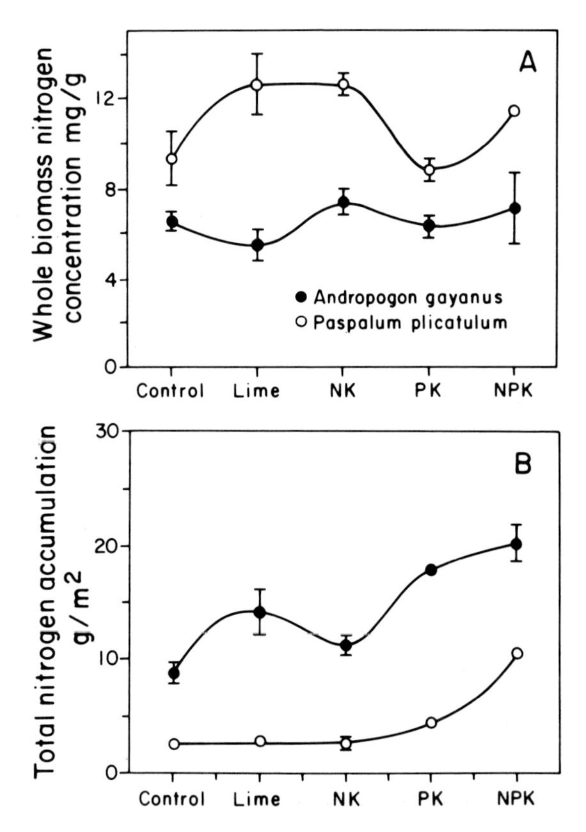

FIG. 3. Nitrogen concentration of total biomass (A) and total nitrogen accumulation (B) at peak biomass development of *Andropogon gayanus* Kunth and *Paspalum plicatulum* Michx. growing in soils with different nutrient supplies.

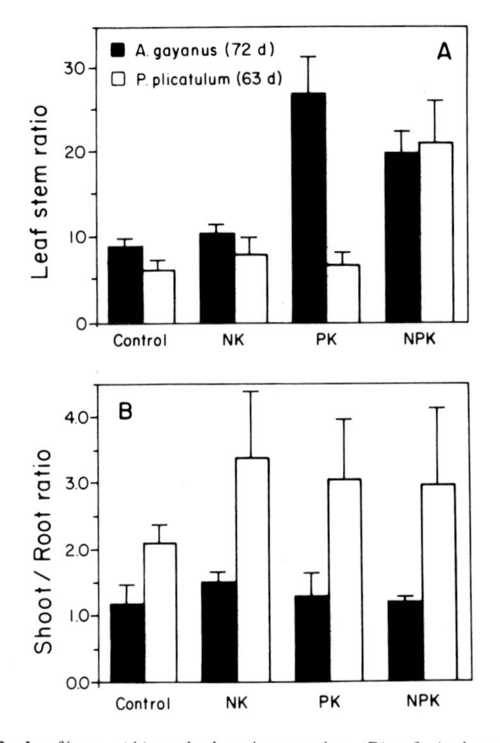

FIG. 2. Leaf/stem (A) and shoot/root ratios (B) of *Andropogon gayanus* Kunth and *Paspalum plicatulum* Michx. grown in soils with different nutrient supplies.

the production of flowering culms. NPK and PK treatments produced the highest values in *A. gayanus* while in *P. plicatulum* only NPK treatment resulted in significant differences with control ($P < 0.05$) (Fig. 2A). This points again to the greater limitation of P for the African grass.

Shoot/root ratios were always higher in *P. plicatulum* and showed little variations with treatments in *A. gayanus* (Fig. 2B). In the former this ratio was significantly higher in NPK treatment.

Nitrogen concentration and accumulation

At the peak of flowering N concentration in biomass was significantly higher for all treatments in *P. plicatulum* as compared with *A. gayanus* ($P < 0.01$) (Fig. 3A). In the latter, no differences in nitrogen concentration between treatments were detected, while in *P. plicatulum* only the PK treatment was similar to the control. Differences in N concentration were similar for all treatments in both species, with the exception of the PK treatment in *A. gayanus*. Total N uptake, however, was higher in *A. gayanus*, as would be expected from its larger biomass production ($P < 0.01$). Differences in N accumulation between PK and NPK treatments were smaller and non-significant ($P < 0.01$) than for *P. plicatulum* (Fig. 3B). In the latter species only the NPK treatment showed significant higher values ($P < 0.001$).

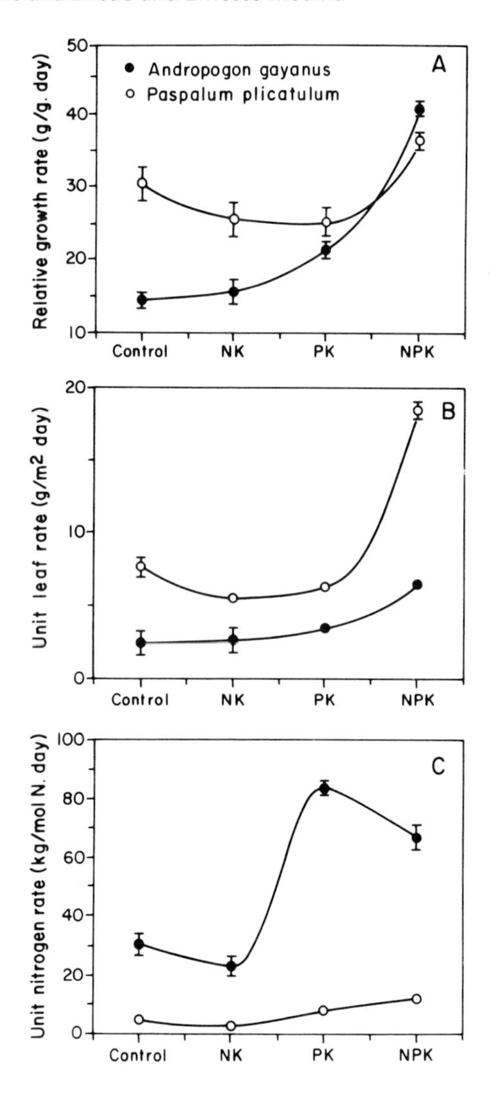

FIG. 4. Growth rates per unit dry weight (RGR) (A), unit leaf area (ULR) (B), and unit nitrogen (UNR) (C) corresponding to the period of peak biomass developent of *Andropogon gayanus* Kunth and *Paspalum plicatulum* Michx. growing in soils with different nutrient supplies.

Growth rates and efficiency of nitrogen use

At the peak of flowering relative growth rate (RGR) increased significantly ($P < 0.01$) in *A. gayanus* with PK and NPK treatments, differences among these treatments being also significant ($P < 0.05$) (Fig. 4A). In *P. plicatulum*, only NPK treatment resulted in significantly higher RGR ($P < 0.01$). However, RGRs in control and NK treatments were significantly higher in *P. plicatulum* than in *A. gayanus* ($P < 0.01$). Unit leaf rates (ULR) behaved similarly (Fig. 4B). *Paspalum plicatulum* showed higher ULRs under all treatments, the differences between PK and NPK treatments were highly significant in this species ($P < 0.01$) but not so in *A. gayanus*. Notice, however, that RGR in the NPK treatment were similar for both species, while this did not hold true for the ULR.

Efficiency of nitrogen use for growth (UNR) was significantly higher for all treatments in *A. gayanus*

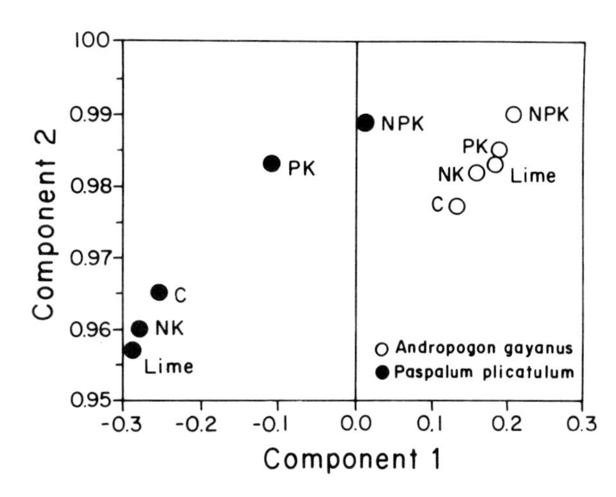

FIG. 5. Species-treatment ordination using biomass, nitrogen content and UNR data. The two axes explained more than 90% of the variance.

($P < 0.01$) (Fig. 4C); UNRs of control and NK treatments were lower than those corresponding to PK treatment (which showed the highest values) and NPK ($P < 0.05$). For *P. plicatulum* UNRs values were homogenously low for all treatments.

Principal components analysis

The behaviour of the two species was clearly differentiated applying a principal component analysis using biomass, nitrogen content, and UNR data (Fig. 5). It is remarkable that even the PK and NPK treatments by *P. plicatulum* did not reach similar values as the control treatment in *A. gayanus*. These differences are probably related to the higher growth potential and lower shoot/root ratio in the latter species.

DISCUSSION

The fertilization experiments described allowed a clear differentiation of the nutritional requirements of *P. plicatulum* and *A. gayanus*. Although potential growth is limited by P supply in both species, reduction of organic matter production is more pronounced in *A. gayanus*. Nitrogen additions did not increase production above the control in both species. This contrasts with earlier reports of fertilization of natural *Trachypogon*-savannas showing a stronger response with N than with P additions (Medina, Mendoza & Montes, 1978). The conclusion is that P requirements in the species analysed are higher than in the dominant grasses of natural savannas, and that P supply is severely limiting in the relatively heavy soils where the experiments were carried out.

Nitrogen concentration in biomass was always higher in *P. plicatulum* indicating a lower efficiency of this species in the utilization of N for growth. However, due to the higher growth potential of *A. gayanus*, the total amount of N extracted from the soil was 2–5 times higher than by *P. plicatulum*. Therefore there was enough N in the soil to support the demands of the latter species, but P limitation and the lower capability for soil exploration, as indicated by the higher shoot/root ratios of this species, limited the expression of its growth potential. Growth indices show that *P. plicatulum* has significantly higher relative growth rates in the control and NK treatments, but not in the PK and NPK treatments, where *A. gayanus* unfolds its stronger growth potential. Unit leaf rates, however, are always higher in *P. plicatulum*, the differences being more pronounced in the NPK treatment, probably as a result of the higher shoot/root ratios of this species.

Results obtained with this pair of species confirm the hypotheses being tested, namely, that the African species is more dependent on P supply for maximal growth, while showing higher nitrogen use efficiency than the South American grass investigated. Indeed, the higher growth potential of *A. gayanus* is based on its higher efficiency of nitrogen utilization for growth, and its capability to exploit scarce P sources in the soil through its relatively higher allocation of biomass for root growth. Comparisons with other species following the methodology applied here should allow the differentiation of savanna grass species according to their P and N requirements for growth, and their capability to exploit soil nutrient sources.

REFERENCES

Baker, H. (1978) Invasion and replacement in Californian and neotropical grassland. *Plant relations in pastures* (ed. by J. R. Wilson), pp. 368–384. CSIRO, Melbourne.

Bogdan, A.V. (1977) *Tropical pastures and fodder plants.* Tropical Agricultural Series. Longman, London.

Coutinho, L.M. (1982) Aspectos ecologicos da sauva do Cerrado. Os murundus de terra, as caracteristicas psamofiticas das species de sua vegetaçao e a sua invasion pelo capim gordura. *Rev. Bras. Bio.* **42**, (1), 147–153,

Cruces, J.M. (1977) *Productividad primaria, fenología y valor nutritivo de la gramínea* Hyparrhenia rufa *(Nees) Stapf, en dos localidades del Edo. Guárico.* Trabajo Especial de Grado. Escuela de Biología. Fac. de Ciencias, UCV, Caracas.

Evans, G.C. (1972) *The quantitative analysis of plant growth.* Blackwell Scientific Publications, Oxford.

Frost, P., Medina, E., Menaut, J.-C. Solbrig, O., Swift, M. & Walker, B. (1986) *Responses of savannas to stress and disturbance.* Biology International. Special Issue No. 10. International Union of Biological Sciences, Paris.

Guzmán-Pérez, J.E. (1986) *Pastos y forrajes de Venezuela. Producción y aprovechamiento.* Espansade Editores, Caracas.

Medina, E., Mendoza, A. & Montes, R. (1978) Nutrient balance and organic matter production in the *Trachypogon* savannas. *Trop. Agric.* **55**, 243–253.

Parsons, J. (1972) Spread of african pasture grasses to the american tropics. *J. Range Managemnt*, **25**, 12–17.

Ramia, M. (1974) *Plantas de las sabanas llaneras.* Monte Avila Editores, Caracas.

Snedecor, G.W. & Cochran, W.G. (1984) *Métodos estadísticos.* Ed. Continental, México.

Soil Taxonomy (1975) *A basic system of soil classification for making and interpreting soil survey.* Soil Conservation Service, Washington.

9/Nutrient cycling in Australian savannas

J. A. HOLT and R. J. COVENTRY *Division of Soils, CSIRO, Davies Laboratory, Private Mail Bag, P.O. Aitkenvale, Townsville, Qld 4814, Australia*

Abstract. Because Australian savannas generally occur on soils that are nutrient deficient, their productivity is dependent upon the rapid recycling of nutrients locked up in the plant biomass. Soil microorganisms play a major role in the decomposition of organic matter and cycling of nutrients in savannas, but their activity is restricted to the wet season.

Termites are widespread in the dry tropics of the world and exert a significant effect on nutrient cycling. The results of studies in the Townsville region of north Queensland suggest that termites might be responsible for up to 20% of organic matter decomposition. The erosional redistribution of termite mound materials is a process which can supply the infertile surface soil with a nutrient enriched soil amendment

Fire is also an active agent in nutrient cycling in some savanna systems. The ecological advantage of this mechanism of nutrient cycling is maintained only if losses by volatilization and atmospheric dispersion are balanced by other inputs.

Key words. Australia, savannas, nutrient cycling, fire, termites.

INTRODUCTION

The productivity of an ecosystem depends on the amounts of nutrient stored in various compartments such as the vegetation, litter, soil, and animal biomass, and on the rates of nutrient transfer among those compartments. The cycling of nutrients in all ecosystems is effected by a combination of biological and physical processes. However, the relative importance of these processes varies considerably between ecosystems as a result of differences between climate, soils, vegetation and management practices. In most Australian savannas the soil fertility status is low (Mott *et al.*, 1985), and the rapid release of nutrients from organic matter is a critical step in ecosystem function. Successful management of nutrient deficient ecosystems therefore requires an understanding of the nature of the cycling process in order that rates of transfer between compartments are optimized.

Physical processes active in nutrient cycling include mineralization of organic matter by fire (both natural and man-managed), losses of nutrients from the soil store by leaching and/or erosion and accessions as a result of atmospheric processes. In undisturbed ecosystems, the gains and losses of nutrients to the system are usually quite small compared to the amounts cycling internally (Swift & Sanchez, 1984). This situation contrasts with some agricultural cropping systems where the large quantities of nutrients that are removed during harvest are replaced to a varying extent by the application of inorganic fertilizers.

Nutrients contained in plant litter are unavailable for uptake by plant roots until they have been mineralized into an inorganic form: for example, the organic nitrogen is transformed into ammonium (NH_4^+) or nitrate (NO_3^-) ions. In most ecosystems, this transformation is completed via the complex interaction of a range of vertebrate and invertebrate organisms, except where organic matter mineralization has occurred as a result of fire. Microorganisms (largely bacteria and fungi) assume a dominant role in this mineralization process, with the invertebrate fauna of the soil and litter generally aiding the process indirectly by contributing to the physical comminution of organic matter. In more arid ecosystems, abiotic factors may also have some influence on the rate of organic matter mineralization.

Studies in the temperate zones of the world have shown that the direct contribution of invertebrates to organic matter decomposition is small (usually <10%) compared with the contribution of soil microbial populations (Peterson & Luxton, 1982; Seastedt, 1984). In tropical semi-arid environments, however, various arthropod detritivores have evolved as important primary decomposers and in some areas have the potential to exert a significant effect in nutrient cycling (Crawford, 1981; Skujins, 1984). Because they possess the ability to modify the microenvironment within their nests and foraging galleries (Luscher, 1961), termites have evolved as the most successful detritivorous insect in the seasonally dry tropics of the world. Their mounds are a conspicuous feature of many northern Australian savannas (Figs. 1a and 1b), particularly in those areas of red and yellow earth soils (Stace *et al.*, 1968; Oxic Paleustalfs and (Oxic) Plinthustalfs of the Soil Survey Staff, 1975) where mound densities of 300 ha^{-1} are common (Lee & Wood, 1971; Holt, Coventry & Sinclair, 1980).

In this paper we discuss certain aspects of nutrient

FIG. 1(a). Termite mounds near Balfes Creek, northeastern Queensland.

cycling in tropical Australian savannas. Because of their importance as primary decomposers, and their widespread occurrence in tropical Australian savannas, special reference is made to the role of termites.

BIOLOGICAL PROCESSES

Rates of carbon dioxide production by decomposer organisms can be used as an index of rates of decomposition (Freckman, Cromack & Wallwork, 1986). This concept has been used by a number of workers to assess the contribution of microorganisms to the decomposition of organic matter in soils. Most studies have shown that the rate of carbon dioxide release by soil microorganisms (that is, their respiration rate) is strongly dependent on soil moisture and temperature (e.g. Wildung, Garland & Buschbom, 1975; Gupta & Singh, 1981; Orchard & Cook, 1983).

In a study of carbon cycling in a savanna woodland with a strongly seasonal rainfall regime, near Townsville in north Queensland, Holt (1987, 1988) also found that soil microbial activity as measured by soil respiration was strongly dependent upon soil moisture. As a consequence, approximately 60% of carbon dioxide production from the soil occurred during the 4-month wet season. Much lower carbon mineralization rates occurred during the 8-month dry season, when low soil moisture limited microbial activity. The total amount of organic carbon mineralized to carbon dioxide by soil microorganisms was estimated to be approximately 2300 kg ha^{-1} yr^{-1}.

In contrast to microbial activity which virtually ceased during the dry season, termite activity at the site near Townsville was continuous throughout the year. Two of the mound-building termite species present, *Amitermes laurensis* Mjoberg and *Nasutitermes longipennis* (Hill), were responsible for the mineralization of approximately 250 kg ha^{-1} yr^{-1} of carbon (Holt, 1988). This conservative estimate represents approximately 10% of the annual carbon turnover at this site. If the contribution of the large, but as yet unquantified, population of subterranean detritivorous termites was also considered, it is possible that the termites (all groups) might be responsible for up to 20% of the carbon mineralized in the ecosystem. The contribution of termites to organic matter decomposition in a tropical savanna is therefore large compared with the direct contribution of all invertebrates to decomposition in temperate ecosystems (usually <10%, Seastedt, 1984).

Although active during the wet season only, soil microorganisms still account for most of the carbon mineralization at the study site. Their inactivity for the major part of the year may be a factor that has allowed termites to exploit a niche which is unavailable in regions where microbial

FIG. 1(b). *Nasutitermes triodiae* (Froggatt) mound near Darwin, Northern Territory.

activity is not moisture limited as it is in the seasonally dry savannas of the tropics (Holt, 1988).

Assuming a functional correspondence between carbon flow and the flow of most other plant nutrients, it is evident that both microorganisms and termites play a significant role in the recycling of nutrients through this savanna ecosystem.

Coventry, Holt & Sinclair (1988) used a different approach to assess the significance of mound-building termites in nutrient cycling in a savanna woodland near Charters Towers, north Queensland. They measured the nutrient content of a range of termite mounds and then used the rate of erosion of these mounds to calculate the amounts of nutrients returned to the soil each year. Given annual erosion losses from the mounds of 3.5% (Bonell, Coventry & Holt, 1986), they estimated that mound-building termites were responsible for the return of at least 24 kg ha^{-1} of organic carbon, 1.8 kg ha^{-1} of nitrogen and 0.015 kg ha^{-1} of weak acid-extractable phosphorus. These estimates did not take into account the return of nutrients as a result of erosion of soil incorporated by termites into their feeding galleries on the soil surface and on trees. The amounts of soil contained in such structures may be of a similar magnitude to the amounts of soil incorporated into termite mounds (Lee & Wood, 1971) and, if taken into account, might well double the above estimates.

Since termites occur in large numbers over the savannas of much of northern Australia, particularly in the more arid regions, their influence on nutrient cycling is widespread. The only areas of semi-arid tropical Australia which do not support large numbers of mound-building termites are the cracking clay soils (Ratcliffe, Gay & Greaves, 1952; Gay & Calaby, 1970). Nevertheless, the biomass of mound-building termites is known for only two locations in northern Australia. Near Townsville, the biomass of one species alone was approximately 25 kg ha^{-1} (Holt, 1988) and at another location near Charters Towers the biomass of five species of termites was estimated at between 40 and 120 kg ha^{-1} (Holt & Easey, 1984). Despite a lack of specific biomass data, observations on the densities of termite mounds at other locations in northern Australia (Williams, 1968; Saunders, 1969; Wood & Lee, 1971; Spain, Okello-Oloya & Brown, 1983; Birkill, 1985) suggest that similar or even higher termite biomasses may be common in many northern Australian savannas.

There are some differences between the termite species assemblages of the Australian region and the African region, where most studies of termites have been conducted. For example, the Hodotermitidae, a family of harvester termites, is not represented in Australia, nor are the fungus cultivating termites (Macrotermitinae) (Gay & Calaby, 1970). The Australian harvester termites are largely

TABLE 1. Liveweight biomass of termites in some African and Australian savannas.

Location	Biomass (kg ha^{-1})	Termite spp.	Author
Transvaal	30	14 spp.	Ferrar (1982)
Nigeria	5	*Macrotermes bellicosus* (Smeathman)	Collins (1981)
Nigeria	31	*Trinervitermes geminatus* (Wasmann)	Ohiagu (1979)
Nigeria	106	12 spp.	Wood & Sands (1978)
Senegal	6	*T. trinervius* (Rambur) *M. bellicosus* (Smeathman)	Lepage (1972)
Australia	25	*Amitermes laurensis* Mjoberg	Holt (1988)
Australia	40–120	4 spp.	Holt & Easey (1984)

represented by the endemic *Drepanotermes* and *Tumulitermes*, both genera occurring with high densities in tropical savannas. Despite these differences, studies conducted in African tropical savannas also show that termites exert a major influence on litter decomposition and nutrient cycling. Although species compositions may be different, the remarkable similarity between the termite biomass of the savannas of Australia and Africa (Table 1), suggests that the functional role of termites in nutrient cycling is similar in both regions.

PHYSICAL PROCESSES

Fire

Much of the savannas of tropical Australia is regularly burnt, both to reduce the bulk of nutritionally poor herbage, and to stimulate the growth of higher protein grass shoots. The passage of fire through savannas causes the rapid mineralization of nutrients contained in organic matter, thereby effectively increasing the nutrient cycling rate. Most of the nutrients thus released are deposited as ash on the soil surface, although some may be dispersed as very fine particles in the atmosphere. In addition to the loss of some particulate matter, losses of nitrogen and sulphur occur by volatilization during most fires. Volatilization of phosphorous only occurs in hot fires when temperatures exceed 600°C (Walker, Raison & Khanna, 1986).

Studies in Australian savannas have shown that tropical grasses are able to relocate nutrients from the above-ground parts to roots at the end of the summer growing season (Norman, 1963; McIvor, 1981). This process effectively reduces the potential loss of nutrients as a result of dry season fires. A similar pattern of nutrient withdrawal from the above-ground parts of tropical grasses was observed by Villecourt, Schmidt & Cesar (1979) in the Ivory Coast. Relatively few studies have quantified nutrient losses to the atmosphere as a direct result of fire. Norman & Wetselaar (1960) found that over 90% of the nitrogen in a native pasture at Katherine, Northern Territory (4.5 kg ha^{-1}), was lost to the atmosphere as a result of burning. They concluded, however, that this relatively small loss of nitrogen was probably offset by annual nitrogen inputs in rainfall and by non-symbiotic nitrogen fixation in the surface soils. Medina (1982) found that sulphur losses caused by fire in a Venezuelan savanna were balanced by sulphur contained in rain, but nitrogen losses were not. However, he suggested that the nitrogen losses might be compensated for by nitrogen fixation on or in the soil.

Annual burning over an extended period of time appears to have had little effect on soil nutrient levels in some tropical and temperate regions. Lamb, Landsberg & Foot (1983) found that the nitrogen levels of a eucalypt forest soils were unaffected despite 30 years of annual burning; a similar result was reported after 10 years from a pine plantation and a eucalypt forest in NSW (Humphreys & Craig, 1981). The results of studies in African savannas also suggest that soil nitrogen and organic matter levels are not significantly affected by annual burning over long periods of time (Moore, 1960; Trapnell *et al.*, 1976). In contrast to those studies which suggest that fire has little effect on soil nitrogen levels, Jones & Richards (1977) found that regular burning of a eucalypt forest caused soil nitrogen levels to decrease. In addition, fire-induced nutrient losses may assume greater significance when expressed as a proportion of plant-available nutrients, rather than as a proportion of the total nutrients contained in an ecosystem (Walker *et al.*, 1986).

Nutrient addition to ecosystems can occur through a variety of mechanisms including rock weathering, atmospheric accession and in the case of nitrogen, biological fixation. To some extent, these mechanisms may serve to offset nutrient losses due to erosion and fire. With the exception of nutrient accession via rainfall however, these processes are particularly difficult to quantify.

There is a strong interaction between fire and termites in savannas, with regular burning of the savanna causing a decrease in numbers of termites. Studies in Ghana have shown that the effect of fire on harvester termite populations is indirect, and occurs as result of a reduction in the availability of dry grass and litter on the soil surface (Benzie, 1986). Observations of termite populations in a savanna near Townsville, north Queensland, suggest that an increase in numbers of mounds of *Nasutitermes longipennis* (Hill) over a period of 3 years may in part be due to the exclusion of fire from the experimental site (J. A. Holt, unpublished information).

Soil erosion

Soil erosion may also result in a substantial loss of nutrients from an ecosystem, with subsequent effects on productivity. Nutrient losses by this process are usually higher in ecosystems where man's activities have led to a greater

TABLE 2. Annual inputs of nutrients (kg ha^{-1} yr^{-1}) in precipitation at five sites in Northern Australia.

Location	Ca	Mg	K	Na	P	N*	S	Source
Townsville (QLD)	4–12	2–3	1–3	10–23	0.2	—	3–6	Probert (1976)
Groote Eylandt (NT)	7	2	11	24	—	1.4	—	Langkamp *et al.* (1982, 1983)
Katherine (NT)	—	0.2	0.3	1.1	—	1.5	—	Wetselaar & Hutton (1963)
Kimberly (WA)	1–3	0.3	1–3	0.2	—	—	1–3	Hingston & Gailitis (1976)
Alligator Rivers Region (NT)	0.5	0.4	1.0	3.4	0.4	2.0	3.2	Noller *et al.* (1985)

* Inorganic N.

degree of disturbance (Saunders & Young, 1983). Natural erosion rates tend to be greater in semi-arid and tropical savanna zones (Saunders & Young, 1983). Overgrazing by domestic herds often exacerbates this problem as the annual rate of soil loss increases dramatically with a decrease in plant cover (Marshall, 1973). Since nutrients lost during soil erosion are not replaced by rock weathering in the short term, this process cannot be considered as recycling.

Nutrient accessions in rainfall

Rainwater samples from several regions of northern Australia contain low but significant levels of nutrients (Table 2). Crutzen *et al.* (1979) suggested that most of the NO_3^-N in the atmosphere is due to the large-scale burning of tropical savannas during the dry season. It is equally likely that some of the other nutrients present in the atmosphere, e.g. Ca, Mg, K, which are subsequently carried to the ground in rainwater, originated from dry season burning. This is also supported by the fact that concentrations of these nutrients in rainwater are generally highest at the beginning of the wet season (Wetselaar & Hutton, 1963; Noller *et al.*, 1985) when the atmosphere is most likely to be more heavily laden with debris from fires and dust storms.

The ratios of nutrients in rainwater at Katherine were found to be similar to the ratios of nutrients in nearby surface soils and vegetation, indicating that the source of these nutrients was terrestrial and local (Wetselaar & Hutton, 1963). Hence the nutrient input via rainfall in the Katherine region was not thought by Wetselaar and Hutton to be 'nutrient accession' in the strict sense, but simply part of a terrestrial cycle. This finding is consistent with the results of studies on the movement of nitrogen compounds in the atmosphere which show that they are returned to the soil close to their source areas (Sanhueza, 1982).

Therefore, there is evidence to suggest that within some tropical savanna systems, nutrient losses caused by burning and the subsequent nutrient gain via rainfall may be closely related. This process might be perceived as a discrete subcycle functioning within the overall nutrient cycle.

CONCLUSION

In tropical savanna ecosystems, microorganisms play a major role in the decomposition of organic matter and cycling of plant nutrients, although their activity is strongly influenced by soil moisture regimes. Termites are the dominant group of soil animals in these ecosystems and they also have a significant effect on nutrient cycling. Fire plays an important role in nutrient cycling by causing the rapid mineralization of organic matter. Nutrient losses may occur with regular burning however, and despite the input of nutrients via atmospheric accession, a net reduction in some nutrients may result. There is some interaction between fire and termites that becomes more significant in those regions where regular burning is conducted.

REFERENCES

Benzie, J.A.H. (1986) The distribution, abundance, and the effects of fire on mound building termites (*Trinervitermes* and *Cubitermes* spp., Isoptera: Termitidae) in northern guinea savanna, West Africa. *Oecologia (Berl.)*, **70**, 559–567.

Birkill, A. (1985) Termite ecology in a tropical savanna grazed by cattle. B.Sc.Hons. thesis, Flinders Univeristy, S.A.

Bonell, M., Coventry, R.J. & Holt, J.A. (1986) Erosion of termite mounds under natural rainfall in semiarid tropical northeastern Australia. *Catena*, **13**, 11–28.

Collins, N.M. (1981) Populations, age structure and survivorship of colonies of *Macrotermes bellicosus* (Isoptera: Macrotermitinae). *J. Anim. Ecol.* **50**, 293–311.

Coventry, R.J., Holt, J.A. & Sinclair, D.F. (1988) Nutrient cycling by mound-building termites in low fertility soils of semi-arid tropical Australia. *Aust. J. Soil Res.* **26**, 375–390.

Crawford, C.S. (1981) *Biology of desert invertebrates.* Springer, Berlin.

Crutzen, P.J., Heidt, L.E., Krasnec, J.P., Pollock, W.H. & Seilor, W. (1979) Biomass burning as a scource of the atmospheric gases CO, H_2, N_2O, NO, CH_3Cl and COS. *Nature*, **282**, 253–256.

Ferrar, P. (1982) Termites of a South African savanna. IV. Subterranean populations, mass determinations and biomass estimations. *Oecologia (Berl.)*, **52**, 147–151.

Freckman, D.W., Cromack, K. & Wallwork, J.A. (1986) Recent advances in quantitative soil biology. *Microfloral and faunal interactions in natural and agro-ecosystems* (ed. by M. J. Mitchel and J. P. Nakas), pp. 399–442. Martinus Nijhoff/Dr W. Junk, Dordrecht.

Gay, F.J. & Calaby, J.H. (1970) Termites of the Australian region. *Biology of termites*, Vol. 2 (ed. by K. Krishna and F. M. Weesner), pp. 643. Academic Press, London.

Gupta, S.R. & Singh, J.S. (1981) Soil respiration in a tropical grassland. *Soil Biol. Biochem.* **13**, 261–268.

Hingston, F.J. & Gailitis, V. (1976) The geographic variation of salt precipitated over Western Australia. *Aust. J. Soil Res.* **14**, 319–335.

Holt, J.A. (1987) Carbon mineralization in semi-arid northeastern Australia: the role of termites. *J. trop. Ecol.* **3**, 255–263.

Holt, J.A. (1988) Carbon mineralization in semi-arid tropical Australia: the role of mound building termites. Ph.D. thesis, University of Queensland, Brisbane, Qld.

Holt, J.A. & Easey, J.F. (1984) Biomass of mound-building termites in a red and yellow earth landscape, north Queensland. *Proc. Nat. Soils Conf. Brisbane, Australia*, p. 363. Aust. Soc. Soil Sci. Inc.

Holt, J.A., Coventry, R.J. & Sinclair, D.F. (1980) Some aspects of the biology and pedological significance of mound-building termites in a red and yellow earth landscape near Charters Towers, north Queensland. *Aust. J. Soil Res.* **18**, 97–109.

Humphreys, F.R. & Craig, F.G. (1981) Effects of fire on soil chemical, structural and hydrological properties. *Fire and the Australian biota* (ed. by A. M. Gill, R. H. Groves and I. R. Noble), pp. 177–200. Aust. Acad. Sci.

Jones, J.M. & Richards, B.N. (1977) Effect of reforestation on turnover of ^{15}N-labelled nitrate and ammonium in relation to changes in soil microflora. *Soil Biol. Biochem.* **9**, 383–392.

Lamb, D., Landsberg, J. & Foot, P. (1983) Effect of prescribed burning on nutrient cycling in *Eucalyptus maculata* forest. *2nd Queensland fire research workshop* (ed. by B. R. Roberts), pp. 21–30. Darling Downs Inst. Adv. Educ. Toowomba.

Langkamp, P.J. & Dalling, M.H. (1983) Nutrient cycling of a stand of *Acacia holosericea* A. Cunn ex Don. III. Calcium, magnesium, sodium and potassium. *Aust. J. Bot.* **31**, 141–149.

Langkamp, P.J., Farnell, G.K. & Dalling, M.J. (1982) Nutrient cycling in a stand of *Acacia holosericea* A. Cunn, ex G. Don. I. Measurements of precipitation interception, seasonal acetylene reduction, plant growth and nitrogen requirement. *Aust. J. Bot.* **30**, 87–106.

Lee, K.E. & Wood, T.G. (1971) *Termites and soils*. Academic Press, London.

Lepage, M. (1972) Recherches ecologiques sur une savane sahelienne du ferlo septentrional, Senegal: donnees preliminaires sur l'ecologie des termites. *La Terre et la Vie*, **26**, 383–409.

Luscher, M. (1961) Air-conditioned termite nests. *Sci. Am.* **205**, 138–145.

Marshall, J.K. (1973) Drought, land use and soil erosion. *The environmental, economic and social significance of drought* (ed. by J. Lovett). Angus and Robertson, Sydney.

McIvor, J.G. (1981) Seasonal changes in the growth, dry matter distribution and herbage quality of three native grasses in northern Queensland. *Aust. J. exp. Agric. Anim. Husb.* **21**, 600–609.

Medina, E. (1982) Physiological ecology of neotropical savanna plants. *Ecology of tropical savannas* (ed. by B. J. Huntly and B. H. Walker), pp. 308–335. Springer, Berlin.

Moore, A.W. (1960) The influence of annual burning on a soil in the derived savanna zone of Nigeria. *Trans. Int. Cong. Soil Sci., 7th Sess., Madison, Wisc.* pp. 257–264.

Mott, J.J., Williams, J., Andrew, M.H. & Gillison, A.N. (1985) Australian savanna ecosystems. *Ecology and management of the worlds savannas* (ed. by J. C. Tothill and J. J. Mott), pp. 56–82. Proc. Int. Savanna Symp., Brisbane, 1984. Aust. Acad. Sci.

Noller, B.N., Currey, N.A., Cusbert, P.J., Tuor, M., Bradley, P. & Harrison, A. (1985) Temporal variability in atmospheric nutrient flux to the Magela and Nourlangie Creek system, Northern Territory. *Proc. Ecol. Soc. Aust.* **13**, 21–31.

Norman, M.J.T. (1963) The pattern of dry matter and nutrient content changes in native pastures at Katherine, N.T. *Aust. J. exp. Agric. Anim. Husb.* **3**, 119–124.

Norman, M.J.T. & Wetselaar, R. (1960) Losses of nitrogen on burning native pasture at Katherine, N.T. *J. Aust. Inst. Agric. Sci.* **26**, 272–273.

Ohiagu, C.E. (1979) Nest and soil populations of *Trinervitermes* spp. with particular reference to *T. geminatus* (Wasmann), (Isoptera), in Southern Guinea savanna near Mokwa, Nigeria. *Oecologia (Berl.)*, **40**, 167–178.

Orchard, V.A. & Cook, F.J. (1983) Relationship between soil respiration and soil moisture. *Soil Biol. Biochem.* **15**, 447–453.

Peterson, H. & Luxton, M. (1982) A comparative analysis of soil fauna populations and their role in decomposition processes. *Oikos*, **39**, 287–388.

Probert, M.E. (1976) The composition of rainwater at two sites near Townsville, Qld. *Aust. J. Soil Res.* **14**, 397–402.

Ratcliffe, F.N., Gay, F.J. & Greaves, T. (1952) *Australian termites*. C.S.I.R.O., Australia.

Sanhueza, E. (1982) The role of the atmosphere in nitrogen cycling. *Plant and Soil*, **67**, 61–71.

Saunders, G.W. (1969) Termites on northern beef properties. *Qld Agric. J.* **95**, 31–36.

Saunders, I. & Young, A. (1983) Rates of surface processes on slopes, slope retreat and denudation. *Earth Surf. Processes Landforms*, **8**, 473–501.

Seastedt, T.R. (1984) The role of microarthropods in decomposition and mineralization processes. *Ann. Rev. Entomol.*, **29**, 25–46.

Skujins, J. (1984) Microbial ecology of desert soils. *Adv. Microbial Ecol.* **7**, 49–91.

Soil Survey Staff, U.S.D.A. (1975) *Soil taxonomy, a basic system of soil classification for making and interpreting soil surveys*. U.S.D.A. Handbook 436. Government Printer, Washington, D.C.

Spain, A.V., Okello-Oloya, T. & Brown, A.J. (1983) Abundances, above-ground masses and basal areas of termite mounds at six locations in tropical north-eastern Australia. *Rev. Ecol. Biol. Sol*, **20**, 547–566.

Stace, H.C.T., Hubble, G.D., Brewer, R., Northcote, K.H., Sleeman, J.R., Mulcahy, M.J. & Hallsworth, E.G. (1968) *A handbook of Australian soils*. Rellim Press, Glenside, S.A.

Swift, M.J. & Sanchez, P.A. (1984) Biological management of tropical soil fertility for sustained productivity. *Nature and Res.* **20**, (4), 2–10.

Trapnell, C.G., Friend, M.T., Chamberlain, G.T. & Birch, H.F. (1976) The effects of fire and termites on a Zambian woodland soil. *J. Ecol.* **64**, 577–588.

Villecourt, P., Schmidt, W. & Cesar, J. (1979) Recherche sur la composition chimique (N, P, K) de la strate herbacee de la savane de Lamto (Cote d'Ivoire). *Rev. Ecol. Biol. Sol*, **16**, 9–15.

Walker, J., Raison, R.J. & Khanna, P.K. (1986) Fire. *Australian soils: the human impact* (ed. by J. S. Russel and R. F. Isbell), pp. 185–216. University of Queensland Press, St Lucia, Qld.

Wetselaar, R. & Hutton, J.T. (1963) The ionic composition of rainwater at Katherine, N.T., and its part in the cycling of plant nutrients. *Aust. J. agric. Res.* **14**, 319–329.

Wildung, R.E., Garland, T.R. & Buschbom, R.L. (1975) The interdependent effects of soil moisture and water content on soil respiration rate and plant decomposition in arid grassland soils. *Soil Biol. Biochem.* **7**, 373–378.

Williams, M.A.J. (1968) Termites and soil development near Brocks Creek, Northern Territory. *Aust. J. Sci.* **31**, 153–154.

Wood, T.G. & Lee, K.E. (1971) The abundance of mounds and competition among colonies of some Australian termite species. *Pedobiologia*, **11**, 341–366.

Wood, T.G. & Sands, W.A. (1978) The role of termites in ecosystems. *Production ecology of ants and termites* (ed. by M. V. Brian), pp. 245–292. I.B.P. 13. Cambridge University Press.

10/Herbivory by insects in Australian tropical savannas: a review

ALAN N. ANDERSEN and W. M. LONSDALE* *Division of Wildlife & Ecology, CSIRO Tropical Ecosystems Research Centre, PMB 44, Winnellie, Northern Territory 0821, Australia, and *Division of Entomology, CSIRO Tropical Ecosystems Research Centre, PMB 44, Winnellie, Northern Territory 0821, Australia*

Abstract. Herbivorous insects are undoubtedly important in savanna ecosystems, but have been largely ignored in studies of herbivory in favour of native ungulates and domestic cattle. In Australia, where the native mammalian herbivore fauna is depauperate, attention has strongly focused on cattle production. In this review we consider three major classes of herbivorous insects, namely grazers, folivores and seed predators, and synthesize information on (1) their composition, diversity and abundance, (2) their ecological effects as herbivores, (3) their importance relative to that of herbivorous mammals, and (4) insect herbivory in Australia compared with that in savannas elsewhere in the world. The most important grazing insects are grasshoppers and harvester termites, although the latter are probably mostly detritivorous. Consumption rates by grasshoppers in African savannas can be comparable to those by large populations of ungulates and cattle; in Australia they are probably the major grazing animals. Folivory and pre-dispersal seed predation by insects are extremely poorly known in Australian savannas, although the results of work on southern species of *Eucalyptus* and *Acacia* (the dominant genera of woody plants throughout Australia) are likely to be at least partly relevant. Harvester ants are the most important post-dispersal seed predators: they consist primarily of omnivorous species of *Monomorium* and *Pheidole*, but also include an endemic radiation of granivorous *Meranoplus*. The overall composition and abundance of herbivorous insects in Australian savannas appears similar to that in other savannas. However, their ecological effects as herbivores are almost totally unknown, and ought to be a priority area for future savanna research.

Key words. Folivory, grazing, herbivory, seed predation, insects, savannas, grasshoppers, termites, Australia, tropics.

INTRODUCTION

'Invertebrates are more important in the maintenance of ecosystems than are vertebrates.'

E. O. Wilson (1987)

'Grasshoppers are responsible for a greater energy turnover (than are cattle) within the (Nylsvley) system.'

M. V. Gandar (1982b)

'Invertebrates are critical to the functional stability of Australian savanna lands.'

J. J. Mott *et al.* (1985)

There is unanimous agreement among savanna ecologists that insects play critical roles in the structure and function of tropical savannas throughout the world. This has meant that savanna insects have been the focus of considerable attention by savanna ecologists, right? Wrong! Despite the great functional importance of savanna insects, they have largely been ignored by savanna ecologists. This is illustrated by the attention given to insects by two recent and influential reviews of savanna ecology. Of the twenty-eight chapters in Huntley & Walker (1982), only four mention insects at all, and just one considers insects in any detail. If all the references to insects are totalled up, they represent less than 2% of the total pages. The situation is even worse in Tothill & Mott (1985), where none of the seventy-three chapters feature insects, and the treatment of insects covers less than 1% of the total pages.

Studies of herbivory in tropical savannas have typically followed one of two approaches. The first is to focus on large, native ungulates, particularly in Africa. The second is to view savannas as places to grow cattle. These approaches reflect traditional mammalian-centred biases more than a balanced consideration of savanna herbivore systems, and promulgate the misleading view that herbivory is synonymous with grazing. In Australia, the second approach to savanna herbivory has predominated, so that Australian savannas have largely been the domain of pasture ecologists.

The obvious consequence of the neglect suffered by insect herbivory in savanna is that little is known about it. The impact of mammalian herbivores on savanna vegetation is well documented (Cumming, 1982; Tainton, 1982;

Milchunas, Sala & Lavenroth, 1988), but in most cases the species composition of insect herbivores is barely known, let alone their ecological effects. Knowledge of insect herbivory is particularly poor in Australia, which means that the best we can do in the following review is to piece together the few fragments of information that are available, rather than offer any comprehensive understanding of the roles of insect herbivores in Australian savanna ecosystems. We define herbivory as the consumption of living plant material by organisms other than pathogens, and identify three major classes of herbivores: (1) grazers, which feed predominantly on grasses and forbs; (2) folivores, which eat the leaves of shrubs and trees; and (3) seed predators, which consume seeds either before or after dispersal.

We begin by summarizing the structure of herbivorous insect communities in tropical savannas. This is based entirely on studies conducted outside Australia, as we know of no relevant data for Australia. By considering each of the major classes of herbivory in turn, we then synthesize information available from Australian savannas, in the context of what is known from savannas elsewhere. Inevitably, the extent of our treatment of each herbivore group reflects the amount of information available on it, and not just its relative importance.

COMMUNITY STRUCTURE OF HERBIVOROUS INSECTS

The community ecology of herbivorous insects has been the focus of considerable study in temperate regions (Strong, Lawton & Southwood, 1984a), but not in tropical savannas. The proportion of total insects that are herbivorous can vary markedly. In a study of six species of broad-leafed trees growing in both Britain and South Africa, Moran & Southwood (1982) found that on average herbivores represented 23% (range 19–27%) of total insect species, 68% (9–93%) of total individuals, and 54% (6—87%) of total insect biomass. The proportion of herbivorous insects was greater on narrower-leafed trees, due mostly to a poorer representation of other guilds. Comparable data are too scanty to make meaningful generalizations about proportional representation in savanna insect communities. In a sweep sample study of four savanna habitats in Costa Rica, counts of insect herbivore species ranged from 54% of total arthropod species in the wettest habitat, to 64% in the driest (Janzen & Schoener, 1968). In African savannas, herbivorous insects often seem to account for about one-third of total insect biomass; this occurred in wet season sweep samples at Lamto, Ivory Coast (Y. Gillon, 1983), in dry season samples at Fete Ole, Senegal (Y. Gillon, 1983), and in wet season light-trap catches at Bauchi, Nigeria (W. M. Lonsdale, unpubl. results).

Herbivorous insect biomass and community structure vary enormously both within and between years. Such year-to-year variation is illustrated by a 5-year study of grasshoppers and leafhoppers in grassland near Nairobi (Denlinger, 1980). The peak yearly totals for both taxa occurred in 1973, with only about half these totals recorded during 1974, and even fewer during the following years. These reductions corresponded with low rainfall. Within any year, herbivorous insects generally reach peak numbers during the west season (e.g. Wolda, 1978, Tanaka & Tanaka, 1982), which is presumably a response to the onset of leaf production.

Insect biomass and community structure are also affected by fire. Regularly burnt savanna has lower arthropod biomass overall than does unburnt savanna, but a variety of responses is shown by different arthropod groups (D. Gillon, 1983). Grasshoppers, for example, were always more abundant in regularly burnt areas (where they fed on fire-induced leaf flush) at Lamto in Ivory Coast, whereas scavenging, detritivorous and predaceous insects were always less abundant.

Studies of phytophagous insects have played a pivotal role in the contemporary debate about the general importance of competition in structuring ecological communities (see Strong *et al.*, 1984b). In many cases it appears that herbivorous insects rarely compete, as their populations are limited by predators or parasites, rather than by food (Lawton & Strong, 1981; Strong, 1984; Strong *et al.*, 1984a). However, competition can bc important in at least some cases (Karban, 1986; Claridge, 1987; Shorrocks & Rosewell, 1987). The role played by competition in structuring communities of savanna insect herbivores is unknown.

INSECT GRAZERS

The low fertility of Australian tropical soils (Mott *et al.*, 1985), and therefore poor nutritional value characteristic of Australian tropical grasses, means that Australian savannas are generally unable to sustain large populations of herbivorous mammals. Kangaroos and other macropods are the dominant native mammalian grazers throughout most of Australia, but generally occur in low densities in tropical savannas (Frith & Calaby, 1969; Calaby, 1980; Newsome, 1983). Indeed, the major grazing mammals in Australian savannas are often feral ungulates, including water buffalo, donkeys, cattle and horses (Freeland, 1990).

Thus insects are probably the dominant native grazers in Australian savannas. Termites and grasshoppers are generally considered to be the most important grazing insects in tropical savannas elsewhere, and, although no comparative data are available, this also seems likely to be the case in Australia. Key (1959), for example claims that grasshoppers alone are probably more important as grazers than all native mammals combined, and are major grazing competitors of domestic stock where they occur. Interestingly, an inverse relationship between the importance of grazing insects on the one hand, and the importance of mammalian grazers on the other, associated with differences in soil fertility, appears to exist within Australian savannas. In a study of thirty sites within the 12,700 km^2 Kakadu National Park in the north of the Northern Territory, termite species richness was found to be positively correlated with grasshopper abundance and richness, but negatively correlated with both soil fertility on the one hand, and the abundance and richness of macropods and rodents on the other (Braithwaite, Miller & Wood, 1988).

The increasing dominance of insects with decreasing soil fertility is therefore not just a direct function of reduced mammalian presence, but also involves a positive response by insects. One potential factor contributing to this response is reduced competition from mammalian herbivores. Herbivores can be food-limited even though they might only consume a small proportion of total primary production (Sinclair, 1975), and Sinclair suggests that competition from ungulates is a major limitation on grasshopper populations in Serengeti long grass systems. However, the generally low abundance of grazing mammals suggests that this is unlikely to be so important in Australian savannas.

An alternative explanation is that sites of low fertility are more favourable for insect grazers, independent of any competition from mammals. One possible mechanism is a reduction in plant chemical defences with reduced soil fertility. According to Rhoades' (1979) 'Optimal Defense Theory', a plant's commitment to defence depends upon its nutrient budget, so that plants growing on infertile soils would be expected to be less well defended against herbivores. Indeed, there are many examples where highest rates of attack by herbivorous insects occur on nutrient-poor soils (reviewed by Mattson & Addy, 1975).

Termites

Grass-eating (harvester) termite mounds are a conspicuous and often spectacular feature of many Australian savanna landscapes, as they are elsewhere in the world. It is tempting to describe them as invertebrate analogues of the ungulate herds that typify African savannas, but this is not strictly correct. Although quantitative data are lacking, Australian harvester termites predominantly forage for dead rather than live grass material (e.g. Watson & Perry, 1981; Watson, Lendon & Low, 1973), as do most harvesters elsewhere (e.g. Sands, 1961; Ohiagu & Wood, 1979), and therefore probably function more as detritivores than as grazers.

The tropical savannas of Africa, Australia and South America each harbour distinctive harvester termite faunas, dominated by endemic genera. In Australia the major harvester genera are *Drepanotermes* (Watson & Perry, 1981; Watson, 1982) and *Tumulitermes* (Lee & Wood, 1971; Holt & Coventry, 1988), although some species of the cosmopolitan *Amitermes* and *Nasutitermes*, which tend to have broader diets, are also important, especially in more mesic savannas (Gay, 1968; Gay & Calaby, 1979; Lee & Wood, 1971). In Africa the most important harvester termites are grass-eating species of *Trinervitermes* and *Hodotermes*, as well as fungus-growing species of *Macrotermes* and *Odontotermes* (Sands, 1961; Lee & Wood, 1971; Wilson, 1971; Josens, 1983). In South America, *Armitermes*, *Syntermes* and other endemic nasute genera predominate. This contrasts with the situation for wood-eaters (e.g. *Coptotermes, Heterotermes, Nasutitermes*) and soil/humus feeders (e.g. *Termes* and allies), where the major genera tend to be widely distributed and often even cosmopolitan. The extent to which the endemism of harvester genera is historical (for example, the Macrotermitinae originated in Africa after it had separated from Australia and South America, which accounts for its absence from the latter continents) rather than ecological is unclear. However, if there are ecological factors predicating against cosmospolitan harvester genera, such as biogeographical differences in the palatability, nutrient content or chemical defences of grasses, then they remain unknown.

Table 1 compares the termites of Kakadu savannas, which are representative of the 'wet' savannas of sub-

TABLE 1. Number of termite species per trophic and taxonomic category in savanna habitats [woodland (W) and open forest (OF)], each summed over nine sites within 12,700 km of Kakadu National Park (data from Braithwaite *et al.*, 1988) compared with those of wooded savannas in Africa (data from Josens, 1983; sampling area and intensity variable).

	Fete Ole, Senegal	Tsavo East, Kenya	Cap Vert, Senegal		Zaria, Nigeria	Mokwa, Nigeria	Lamto, Ivory Coast	Kakadu NP, Australia	
								W	OF
Latitude	16°N	3.5°S	15°N	15°N	11°N	9°N	6°N	12–13°S	
Longitude	15°W	39°E	17°W	17°W	7.5°E	5°E	5°W	132–3°E	
Mean annual rainfall (mm)	375	400	500	575	1170	1175	1290	1300–1600	
Feeding predominantly on:									
Grass/grass litter	8	8	6	6	11	10	10	5	6
Wood/leaf litter	11	17	18	11	16	18	21	20	17
Humus/soil	3	3	8	10	9	9	13	11	12
Family/sub-family									
Mastotermitidae	0	0	0	0	0	0	0	1	1
Kalotermitidae	0	4	1	0	0	0	2	0	1
Rhinotermitidae	3	1	3	0	0	1	1	5	6
Termitinae	8	9	12	12	10	9	10	20	20
Apicotermitinae	0	0	2	3	3	3	4	0	0
Macrotermitinae	3	8	6	6	11	11	9	0	0
Nasutitermitinae	5	3	4	4	5	7	10	9	8
Total species	19	25	28	25	29	31	36	35	36

TABLE 2. Mound densitities (no. ha^{-1}) of grass-eating termite species in tropical savannas of Africa and Australia.

	Location	Density	Reference
Africa			
Trinervitermes trinervoides	S. Africa	534	Murray, 1938
Trinervitermes spp.	Ivory Coast	9–72	Bodot, 1967; Josens, 1983
	Nigeria	70–755	Sands, 1965
	Upper Volta	800–1300	Roose, 1976
Macrotermes bellicosus (Smeathman)	Congo	2–3	Bouillon & Kidieri, 1964
	Nigeria	2–25	Sands, 1965
Macrotermes spp.	E. Africa	3–4	Hesse, 1955
Odontotermes spp.	Kenya	5–7	Glover *et al.*, 1964
	Nigeria	0–15	Sands, 1965
Australia			
**Amitermes laurensis* Mjöberg	N. Australia	28–210	Lee & Wood, 1971
	Townsville area, Qld	72–384	Spain & Brown, 1970
		66–348	Spain *et al.*, 1983
**A. vitiosus* Hill	Pine Creek area, NT	60–268	Lee & Wood, 1971
	nr Charters Towers, Qld	163–688	Spain & Brown, 1979
		284	Holt *et al.*, 1980
	Townsville area, Qld	64–693	Spain *et al.*, 1983
	Musgrave area, Qld	0–12	Spain *et al.*, 1983
Drepanotermes spp.	Larrimah, NT	354	Lee & Wood, 1971
	Central Australia	max. 350,	
		commonly 50–100	Watson *et al.*, 1973
		224–450	Holt *et al.*, 1980
	nr Charters Towers, Qld	0–40	Spain & Brown, 1979
		224–450	Holt *et al.*, 1980
	Townsville area, Qld	0–72	Spain *et al.*, 1983
Nasutitermes triodiae (Froggatt)	N. Australia	3–7	Lee & Wood, 1971
	Musgrave area, Qld	0–9	Spain *et al.*, 1983
Tumulitermes spp.	Pine Creek area, NT	180–500	Lee & Wood, 1971
	nr Charters Towers, Qld	0–43	Spain & Brown, 1979

* General foragers rather than specialist harvesters.

coastal northern Australia, with those of various wooded savannas in Africa. Compared to Africa, Kakadu appears to have a normal complement of non-harvesting species, but seems to be depauperate in harvesters. The occurrence of five or six harvester species in any one locality appears to be typical for Australian savannas (e.g. Spain & Brown, 1979; Holt, Coventry & Sinclair, 1980; Spain, Okello-Oloya & Brown, 1983; Braithwaite *et al.*, 1988). The composition of the termite fauna in the two major savanna habitats of Kakadu, woodland and open forest, is very similar.

Densities of harvester termite mounds in Australian savannas seem to be broadly similar to those in Africa (Table 2). The giant mounds of *Macrotermes* spp. (Africa) and *Nasutitermes triodiae* (Froggatt) (Australia), which often exceed 5 m in height, occur at low densities, typically less than 5 ha^{-1}. However, small mounds such as those of *Trinervitermes* (Africa) and *Tumulitermes* (Australia) can be extremely numerous, and commonly occur at densities of many hundreds per hectare. The composition and abundance of harvester termite mounds vary markedly with soil type (Josens, 1983). In the Northern Territory, for example, greatest mound densities (>1000 ha^{-1}) occur on 'earth' soils, which support several common species. Mound density (240 ha^{-1}) and species richness (only two common species) are much lower on shallow, gravelly loams (Lee & Wood, 1971). Mound density obviously also varies with grass biomass and floristics (Lee & Wood, 1971; Josens, 1983).

There are several reasons why one should approach comparisons based on termite mound density with a great deal of caution. In the first place, figures reported in the literature tend to be maxima rather than means, as the sites with highest densities naturally attract most interest from ecologists (and journal editors). The extent to which this might distort the broader picture of termite abundance is illustrated by a study of sites located randomly within savanna habitats of northern Ghana, West Africa (Benzie, 1986). Mean densities of *Trinervitermes* spp. mounds were 69 ha^{-1} in guinea savanna, 73 ha^{-1} in open woodland, and 30 ha^{-1} in open grassland, figures that are far lower than most of those given for *Trinervitermes* in Table 2. Similarly, in a study of sites located randomly within 700 km^2 of Kakadu, total density of harvester termite mounds averaged only 4, 5 and 12 ha^{-1} in grassland, open forest and woodland respectively (M. Hodda, unpubl. data). Interestingly, Hodda found a strong positive correlation between mound density and the biomass of perennial grasses. The dominant grasses throughout much of the Kakadu region are in fact annual *Sorghum* spp. In Queensland, by contrast, the dominant grasses tend to be perennial species of *Heteropogon, Themeda* and *Bothriochloa* (Mott *et al.*, 1985), and harvester mound densities seem to be much higher than in Kakadu (Table 2).

Another reason why comparisons of mound densities can be unreliable is that many mounds no longer contain termites, and the proportion of abandoned mounds varies with both site and species. For example, Sands (1965) found that from 8% to 47% of *Trinervitermes geminatus* (Wasmann) mounds are abandoned in Nigeria, and that this sometimes rose to 100% in *Macrotermes bellicosus* (Smeathman). Similarly, in the Northern Territory of Australia, 36% of *Tumulitermes* mounds were abandoned at Larrimah, compared with 65% at Brocks Creek (Lee & Wood, 1971). Further, termites often occupy mounds built by different species, and many species do not occupy mounds at all. Within a single harvesting species of *Drepanotermes*, for example, a colony may build its own mound, occupy another species' mound, or have an entirely subterranean nest (Watson & Perry, 1981). Finally, counts of termite mounds provide questionable quantitative information because there is no simple relationship between mound density and termite population size, which varies with species, mound size, site and season (Josens, 1981).

Actual data on population sizes and biomasses of harvester termites in tropical savannas are very sparse, but using information on colony sizes and mound densities from Nigeria (Sands, 1965) and the Ivory Coast (Bodot, 1967), Lee & Wood (1971) estimated that population sizes range from 1000 to 10 000 m^{-2}, representing a biomass range from 5 to 50 g m^{-2}. We know of no comparable data from Australia.

As mentioned previously, it is likely that harvester termites function primarily as detritivores, rather than strictly as grazers, because they feed predominantly on dead grass material. There is a wealth of information on the effects of termites on soil structure and fertility (reviewed by Lee & Wood, 1971), which often can have dramatic effects on vegetation (e.g. Glover, Trump & Wateridge, 1964; Harris, 1966). More recently, the pedological effects of mound-building termites in the savannas of northeastern Australia have been extensively studied. This includes work on the structure and chemistry of mound soil (Spain & Brown, 1979; Holt *et al.*, 1980; Spain, John & Okello-Oloya, 1983; Okello-Oloya, Spain & John, 1985) and its effects on associated vegetation (Spain & McIvor, 1988), and on nutrient cycling (Coventry, Holt & Sinclair, 1988) and carbon mineralization (Holt, 1987). Harvester termites can have other indirect ecological effects, such as accelerating soil erosion by denuding vegetation (Murray, 1938; Watson *et al.*, 1973). However, little is known about the direct impact of harvester termites as grazers. Circumstantial evidence suggests that species of *Trinervitermes* might influence grass species composition in West Africa because of species preferences (Sands, 1961), but no such information is available from Australia.

Grasshoppers

Grasshoppers differ from termites as grazers in two important respects. First, they almost exclusively eat live plant (predominantly leaf) material (Key, 1959; Gandar, 1982a). Second, their roles as grazers in tropical savannas have been intensively studied, if only in a few cases. Grasshoppers are probably the most important grazing insects in the majority of tropical savannas. At Nylsvley in South Africa, for example, they represent 76% of the total biomass of phytophagous insects in *Burkea* savanna, and 93% in *Acacia* savanna (Gandar, 1982a). Elsewhere in Africa, grasshoppers have been found to represent between 40% and 50% of total arthropod biomass in the grass layer (Y. Gillon, 1983). They can occur at average densities of more than 10 m^{-2}, representing a biomass of 2.3 kg ha^{-1} (Gandar, 1982a), and can consume several hundred kilograms of grass per hectare each year (Table 3). From studies of growth rates, respiration rates and assimilation efficiency, Gandar (1982a) calculated that grasshoppers consume about one third of their body weight per day. Grasshoppers are major grazers even in savannas supporting large populations of herbivorous mammals (Table 3).

Australia has a diverse grasshopper fauna, with over 800 acridoid species, belonging to predominantly autochthonous genera (Key, 1959, 1970). Although the biology of some species (particularly pest locusts) is well known, the ecology of Australian grasshoppers is generally poorly known, and estimates of densities are unavailable. There has been only one ecological study of a savanna fauna, that of Friend (1985) in Kakadu National Park. Friend recorded a total of eighty-nine species in the Park, with savanna habitats (woodlands and open forests) containing about twice as many species as did others. Seventy-three (84%) species

TABLE 3. Grass consumption by grasshoppers in African savannas, compared with consumption by mammalian herbivores. The Serengeti supports large populations of ungulates, and also herbivorous rodents. Nyslsvley supports domestic cattle as well as native ungulates.

	Serengeti, Tanzania (Sinclair, 1975)			Nylsvley, South Africa (Gandar, 1982a, b)	
	Long grassland	Short grassland	Kopjes	*Burkea* savanna	*Acacia* savanna
Biomass consumed (kg ha^{-1} yr^{-1})	456	194	484	130	406
Per cent total herbivore consumption	28	11	56	40	73

belonged to the family Acrididae, which is similar to the acridid dominance of African savanna faunas (Gandar, 1982a; Y. Gillon, 1983). Individual woodland sites supported an average of eighteen species during the west season compared with thirteen in open forests, and grasshopper abundance was also higher in woodlands. Grasshopper richness and abundance were both far higher during the wet season than the dry season, which is also the case in African savannas (Gandar, 1982a; Y. Gillion, 1983).

Other grazing insects

Many other groups of phytophagous insects occur in the grass layer of tropical savannas, including caterpillars, phasmids, beetles, and various hemiptera (Sinclair, 1975; Y. Gillon, 1983). However, apart from irregular episodes of severe defoliation by caterpillars (e.g. Bucher, 1987), they tend to be of minor importance compared with grasshoppers. Little is known of these other insect groups in Australian savannas.

FOLIVORY

Insect folivores, principally caterpillars, chrysomelid beetles, sucking bugs and sawfly larvae (Hodkinson & Hughes, 1982; Strong *et al.*, 1984a), are almost universally more important than mammalian folivores. Janzen (1988) has claimed that caterpillars alone consume more living leaves in most forests of the world than all other animals combined. At Nylsvley in South Africa, for example, insect folivores removed 62% of the total leaf material consumed by herbivores (Ganda, 1982b). Of the insect consumption, 74% was by caterpillars, 23% by hemipterans and 3% by beetles. Leaf-cutting ants (Wilson, 1971; Wheeler, 1973) are important defoliators (the leaves are used as a medium for growing fungus, rather than for direct consumption) in Central America, but do not occur on other continents.

We know of no studies of insect folivory in Australian savannas. However, it has been argued that insect attack on eucalypts elsewhere in Australia is high in comparison with forests overseas (Morrow, 1977; Fox & Morrow, 1983, 1986; Morrow & Fow, 1989) (Fig. 1). This view has been opposed by Ohmart, Stewart & Thomas (1983a, b) on the grounds that many of the Australian observations were carried out under unrepresentative conditions (e.g. during periods of insect outbreak), and that the level of herbivory in eucalypt forests are similar to that in comparable forests in the northern hemisphere. Lowman (1985), too, has argued that for tropical rainforests differences in defoliation levels between Australia and the rest of the world are probably negligible. As yet the evidence is not good enough to make a useful comparison, especially as the commonly used technique of measuring missing leaf area on single occasions has been shown to underestimate the amount of herbivory by up to 5 times (Lowman, 1984).

Defoliation by insects is notoriously patchy in space and time (Crawley, 1983). For example, although woody plants in a Costa Rican dry forest commonly lose 1–20% of total leaf area, mostly to caterpillars, individual plants are sometimes totally defoliated, generally by a single generation of

FIG. 1. An example of severe defoliation of a *Eucalyptus* sapling by insects in an Australian savanna. (Photo: A. Andersen.).

one or two caterpillar species (Janzen, 1988). Some insects (e.g. sphingid caterpillars and chrysomelid beetle larvae in Costa Rican dry forest) have outbreak years in which they defoliate all members of their host plant population (Janzen, 1981). At Nylsvley in South Africa, *Vitex rehmanii* Gürake was suddenly stripped by caterpillars in 1977, while *Burkea africana* Hook is attacked and often denuded by *Cerina* caterpillars early in the wet season (Gandar, 1982b). Janzen (1981) saw twenty-four instances of more than 90% defoliation of woody plants at Santa Rosa National Park in Costa Rica during two consecutive growing seasons. Caterpillars were largely responsible. Damage intensities ranged from defoliation of entire populations, a fairly rare occurrence, to the defoliation of scattered individuals only, a more widespread event.

While scattered defoliation is unlikely to affect the overall population size of the host plant, it will clearly influence the success of individual plants. Rockwood (1973) showed that artificial defoliation of six dry forest species in Costa Rica drastically reduced seed production in the following year, with 80% of the test plants producing no seeds whatsoever. *Randia* spp. trees defoliated by caterpillars bore no fruits in the following rainy season (Janzen, 1985).

In addition, once defoliated, a tree is likely to accrue even more damage, as insects are attracted to the new leaf flush produced in response to the previous defoliation (Rockwood, 1974; Lieberman & Lieberman, 1984). Alternatively, trees may not produce a new leaf crop until a year later, thus losing a year's growth (cf. Janzen, 1988).

Phytophagous insects are often highly host specific, primarily because specialization is required to overcome plant chemical and physical defences (Crawley, 1983; Strong *et al.*, 1984a). In addition, like vertebrate herbivores, insects frequently feed preferentially on new leaves (Coley, 1982), which have higher protein levels than do old ones (Sinclair, 1975). In a Ghanaian dry forest, insects (mostly caterpillars) attacked fifteen out of fifty-nine woody species, damaging only the new leaf flushes of those species (Lieberman & Lieberman, 1984). Species known to be rich in secondary compounds, or having hairy leaves, tended to suffer less damage. Rockwood (1974), working in Costa Rican dry forest, found that the flea beetle *Oedionychus* sp. attacked 100% of new leaves of *Crescentia alata* H.B.K. produced after experimental defoliation, and 82% of a second flush of leaves, but only 29% of mature leaves on control trees.

SEED PREDATION

Of the three classes of herbivory, seed predation is possibly the most poorly known in tropical savannas. This contrasts with the situation for neotropical rainforests, which have been the focus of considerable research on the topic (Janzen, 1975; Heithaus & Anderson, 1982; de Steven & Putz, 1984; Roberts & Heithaus, 1986; Schupp, 1988a, b). Seed predators in these forests consume a large proportion of total seed production (Janzen, 1971a), and, over evolutionary time, are believed to have led to widespread modifications to the size, morphology and chemistry of seeds (Janzen, 1969) and to patterns of seed production (Janzen, 1971b). Seed predators possibly also contribute to the high tree diversity characteristic of tropical rainforests (Janzen, 1970; Connell, 1971; but see Hubbell, 1980). Pre-dispersal seed predation in these forests is predominantly by insects, with a noteworthy radiation of bruchid beetles specializing on the seeds of legumes (Janzen, 1969, 1975; Heithaus & Anderson, 1982; including those of deciduous forests, Janzen, 1980), whereas both insects, predominantly ants (Risch & Carroll, 1986; Roberts & Heithaus, 1986), and mammals (de Steven & Putz, 1984; Schupp, 1988a, b) are important post-dispersal seed predators.

Pre-dispersal

We know of no studies of pre-dispersal seed predation in Australian savannas. However, many plant species common in savannas belong to genera that are widely distributed in Australia, and have been the subject of seed predation studies elsewhere. This particularly applies to *Eucalyptus* and *Acacia*, which together dominate the woody component of most Australian vegetation types, including tropical savannas. In all taxa studied, insects are by far the most important pre-dispersal seed predators, and this seems certain to be true for tropical species.

Species of *Eucalyptus* have woody, capsular fruit that in many cases can persist unopened for several years (Andersen, 1989). The major seed-eating insects supported by southern species are anobiid beetles of the genus *Dryophilodes*, and chalcidoid wasps of the genera *Megastigmus* (Torymidae) and *Eurytoma* (Eurytomidae) (Andersen & New, 1987). The predominance of *Dryophilodes* as a seed predator is noteworthy in that most other anobiid beetles are wood-borers rather than granivores (Britton, 1970). Its extensive radiation inside the woody fruits of *Eucalyptus* and other Myrtaceae (Andersen & New, 1987) parallels that of bruchid beetles in neotropical legumes. An exclusion experiment on *Eucalyptus baxteri* (Benth.) Maiden & Blakely, showed that as well as directly consuming seeds, insects reduced total numbers of seeds per fruit and caused large reductions in seed viability, resulting in total seed losses of at least 66% (Andersen, 1988a). The fruits of most tropical eucalypts are far less persistent than those of southern species, so that it is possible that *Dryophilodes* is not so prominent in them.

Pre-dispersal seed predation has been studied in many southeastern Australian species of *Acacia* (Auld, 1983, 1986; New, 1983), and in each case the major predators are weevils, particularly of the genus *Melanterius*. This contrasts with the situation for acacias in Africa and the Middle East (New, 1983), as well as those in the neotropics, which are all attacked by bruchids. Infestation levels of *Bruchidius spadicens* (Fabr.) in seeds of *Acacia tortilis* (Forssk.) Hayne range from 5.1% (Pellew & Southgate, 1984) to 99.6% (Lamprey, Halery & Makacha, 1974). It should be noted, however, that in the latter case seed samples were stored for a year prior to analysis, possibly enabling the bruchids to complete several generations in them. For the leguminous shrub *Crotalaria pallida* Ait in Tanzania, predation intensity by insects averaged 49%, ranging from less than 20% to nearly 100% for individual plants (Moore, 1978). The proportion of seeds of Australian species lost to weevils varies considerably, but can exceed 50% (Auld, 1986).

Post-dispersal

Ants are dominant post-dispersal seed predators throughout most of Australia (Morton, 1982; Andersen & Ashton, 1985; Andersen, 1987, 1988b), and this certainly is true for its tropical savannas. Other insects feeding on fallen seeds in Australian savannas include omnivorous species of gryllid crickets and tenebrionid (helaeine) beetles (A. N. Andersen, unpubl. obs.), and probably also harvester termites (cf. Sands, 1961) and lygaeid bugs (Andersen, 1985). However, these are probably of negligible importance compared to ants. Australia has an impoverished fauna of granivorous mammals (Morton, 1979) and, aside from ants, granivorous birds (e.g. Morton & Davies, 1983) are probably the only other taxa consuming substantial numbers of fallen seeds.

The most important harvester ants in Australian savannas are omnivorous species of *Monomorium* (especially of the *rothsteini* Forel group) and *Pheidole* (Greenslade & Mott, 1979; A. N. Andersen, unpubl. data), genera that are widely distributed throughout Australia and the rest of the world. The specialist granivorous genera of Africa and America, such as *Messor*, *Pogonomyrmex* and *Veromessor* (Brown, Reichman & Davidson, 1979; Abramsky, 1983; Lévieux, 1983), do not occur in Australia. Species of *Rhytidoponera* are important seed predators in southern Australia (Majer,

FIG. 2. *Meranoplus* sp. (*diversus* F. Smith group), with midden of *Sorghum intrans* F. Muell. ex Benth. seed husks. (Photo: A. Andersen.)

1982; Andersen & Ashton, 1985; Andersen, 1988b) but, although many *Rhytidoponera* species occur in Austalian savannas, none is known to eat seeds. The same is true for *Melophorus*, which includes several radiations of granivorous species in arid central and semi-arid southern Australia (Greenslade, 1982).

Despite the overall preponderance of omnivorous species, probably the most notable group of harvester ants in Australian savannas is an endemic radiation of *Meranoplus* (*diversus* F. Smith group) that appears to be exclusively granivorous (Fig. 2). The group is represented throughout most of nothern Australia, and consists of very many (mostly undescribed) species which appear to specialize on the seeds of only one or a few plant (mostly grass) species (A. N. Andersen, unpubl. obs.). One of these subsists entirely on the seeds of *Sorghum intrans* F. Muell. ex Benth., the dominant annual grass in the Kakadu region, and has been studied by Andrew (1986). Colony activity is timed to seed availability, with foragers first becoming active at the end of the wet season, when seedfall occurs, and then becoming dormant late during the dry season, when seed supplies have been depleted. Andrew estimated that the ants consume about one third of total seed production, but concluded that this has little or no effect on the persistence of *Sorghum* populations because of compensatory, density-dependent seed production. However, population models of *Sorghum* suggest that high densities of *Meranoplus* can reduce plant density by up to an order of magnitude (Watkinson, Lonsdale & Andrew, 1989).

One of us (A.N.A.) has made detailed studies of the harvester ant communities of two 30×30 m plots (separated by 50 m) in savanna woodland at Kakadu (Table 4). A total of eighteen species were recorded, including seven species each of *Pheidole* and *Monomorium*, and three species of the *diversus* group of *Meranoplus*. This is an exceptionally rich harvester ant fauna by world standards, but appears to be typical for Australian savannas, and reflects the extraordinary richness of the ant fauna in general (A. N. Andersen, unpubl. data). Total harvester nest density was 0.04 m^{-2} in one plot and 0.17 m^{-2} in the other, with this difference due mostly to a single species of *Pheidole*, whose nests were exceptionally abundant in the second plot. However, even here the most common harvesters were species of the *rothsteini* group of *Monomorium*, which represented 19% of all ants collected in pitfall traps compared with only 2% for *Pheidole* (A. N. Andersen, unpubl. data). Species of the *rothsteini* group of *Monomorium* are important granivores throughout arid, semi-arid, and seasonally arid Australia (e.g. Briese, 1982; Davison, 1982). In the Kakadu plots they are responsible for most of the harvesting of seeds of *Eucalyptus tetrodonta* F. Muell, the dominant tree species. Seed depot experiments indicate that 90% of these seeds are harvested by ants within 2 days (A. N. Andersen, unpubl. data), which is similar to the removal rates recorded for eucalypt seeds in southern Australia (Andersen & Ashton, 1985; Andersen, 1987).

TABLE 4. Number of harvester ant nests in two 30×30 m plots (separated by 50 m) in savanna woodland at Kakadu National Park, Northern Territory (A. N. Andersen, unpubl. data). Figures in parentheses are percentages of total nests.

	Plot 1	Plot 2
Pheidole (7 spp.)	10 (26)	115 (76)
Meranoplus diversus gp (3 spp.)	25 (64)	27 (18)
Monomorium rothsteini gp (2 spp.)	3 (8)	4 (3)
Others* (6 spp.)	1 (3)	5 (3)
Total	39 (100)	151 (100)

* Other species of *Monomorium* and *Meranoplus*.

CONCLUSIONS

Insects are probably the dominant native grazers, folivores and seed predators in Australian savannas, and apparently have been throughout their evolutionary history. This puts Australian savannas in a unique position, as savannas elsewhere either currently support a diverse and abundant fauna of large mammalian herbivores (e.g. Africa), or did so in their recent past (e.g. South America; Bucher, 1987).

Has the paucity of mammalian herbivores led to a measurable response by Australian savanna insects in terms of their diversity or biomass? If so, this might have important implications for community structure at higher trophic levels. For example, the extraordinary diversity of lizards in the Australian arid zone has been linked to an unusually abundant and diverse termite food resource (Morton & James, 1988), and Braithwaite *et al.* (1988) found a positive correlation between the diversity of termites in Kakadu, and the diversity of insectivorous reptiles, birds and mammals. However, although the data base is poor, the overall composition and abundance of herbivorous insects in Australian savannas appears similar to that in savannas elsewhere. This parallels the situation for arid zone granivory, where the structure of harvester ant communities in Australia is similar to that elsewhere in the world, despite Australian deserts having a depauperate rodent fauna (Morton, 1982; Morton & Davidson, 1988).

Whereas data on the composition and abundance of herbivorous insects in Australian savannas is scant, the key question 'What ecological effects do insect herbivores have?' remains totally unanswered. These effects may be manifested in terms of the growth and reproduction of individual plants, in the dymanics of plant populations, or in the structure of plant communities. Even where data are available, it has become increasingly clear that the most commonly used techniques for measuring insect herbivory are seriously flawed (Lowman, 1984; Andersen, 1988a). The message from this paper is obvious: the ecology of savanna insects ought to be a priority for future savanna research. We hope this paper will contribute to a change of emphasis in savanna ecology, towards 'the little things that run the world' (Wilson, 1987), and away from the big things that merely catch our eye.

ACKNOWLEDGMENTS

We are grateful to G. Friend and M. Hodda for providing us with their unpublished data, and to J. Calaby, J. Holt, L. Miller and J. A. L. Watson for their advice during the preparation of the manuscript. R. W. Braithwaite, A. R. E. Sinclair and C. G. Wilson made valuable comments on the manuscript. This is contribution No. 636 of the CSIRO Tropical Ecosystems Research Centre.

REFERENCES

Abramsky, Z. (1983) Experiments on seed predation by rodents and ants in the Israeli desert. *Oecologia*, **57**, 328–332.

Andersen, A.N. (1985) Seed-eating bugs (Hemiptera: Heteroptera: Lygaeidae) at Wilson's Promontory. *Vict. Nat.* **102**, 200–204.

Andersen, A.N. (1987) Effects of seed predation by ants on seedling densities at a woodland site in S.E. Australia. *Oikos*, **48**, 171–174.

Andersen, A.N. (1988a) Insect seed predators may cause far greater losses than they appear to. *Oikos*, **52**, 337–340.

Andersen, A.N. (1988b) Immediate and longer-term effects of fire on seed predation by ants in sclerophyllous vegetation in southeastern Australia. *Aust. J. Ecol.* **13**, 285–293.

Andersen, A.N. (1989) Impact of insect predation on ovule survivorship in *Eucalyptus baxteri*. *J. Ecol.* **77**, 62–69.

Andersen, A.N. & Ashton, D.H. (1985) Rates of seed removal by ants at heath and woodland sites in southeastern Australia. *Aust. J. Ecol.* **10**, 381–390.

Andersen, A.N. & New, T.R. (1987) Insect inhabitants of fruits of *Eucalyptus*, *Leptospermum* and *Casuarina* in southeastern Australia. *Aust. J. Zool.* **35**, 327–336.

Andrew, M.H. (1986) Granivory of the annual grass *Sorghum intrans* by the harvester ant *Meranoplus* sp. in tropical Australia. *Biotropica*, **18**, 344–349.

Auld, T.D. (1983) Seed predation in native legumes of southeastern Australia. *Aust. J. Ecol.* **8**, 367–376.

Auld, T.D. (1986) Variation in predispersal seed predation in several Australian *Acacia* spp. *Oikos*, **47**, 319–326.

Benzie, J.A.H. (1986) The distribution, abundance, and the effects of fire on mound building termites (*Trinervitermes* and *Cubitermes* spp., Isoptera: Termitidae) in northern guinea savanna West Africa. *Oecologia*, **70**, 559–567.

Bodot, P. (1967) Étude écologique des termites des savanes de Basse Cote d'Ivoire. *Insectes Soc.* **14**, 229–258.

Bouillon, A. & Kidieri, S. (1964) Répartition des termitieres de *Bellicositermes bellicosus rex* Grassé et Noirot dan l'Ubangi, d'après les photos aeriennes. Corrélations écologiques qu'elle révèle. *Études sur les Termites Africains* (ed. by A. Bouillon), pp. 373–376. Leopoldville University, Leopoldville.

Braithwaite, R.W., Miller, L. & Wood, J.T. (1988) The structure of termite communities in the Australian tropics. *Aust. J. Ecol.* **13**, 375–391.

Briese, D.T. (1982) Relationship between the seed-harvesting ants and the plant community in a semi-arid environment. *Ant–plant interactions in Australia* (ed. by R. C. Buckley), pp. 11–24. W. Junk Press, The Hague.

Britton, E.B. (1970) Coleoptera. *Insects of Australia*, pp. 495–621. Melbourne University Press, Melbourne.

Brown, J.H., Reichman, O.J. & Davidson, D.W. (1979) Granivory in desert ecosystems. *Ann. Rev. Ecol. Syst.* **10**, 201–227.

Bucher E.H. (1987) Herbivory in arid and semi-arid regions of Argentina. *Revista Chilena de Historia Natural*, **60**, 265–273.

Calaby, J. (1980) Ecology and human use of the Australian savanna environment. *Human ecology in savanna environments* (ed. by D. R. Harris), pp. 321–338. Academic Press, London.

Claridge, M.F. (1987) Insect assemblages – diversity, organization, and evolution. *Organization of communities past and present* (ed. by J. H. R. Gee and P. S. Giller), pp. 141–162. Blackwell Scientific Publications, Oxford.

Coley, P.D. (1982) Rates of herbivory on different tropical trees. *The ecology of a tropical forest* (ed. by E. G. Leight, A. S. Rand and D. M. Windsor), pp. 123–132. Smithsonian Institute Press, Washington, D.C.

Connell, J.H. (1971) On the role of natural enemies in preventing competitive exclusion in some marine animals and in rain forest trees. *Dynamics of populations* (ed. by P. J. Den Boer and G. R. Girndwell), pp. 298–310. Centre for Agricultural Publishing and Documentation, Wageningen.

Crawley, M.J. (1983) *Herbivory: the dynamics of animal–plant interactions*. Blackwell Scientific Publications, Oxford.

Cumming, D.H.M. (1982) The influence of large herbivores on savanna structure in Africa. *Ecology of tropical savannas* (ed. by B. J. Huntley and B. H. Walker), pp. 217–245. Springer, Berlin.

Coventry, R.J., Holt, J.A. & Sinclair, D.F. (1988) Nutrient cycling by mound-building termites in low-fertility soils of semi-arid tropical Australia. *Aust. J. Soil Res.* **26**, 375–390.

Davison, E.A. (1982) Seed ultilization by harvester ants. *Ant–plant interactions in Australia* (ed. by R. C. Buckley), pp. 1–6. W. Junk Press, The Hague.

Denlinger, D.L. (1980) Seasonal and annual variation of insect abundance in the Nairobi National Park, Kenya. *Biotropica*, **12**, 100–106.

de Steven, D. & Putz, F. E. (1984) Impact of mammals on early recruitment of a tropical canopy tree, *Dipteryx panamensis*, in Panama. *Oikos*, **43**, 207–216.

Fox, L.R. & Morrow, P.A. (1983) Estimates of damage by herbivorous insects on *Eucalyptus* trees. *Aust. J. Ecol.* **8**, 139–147.

Fox, L.R. & Morrow, P.A. (1986) On comparing herbivore damage in Australian and north temperature systems. *Aust. J. Ecol.* **11**, 387–393.

Freeland, W.J. (1990) Large herbivorous mammals: exotic species in northern Australia. *J. Biogeogr.* **17**, 445–449.

Friend, G.R. (1985) Grasshoppers. *Kakadu fauna survey – final report*, Vol. 2 (ed. by R. W. Braithwaite), pp. 430–453. CSIRO Division of Wildlife and Rangelands Research, Canberra.

Frith, H.J. & Calaby, J.H. (1969) *Kangaroos*. F. W. Chesire, Melbourne.

Gandar, M.V. (1982a) The dynamics and trophic ecology of grasshoppers (Acridoidea) in a South African savanna. *Oecologia*, **54**, 370–378.

Gandar, M.V. (1982b) Trophic ecology and plant/herbivore energetics. *Ecology of tropical savannas* (ed. by B. J. Huntley and B. H. Walker), pp. 514–543. Springer, Berlin.

Gay, F.J. (1968) A contribution to the systematics of the genus *Amitermes* (Isoptera: Termitidae) in Australia. *Aust. J. Zool.* **16**, 405–457.

Gay, F.J. & Calaby, J.H. (1970) Termites from the Australian region. *Biology of termites*, vol. 2 (ed. by K. Drishna and F. M. Weesner), pp. 393–448. Academic Press, New York.

Gillon, D. (1983) The fire problem in tropical savannas. *Ecosystems of the world 13: tropical savannas* (ed. by F. Bourliere), pp. 617–641. Elsevier, Amsterdam.

Gillon, Y. (1983) The invertebrates of the grass layer. *Ecosystems of the world 13: Tropical savannas* (ed. by F. Bourliere), pp. 289–311. Elsevier, Amsterdam.

Glover, P.E., Trump, E.C. & Wateridge, L.E.D. (1964) Termitaria and vegetation patterns on the Loita plains of Kenya. *J. Ecol.* **52**, 365–377.

Greenslade, P.J.M. (1982) Diversity and food specificity of seed-harvesting ants in relation to habitat and community structure. *Proceedings of the 3rd Australasian conference on grassland invertebrate ecology* (ed. by K. E. Lee), pp. 227–233. S.A. Government Printer, Adelaide.

Greenslade, P.J.M. & Mott, J.J. (1979) Ants of native and sown pastures in the Katherine area, Northern Territory, Australia (Hymenoptera: Formicidae). *Proceedings of the 2nd Australasian conference on grassland invertebrate ecology* (ed. by T. K. Crosby and R. P. Pottinger), pp. 153–156. Government Printer, Wellington.

Harris, M.V. (1966) The role of termites in tropical forestry. *Insectes Soc.* **13**, 255–266.

Hesse, P.R. (1955) A chemical and physical study of the soils of termite mounds in East Africa. *J. Ecol.* **43**, 449–461.

Heithaus, E.R. & Anderson, P.K. (1982) Cumulative effects of plant-animal interactions on seed production by *Bauhinia ungulata*, a neotropical legume. *Ecology*, **63**, 1294–1302.

Hodkinson, I.D. & Hughes, M.K. (1982) *Insect herbivory*. Chapman and Hall, London.

Holt, J.A. (1987) Carbon mineralization in semi-arid northeastern Australia: the role of termites. *J. Trop. Ecol.* **3**, 255–263.

Holt, J.A. & Coventry, R.J. (1988) The effects of tree clearing and pasture establishment on a population of mound-building termites (Isoptera) in North Queensland. *Aust. J. Ecol.* **13**, 321–325.

Holt, J.A., Coventry, R.J. & Sinclair, D.F. (1980) Some aspects of the biology and pedological significance of mound-building termites in a red and yellow earth landscape near Charters Towers, north Queensland. *Aust. J. Soil Res.* **18**, 97–109.

Hubbell, S.P. (1980) Seed predation and the coexistence of tree species in tropical forests. *Oikos*, **35**, 214–229.

Huntley, B.J. & Walker, B.H. (eds) (1982) *Ecology of tropical savannas*. Springer, Berlin.

Janzen, D.H. (1969) Seed-eaters versus seed size, number, toxicity and dispersal. *Evolution*, **23**, 1–27.

Janzen, D.H. (1970) Herbivores and the number of tree species in tropical forests. *Am. Nat.* **104**, 501–528.

Janzen, D.H. (1971a) Seed predation by animals. *Ann. Rev. Ecol. Syst.* **2**, 465–492.

Janzen, D.H. (1971b) Escape of *Cassia grandis* L. beans from predators in time and space. *Ecology*, **52**, 964–979.

Janzen, D.H. (1975) Intra- and interhabitat variations in *Guazuma ulmifolia* (Sterculiaceae) seed predation by *Amblycertus cistelinus* (Bruchidae) in Costa Rica. *Ecology*, **56**, 1009–1013.

Janzen, D.H. (1980) Specificity of seed-attacking beetles in a Costa Rican deciduous forest. *J. Ecol.* **68**, 929–952.

Janzen, D.H. (1981) Patterns of herbivory in a tropical deciduous forest. *Biotropica*, **13**, 271–282.

Janzen, D.H. (1985) A host plant is more than its chemistry. *Ill. Nat. Hist. Surv. Bull.* **33**, 141–174.

Janzen, D.H. (1988) Ecological characterization of a Costa Rican dry forest caterpillar fauna. *Biotropica*, **20**, 120–135.

Janzen, D.H. & Schoener, T.W. (1968) Differences in insect abundance and diversity between wetter and drier sites during a tropical dry season. *Ecology* **49**, 96–110.

Josens, G. (1983) The soil fauna of tropical savannas. III. The termites. *Ecosystems of the world 13: Tropical savannas* (ed. by F. Bouliere), pp. 505–524. Elsevier, Amsterdam.

Karban, R. (1986) Interspecific competition between folivorous insects on *Erigeron glaucus*. *Ecology*, **67**, 1063–1072.

Key, K.H.L. (1959) The ecology and biogeography of Australian grasshoppers and locusts. *Biogeography and ecology in Australia* (ed. by A. Keast, R. L. Crocker and C. S. Christian), pp. 192–210. W. Junk Press, The Hague.

Key, K.H.L. (1970) Orthoptera. *Insects of Australia*, pp. 323–347. Melbourne University Press, Melbourne,

Lamprey, H.F., Halevy, G. & Makacha, S. (1974) Interactions between *Acacia* bruchid seed beetles and large herbivores. *E. Afr. Wildl. J.* **12**, 81–85.

Lawton, J.H. & Strong, D.R. (1981) Community patterns and competition in folivorous insects. *Am. Nat.* **118**, 317–338.

Lee, K.E. & Wood, T.G. (1971) *Termites and soils*. Academic Press, London.

Lévieux, J. (1983) The soil fauna of tropical savannas. IV. The ants. *Ecosystems of the world 13: Tropical savannas* (ed. by F. Bourliere), pp. 525–540. Elsevier, Amsterdam.

Lieberman, D. & Lieberman, M. (1984) The causes and consequences of synchronous flushing in a tropical dry forest. *Biotropica*, **16**, 193–201.

Lowman, M.D. (1984) An assessment of techniques for measuring herbivory: is rainforest defoliation more intense than we thought? *Biotropica*, **16**, 264–268.

Lowman, M.D. (1985) Insect herbivory in Australian rainforests – is it higher than in the Neotropics? *Proc. Ecol. Soc. Aust.* **14**, 109–119.

Majer, J.D. (1982) Ant–plant interactions in the Darling Botanical

District of Western Australia. *Ant–plant interactions in Australia* (ed. by R. C. Buckley), pp. 45–61. W. Junk Press, The Hague.

Mattson, W.J. & Addy, N.D. (1975) Phytophagous insects as regulators of forest primary production. *Science*, **190**, 515–522.

Milchunas, D.G., Sala, O.E. & Lavenroth, W.K. (1988) A generalized model of the effects of grazing by large herbivores on grassland community structure. *Am. Nat.* **132**, 87–106.

Moore, L.R. (1978) Seed predation in the legume *Crotalaria*. I. Intensity and variability of seed predation in native and introduced populations of *C. pallida* Ait. *Oecologia*, **34**, 185–202.

Moran, V.C. & Southwood, T.R.E. (1982) The guild composition of arthropod communities in trees. *J. Anim. Ecol.* **51**, 289–306.

Morrow, P.A. (1977) The significance of phytophagous insects in the eucalypt forests of Australia. *The role of arthropods in forest ecosystems* (ed. by W. J. Mattson), pp. 19–30. Springer, Berlin.

Morrow, P.A. & Fox, L.R. (1989) Estimates of pre-settlement insect damage in Australian and North American forests. *Ecology*, **70**, 1055–1060.

Morton, S.R. (1979) Diversity of desert-dwelling mammals: a comparison of Australia and North America. *J. Mammal.* **60**, 253–264.

Morton, S.R. (1982) Granivory in the Australian arid zone: diversity of harvester ants and structure of their communities. *Evolution of the flora and fauna of arid Australia* (ed. by W. R. Barker and P. J. M. Greenslade), pp. 257–262. Peacock Publications, South Australia.

Morton, S.R. & Davidson, D.W. (1988) Comparative structure of harvester ant communities in arid Australia and North America. *Ecol. Monogr.* **58**, 19–38.

Morton, S.R. & Davies, P.H. (1983) Food of the zebra finch (*Poephila guttata*), and an examination of granivory in birds of the Australian arid zone. *Aust. J. Ecol.*, **8**, 235–243.

Morton, S.R. & James, C.D. (1988) The diversity and abundance of lizards in arid Australia: a new hypothesis. *Am. Nat.*, **132**, 237–256.

Mott, J.J., Williams, J., Andrew, M.H. & Gillison, A.N. (1985) Australian savanna ecosystems. *Ecology and management of the world's savannas* (ed. by J. C. Tothill and J. J. Mott), pp. 56–82. Australian Academy of Science, Canberra.

Murray, J.M. (1938) An investigation of the interrelationships of the vegetation, soils and termites. *S. Afr. J. Sci.* **35**, 288–297.

New, T.R. (1983) Seed predation of some Australian Acacias by weevils (Coleoptera: Curculionidae). *Aust. J. Zool.* **31**, 345–352.

Newsome, A.E. (1983) The grazing Australian marsupials. *Ecosystems of the world 13: tropical savannas* (ed. by F. Bourliere), pp. 441–461. Elsevier, Amsterdam.

Ohiagu, C.E. & Wood, T.G. (1979) Grass production and decomposition in Southern Guinea Savanna, Nigeria. *Oecologia*, **40**, 155–165.

Ohmart, C.P., Stewart, L.G. & Thomas, J.R. (1983a) Leaf consumption by insects in three *Eucalyptus* forest types in south eastern Australia and their role in short-term nutrient cycling. *Oecologia*, **59**, 322–330.

Ohmart, C.P., Stewart, L.G. & Thomas, J.R. (1983b) Phytophagous insect communities in the canopies of three *Eucalyptus* forest types in south eastern Australia. *Aust. J. Ecol.* **8**, 395–403.

Okello-Oloya, T., Spain A.V. & John, R.D. (1985) Selected chemical characteristics of the mounds of two species of *Amitermes* (Isoptera, Termitinae) and their adjacent surface soils from northeastern Australia. *Rev. Écol. Biol. Sol*, **22**, 291–311.

Pellew, R.A. & Southgate, B.J. (1984) The parasitism of *Acacia tortilis* seeds in the Serengeti. *Afr. J. Ecol.* **22**, 73–75.

Rhoades, D.F. (1979) Evolution of plant chemical defense against herbivores. *Herbivores: their interaction with secondary plant metabolites* (ed. by G. A. Rosenthal and D. H. Janzen), pp. 3–54. Academic Press, New York.

Risch, S.J. & Carroll, C.R. (1986) Effects of seed predation by a tropical ant on competition among weeds. *Ecology*, **67**, 1319–1327.

Roberts, J.T. & Heithaus, E.R. (1986) Ants rearrange the vertebrate-generated seed shadow of a neotropical fig tree. *Ecology*, **67**, 1046–1051.

Rockwood, L.L. (1973) The effect of defoliation on seed production of six Costa Rican tree species. *Ecology*, **54**, 1363–1369.

Rockwood, L.L. (1974) Seasonal changes in the susceptibility of *Crescentaia alata* leaves to the flea beetle, *Oedionychus* sp. *Ecology*, **55**, 142–148.

Roose, E.J. (1976) Contribution à l'étude de l'influence de la mèsofaune sur la pédogénèse actuelle en milieu tropical. Rapport ORSTOM. Centre d'Adiopodoumé, Ivory Coast, 56pp.

Sands, W.A. (1961) Foraging behaviour and feeding habits in five species of *Trinervitermes* in West Africa. *Entomologia exp. appl.* **4**, 277–288.

Sands, W.A. (1965) Termite distribution in man-modified habitats in West Africa, with special reference to species segregation in the genus *Trinervitermes* (Isoptera, Termitidae, Nasutitermitinae). *J. Anim. Ecol.* **34**, 557–571.

Schupp, E.W. (1988a) Factors affecting post-dispersal seed survival in a tropical forest. *Oecologia*, **76**, 525–530.

Schupp, E.W. (1988b) Seed and early seedling predation in the forest understorey and in treefall gaps. *Oikos*, **51**, 71–78.

Shorrocks, B. & Rosewell, J. (1987) Spatial patchiness and community structure: coexistence and guild size of drosophilids on ephemeral resources. *Organization of communities past and present* (ed. by J. H. R. Gee and P. S. Giller), pp. 29–51. Blackwell Scientific Publications, Oxford.

Sinclair, A.R.E. (1975) The resource limitation of trophic levels in tropical grassland ecosystems. *J. Anim. Ecol.* **44**, 497–520.

Spain, A.V. & Brown A.J. (1979) Aspects of the biology of harvester termites in a grassland environment (Isoptera). *Proceedings of the 2nd Australasian conference on grassland invertebrate ecology* (ed. by T. K. Crosby and R. P. Pottinger), pp. 141–145. Government Printer, Wellington.

Spain, A.V., John, R.D. & Okello-Oloya, T. (1983) Some pedological effects of selected termite species at three locations in north-eastern Australia. *Proceedings of the VIII international colloquium of soil zoology*, pp. 143–149.

Spain, A.V. & McIvor, J.G. (1988) The nature of herbaceous vegetation associated with termitaria in north-eastern Australia. *J. Ecol.* **76**, 181–191.

Spain, A.V., Okello-Oloya, T. & Brown, A.J. (1983) Abundances, above-ground masses and basal areas of termite mounds at six locations in tropical north-eastern Australia. *Rev. Écol. Biol. Sol*, **20**, 547–566.

Strong, D.R. (1984) Exorcising the ghost of competition past: phytophagous insects. *Ecological communities: conceptual issues and the evidence* (ed. by D. R. Strong, D. Simberloff, L. G. Abele and A. B. Thistle), pp. 28–41. Princeton University Press.

Strong, D.R., Lawton, J.H. & Southwood, T.R.E. (1984a) *Insects on plants*. Blackwell Scientific Publications, Oxford.

Strong, D.R., Simberloff, D., Abele, L.G. & Thistle, A.B. (eds) (1984b) *Ecological communities: conceptual issues and the evidence*. Princeton University Press.

Tainton, N.M. (1982) Response of the humid subtropical grasslands of South Africa to defoliation. *Ecology of tropical savannas* (ed. by B. J. Huntley and B. H. Walker), pp. 405–414. Springer, Berlin.

Tanaka, L.K. & Tanaka, S.K. (1982) Rainfall and seasonal changes in arthropod abundance on a tropical oceanic island. *Biotropica*, **14**, 114–123.

Tothill, J.C. & Mott, J.J. (eds) (1985) *Ecology and management of the world's savannas*. Australian Academy of Science, Canberra.

Watkinson, A.R., Lonsdale, W.M. & Andrew, M.H. (1989) Modelling the population dynamics of an annual plant: *Sorghum intrans* in the Wet–Dry tropics. *J. Ecol.* **77**, 162–181.

Watson, J.A.L. (1982) Distribution, biology and speciation in the Australian harvester termites, *Drepanotermes* (Isoptera: Termitinae). *Evolution of the flora and fauna of arid Australia* (ed. by W. R. Barker and P. J. M. Greenslade), pp. 263–265. Peacock, Adelaide.

Watson, J.A.L., Lendon, C. & Low, B.S. (1973) Termites in mulga lands. *Tropical Grasslands*, **7**, 121–126.

Watson, J.A.L. & Perry D.H. (1981) The Australian harvester termites of the genus *Drepanotermes* (Isoptera: Termitinae). *Aust. J. Zool. Suppl. Ser.* **78**, 1–153.

Wheeler, W.M. (1973) *The fungus-growing ants of North America*. Dover Publications, New York.

Wilson, E.O. (1971) *The insect societies*. Belknap Press, Cambridge, Mass.

Wilson, E.O. (1987) The little things that run the world. *Conservation Biology*, **1**, 344–346.

Wolda, H. (1978) Fluctuations in abundance of tropical insects. *Am. Nat.* **122**, 1017–1045.

Wolda, H. (1980) Seasonality of tropical insects. I. Leafhoppers (Homoptera) in Las Cumbres, Panama. *J. Anim. Ecol.* **49**, 277–290.

11/Large herbivorous mammals: exotic species in northern Australia

W. J. FREELAND *Conservation Commission of the Northern Territory, P.O. Box 496, Palmerston, Northern Territory 0831, Australia*

Abstract. Northern Australia lacks an extant, diverse fauna of native, large herbivorous mammals. The native large mammal fauna became extinct during the Pleistocene, and has been replaced by introduced ungulates during the past 150 years. The introduced fauna is as species rich as many native tropical savanna megafaunas. Carrying capacities for the introduced species are greater than would be expected from their body sizes, and greater than the species achieve in their native habitats. Available evidence suggests that the feral ungulates lack significant impact from predators and pathogens, and that populations are either density independent, or regulated by a mechanism of food limitation unlikely to occur among species in their native habitats. Control of these populations is central to the conservation of northern Australian savannas, and may be best achieved by the introduction of pathogens.

Key words. Herbivore, large mammals, carrying capacity, population density, population regulation, body size, chemical defence, mineral nutrition, parasites, predators, feral, ungulates, introduced species.

INTRODUCTION

Compared to the tropical savannas of Asia and Africa, the northern Australian savannas are depauperate in native species of large herbivorous mammal. In the Northern Territory north of 18°S there are only six species of macropod marsupial that could qualify as large herbivorous mammals. It is unusual to find more than three or possibly four of these species living in the same geographic area, and even then the species appear to exhibit different patterns of habitat choice. When compared to the Asian and African savannas which have at least six to eight species of large herbivore (McKay & Eisenberg, 1974; Eisenberg & Seidensticker, 1976), the northern Australian region is clearly depauperate.

The paucity of large herbivorous mammals in northern Australia is due to extinction of large herbivorous marsupials rather than the habitat being in some way unsuitable for large grazing or browsing mammals. Not only do the northern Australian savannas bear a strong structural resemblance to those from other parts of the tropics, but over two-thirds of the vascular plant genera present in the Northern Territory between 11 and 16°S have global distributions including other areas of the world's tropics. Only 15% of genera have distributions restricted to Australia (Bowman, Wilson & Dunlop, 1988).

Up until the Pleistocene Australia had a fauna of large herbivorous marsupials, as well as large ratite birds, and large marsupial and varanid predators (Hope, 1984). Man's first arrival on the continent may have caused the extinction either through hunting or indirectly by habitat modifications wrought by the use of fire, or climate change may have played the major role (Horton, 1980). Whatever the cause of the extinctions, it is clear that when European man arrived in northern Australia approximately 150 years ago the habitat was underutilized by herbivorous mammals, and had a flora with a long evolutionary history of herbivory from large mammals.

Since the arrival of European man the extinct megafauna has been replaced by large ungulates introduced from Asia, Europe and Africa. A total of eight species of ungulate (and one size variant, the Timor Pony) have been introduced to various places in the Northern Territory. Some species are now widely spread throughout the region (e.g. *Bos taurus* and *Equus caballus*), others have retained a localized distribution (e.g. *Bos banteng* and *Cervus unicolor*), and some are still expanding their ranges (e.g. *Camelus dromedarius*) (Bayliss & Yeomans, 1989a, b). The species are either entirely feral, primarily feral, or maintain feral populations together with minimally managed harvested stock. Expansion of the feral and the near feral herds has resulted in overgrazing and alteration of native habitats (Letts, Bassingthwaight & deVos, 1979; Braithwaite *et al.*, 1984; Bowman & Panton, 1989).

In northern Australia it is assumed that there is something unusual or undesirable about habitat change caused by the feral herds. Undesirable effects include loss of habitat for native species and soil erosion (e.g. Letts *et al.*, 1979; Bowman & Panton, 1989). It is open to question whether the impact of the feral herds is in any way greater than or in some way different from that which occurs in the savannas of Asia or Africa.

If the impact of the feral herds on savanna vegetation in northern Australia is greater than occurs with native herds, one possible explanation is that feral herds exist at densities greater than those achieved in their native habitats. Data on the densities of introduced herbivorous mammals in the Australasian region are summarized and compared to expectations of population density based on body size (Damuth, 1981), and densities particular species of introduced herbivore achieve in their native habitats. Results are discussed in relation to what is known of factors likely to influence the size of Australian and native populations.

METHODS

Data on the population densities of ten species of herbivorous mammal in Australasia, and seven species in both Australasian and native habitats, have been extracted from a variety of sources and are available on request. All species and populations considered have been in their introduced habitats for greater than 100 years. Data for the large mammals are taken primarily from aerial survey results, whereas data for the smaller species are based on capture–recapture studies. Data on the introduced populations are compared to the data set assembled by Damuth (1981) in his study of the relationship between mammalian herbivore body size and population density. Some of the variance in his data set was due to inclusion of data from populations introduced to alien environments. All such data were removed from the data set and body size–population density relationship recalculated for comparison with the data on introduced species. The adjusted body size–population density relationship is described by:

$$\log_{10}\text{Density}=4.196-0.74(\log_{10}W),$$

where Density is km^{-2}, and W=body mass in g (r^2=0.73, N=319).

TABLE 1. Ecological densities of herbivorous mammals as predicted from the Damuth relationship, as observed in their native habitats and as observed in Australasia. Densities are in km^{-2}, – indicates missing data, and * indicates species for which there are data from northern Australia. The native density for *E. asinus* is that of *E. heminonus*. Species have been arranged in ascending order of body size.

Species	Population density			
		Native	Australasia	
	Predicted	habitat	Mean	Maximum
Mus musculus	1842.00	–	31769.00[1]	–
Rattus exulans	890.00	1073.00[2]	8500.00[3]	18500.00[3]
Rattus rattus	446.00	–	4500.00[4]	6400.00[4]
Oryctolagus cuniculus	59.00	–	135.00[5]	–
*Sus scrofa**	4.78	0.79[6]	6.61[7]	12.61[7]
Capra hircus	3.81	3.10[8]	5.00[9]	–
*Bos banteng**	1.50	0.85[10]	2.46[10]	–
*Equus asinus**	1.35	0.33[12]	2.20[13]	10–15[13]
*E. caballus**	1.10	–	1.40[13,14]	7.20[13,14]
*Bubalus bubalis**	1.08	0.83[6]	3.30[14]	25.20[14]

1=Newsome (1969); 2=Harrison (1969) and Dwyer (1978); 3=Strecker (1962) and Wirtz (1972); 4=Tamarin & Malecha (1971); 5=Dunsmore (1974) and Wood (1980); 6=Eisenberg & Seidensticker (1976); 7=Hone (1986); 8=Damuth (1982); 9=Henzell & McCloud (1984); 10=Hoogerwerf (1970); 11=Bayliss & Yeomans (1989a); 12=Gee (1963) and Wolfe (1979); 13=Graham *et al.* (1982); 14=Bayliss & Yeomans (1989b).

RESULTS AND DISCUSSION

The phenomenon

Herbivorous mammals introduced to Australasia exist at higher population densities than those predicted from knowledge of their body sizes or that occur in the species' native habitats. All ten introduced species for which there are data have ecological densities greater than predicted by the Damuth relationship (Table 1). This frequency of occurrence of densities greater than predicted is higher than expected by chance (binomial probability=0.00098). As five of the six species for which there are data in both the Australasian and native habitats have observed native densities less than predicted by the Damuth relationship (Table 1), the observed high densities in Australasia are not due to the species being a biased selection that has higher than usual natural densities. In all cases for which there are data, the mean Australasian density is greater than is observed in the native habitat (Table 1). Maximum densities over minimum areas of at least 100 km^2 in Australasia for species surveyed from the air are 16–45 times densities recorded in native habitats.

Possible explanations for the unusually high population densities of introduced herbivores in the Australasian environment include: (a) an absence of competition from species rich herbivore communities, (b) a paucity of potential predators, (c) a paucity of parasites and diseases, and (d) an absence of allelochemical/physical defences capable of protecting Australasian plants against introduced herbivores.

Competition

Lack of interspecific competition may contribute to the high densities of introduced herbivores in Australia.

Species richness of mammalian herbivores (native and introduced) in northern Australia compares favourably with that for savannas in other parts of the tropics. Large mammal species richness for areas in the Australian tropical savannas from which some of the density data were obtained range from seven to ten. Examples include the Cobourg Peninsula (nine species) and the Victoria River District (ten species) (Table 2). These areas were selected because together they embrace the majority of the latitudinal moisture gradient present in the Northern Territory, and be-

TABLE 2. Species present in two communities of large mammalian herbivores in the Northern Territory, Australia. (a)=horse, (b)=Timor Pony.

Species	Cobourg Peninsula	Victoria River District
Native		
Macropus robustus		*
M. antilopinus	*	*
M. agilis	*	*
Onychogalea ungifera	*	*
Introduced		
Bos banteng	*	
B. taurus		*
Bubalus bubalis	*	*
Cervus unicolor	*	
Camelus dromedarius		*
Equus caballus (a)	*	*
E. caballus (b)	*	
E. asinus		*
Sus scrofa	*	*
Total species	9	10

cause both areas have been used to derive some of the population density estimates. These levels of species richness compare favourably with those of the Asian (six to eight species) and many African savannas (McKay & Eisenberg, 1974; Eisenberg & Seidensticker, 1976).

While species richness may not be a significant factor in any possible absence of competition among the feral and native populations in northern Australia, the combinations of species involved differs greatly from that of natural communities, and this may influence the types and effects of competitive interactions. The major differences between the Australian man-made communities and those of Asia/Africa are: (a) the absence of any extensive period during which the Australian herbivores could have undergone co-evolution resulting in a minimization of competitive interactions, (b) the absence from Australia of exceedingly large herbivores such as elephants and rhinoceros, and (c) the absence from Australia of browsing species. The first and last factors result in an expectation of more intense competitive interactions in Australian as opposed to natural communities, while any possible effects of the second are difficult if not impossible to predict.

A reduction of more than 50% of the feral donkeys (*E. asinus*) in an area in the Victoria River District appeared to result in a significant increase in the population of feral horses (*E. caballus*) (Freeland & Choquenot, 1990). Competition clearly influences patterns of relative abundance of large herbivorous mammals in northern Australia, but it is not known whether this has any greater or lesser effect in Australian as opposed to native communities. The interaction between donkeys and horses may be a consequence of the haphazard man-made nature of the community resulting in sympathy of species lacking a co-evolutionary background. Natural communities do not usually have more than one species of wild equid.

Available information on levels of biomass per unit of land area suggest that northern Australia ungulates may experience more inter-specific competition than do ungulate populations in southern Asia. Data from southern Asia (Eisenberg & Seidensticker, 1976) suggest a mean biomass of native ungulates of 1153 kg km^{-2} (standard deviation =865, N=8) significantly lower than is found in introduced ungulates in parts of northern Australia (water buffalo, horses, cattle and donkeys only) (mean=2225 kg km^{-2}, standard deviation=767, N=6 areas (9000–17,700 km^{2})), that have been colonized by introduced species for >50 years (Bayliss & Yeomans, 1989b). Because the introduced ungulates of northern Australia have ecological equivalents (in some cases the same species) in Asia, it might be taken that levels of competition could be higher in Australia. This is especially so given the absence of introduced leaf-eating specialists from northern Australia, and the absence of feral pigs and native macropods from the Australian data. Any conclusion drawn from these data requires that soils, rainfall and survey methods have not significantly biased the results.

Australian levels of biomass are in one case equivalent to that of Africa, and in another the Australian biomass is much lower. The pronounced biological differences between the African and introduced Australian faunas make interpretation of these patterns difficult if not impossible. The Australian biomass mean is equivalent to that for ungulates in Rwenzori National Park, Uganda (2696 kg km^{-2}), even though the Australian figure does not include data for native species or feral pigs. The Australian estimate is far less than the reported 8427 kg km^{-2} for Transvaal lowveld (Damuth, 1982). While the Australian figure does not include pigs or the large native herbivores, it seems unlikely that they could account for the difference between Australia and the lowveld. The absence from Australia of functional equivalents to the giraffe (40% of the lowveld biomass), the leaf eating kudu (6% of biomass) and a migratory equivalent to the wildebeeste (26% of biomass) make direct comparisons difficult at best. Migratory species can have 11 or more times the population sizes of non-migratory species (Fryxell, Greever & Sinclair, 1988). These biological differences are likely to have a major impact on any differences between the areas' potentials for interspecific competition. Certainly the levels of biomass recorded from northern Australia are substantial, and given the haphazard nature of the community composition and the absence of browsing species, interspecific competition may be unusually high, rather than unusually low.

Resolution of the problem requires more detailed data on total community biomass (rather than species specific population sizes) and species richness in relation to rainfall and soil fertility, as well as manipulative experiments in the Australian and native environments. Account needs also to be taken of any effects that wildlife management (or its absence) and fencing may play in biasing some of the African results.

Predators and pathogens

Low levels of predation and low pathogen loads are likely to contribute towards the maintenance of high densities of introduced mammalian herbivores in Australia.

Predation can have a marked impact on juvenile survivorship of large herbivorous mammals (e.g. Beasom, 1974), and Australia is depauperate in large predators. The only predator potentially capable of consuming any of the larger species of introduced herbivore is the dingo (*Canis familiaris dingo*) (i.e. the domestic dog, introduced by man). Although the dingo's effect on prey populations remains unquantified, its effect is unlikely to equate with that of species rich natural communities of predators that have species with much larger body sizes (e.g. lions, tigers, hyaenas).

Introduced ungulates in northern Australia experience a reduced pathogen load. Native Australian herbivores are marsupials and rodents, whereas the introduced species are primarily ungulate. Relatively few of the native pathogens are capable of infesting the introduced species (e.g. Freeland, 1983). Relatively few pathogens accompanied introduction of the herbivores, and virtually none of the major ungulate pathogens is present in Australia (e.g. African horse sickness, anthrax, foot and mouth disease, rinderpest, blue tongue, equine influenza) (Seddon, 1953; Meischke & Geering, 1985). Disease is a major cause of mortality in natural populations of herbivorous mammals (Dobson & May, 1982), and is likely to be less significant among Australia's feral herds.

Plant defences

Plants in the northern Australian savannas are unlikely to lack defences against, or have less impact on mammalian populations than do plants in other savannas. Nor should any such effects be expected given the long plant–mammalian herbivore coevolution of Australian environments, and the great similarity between Australian and pantropical savanna floras (Bowman *et al.*, 1988). What might be expected is that because of reduced impact from predators and pathogens, populations of feral species in Australia exceed bounds imposed in native environments, only to have imposed on them limits not usually observed in natural situations.

Feral donkey populations in northern Australia are limited primarily by poor juvenile survivorship at carrying capacity densities (Choquenot, 1988). Poor juvenile survivorship is linked to the apparent inability of females to lactate when depleted of mineral nutrients (Na, Ca, P) (Freeland & Choquenot, 1990). Females at carrying capacity are depleted of these nutrients relative to females in growing populations. Low mineral nutrient status is caused by a lower level of mineral intake, and fecal loss of mineral nutrients (apparently caused by the high fibre, species poor diet eaten when at carrying capacity density) (Freeland & Choquenot, 1990).

If, as seems likely, predators and pathogens have significant impacts on populations of mammalian herbivores in natural settings, population limitation as observed in feral donkeys may only occur in feral or man-disturbed communities.

CONCLUSIONS

Species of large mammalian herbivore introduced to northern Australia exist at densities greater than those achieved in the species' native habitats. These high density populations appear to result from the absence of significant predation and/or significant pathogen loads. In at least one case this has resulted in population regulation imposed by grazing caused changes in food eaten. The northern Australian savannas may be subject to unusually high levels of degradation from large mammals because of the absence of factors that regulate natural populations of ungulates. This expectation is in accordance with reports of severe habitat alteration (e.g. Letts *et al.*, 1979). Conservation of the northern Australian environment requires the control of feral ungulates. This may be achieved at least cost and least environmental disturbance by the introduction of pathogens, rather than physical intervention by man or the introduction of predators.

ACKNOWLEDGMENTS

D. Bowman, R. Braithwaite, S. Tidemann, P. Whitehead and J. Woinarski are thanked for their critical comments. J. Damuth is thanked for making his 1981 data set available.

REFERENCES

Bayliss, P. & Yeomans, K.M. (1989a) Correcting bias in aerial survey population estimates of feral livestock in Northern Australia using the double count technique. *J. appl. Ecol.* **26**, 925–934.

Bayliss, P. & Yeomans, K.M. (1989b) The distribution and abundance of feral livestock in the 'Top End' of the Northern Territory (1985–1986), and their relation to population control. *Aust. Wildl. Res.* **16**, 651–676.

Bayliss, P. & Yeomans, K.M. (1989c) The distribution and abundance of Bali Cattle and other feral ungulates on Cobourg Peninsular, Northern Territory, 1985. *Aust. Wildl. Res.* (in press).

Beasom, S.L. (1974) Relationships between predator removal and white-tailed deer net productivity. *J. Wildl. Managemnt*, **38**, 854–859.

Bowman, D.M.J.S. & Panton, W. (1989) Banteng (*Bos javanicus*) and pig (*Sus scrofa*) habitat impact, Cobourg Peninsular, Northern Australia. *Aust. J. Ecol.* (in press).

Bowman, D.M.J.S., Wilson, B.A. & Dunlop, C.R. (1988) Preliminary biogeographic analysis of the Northern Territory vascular flora. *Aust. J. Bot.* **36**, 503–517.

Braithwaite, W.R., Dudzinski, M.L., Ridpath, M.G. & Parker, B.S. (1984) The impact of water buffalo on the monsoon forest ecosystem in Kakadu National Park. *Aust. J. Ecol.* **9**, 309–322.

Choquenot, D. (1988) Feral donkeys in northern Australia: population dynamics and the cost of control. M.Sc. dissertation, Canberra College of Advanced Education.

Damuth, J. (1981) Population density and body size in mammals. *Nature*, **290**, 699–700.

Damuth, J. (1982) The analysis of the degree of community structure in assemblages of fossil mammals. Ph.D. dissertation, University of Chicago.

Dobson, A.P. & May, R.M. (1982) Disease and conservation. *Animal disease in relation to animal conservation* (ed. by M. A. Edwards and U. McDonnell), pp. 345–365. Symposia of the Zoological Society of London, No. 50. Academic Press, London.

Dunsmore, J.D. (1974) The rabbit in subalpine southeastern Australia. I. Population structure and productivity. *Aust. Wildl. Res.* **1**, 1–16.

Dwyer, P.D. (1978) A study of *Rattus exulans* (Peale) (Rodentia: Muridae) in the New Guinea highlands. *Aust. Wildl. Res.* **5**, 221–248.

Eisenberg, J.F. & Seidensticker, J. (1976) Ungulates in Southern Asia: a consideration of biomass estimates for selected habitats. *Biol. Conserv.* **10**, 293–308.

Freeland, W.J. (1983) Parasites and the co-existence of animal host species. *Am. Nat.* **121**, 223–236.

Freeland, W.J. & Choquenot, D. (1990) Determinants of herbivore carrying capacity: plants, nutrients and *Equus asinus* in northern Australia. *Ecology* (in press).

Fryxell, J.M., Greever, J. & Sinclair, A.R.E. (1988) Why are migratory ungulates so abundant? *Am. Nat.* **131**, 781–798.

Gee, E.P. (1963) The Indian Wild Ass (a survey). *Oryx*, **7**, 9–21.

Graham, A., Raskin, S., McConnell, M. & Begg, R. (1982) An aerial survey of feral donkeys in the V.R.D. Conservation Commission of the Northern Territory, Technical Report.

Harrison, J.L. (1969) The abundance and population densities of mammals in Malayan lowland forests. *Malayan Nature J.* **22**, 174–178.

Henzell, R.P. & McCloud, P.I. (1984) Estimation of the density of feral goats in part of arid South Australia by means of the Petersen Estimate. *Aust. Wildl. Res.* **11**, 93–102.

Hone, J. (1986) An evaluation of helicopter shooting of feral pigs. Report to the Conservation Commission of the Northern Territory.

Hoogerwerf, A. (1970) *Udjung Kulon: the land of the last Javan rhinoceros*. Brill, Leiden.

Hope, J. (1984) The Australian Quaternary. *Vertebrate zoogeography and evolution in Australasia* (ed. by M. Archer and G. Clayton), pp. 69–81. Hesperian Press, Victoria Park.

Horton, D.R. (1980) A review of the extinction question: Man, climate and megafauna. *Archaeol. Phys. Anthropol. Oceania*, **15**, 86–97.

Letts, G.A., Bassingthwaight, A. & deVos, W.L. (1979) Feral animals in the Northern Territory: report of a board of inquiry. Northern Territory Government Printer, Darwin.

McKay, G.M. & Eisenberg, J.F. (1974) Movement patterns and habitat utilization of ungulates in southeastern Ceylon. In: *The behaviour of ungulates and its relation to management* (ed. by V. Geist and F. Walther). IUCN Publ. No. 24.

Meischke, H.R.C. & Geering, W.A. (1985) Exotic animal diseases. *Pests and parasites as migrants* (ed. by A. Gibbs and R. Meischke), pp. 23–27. Cambridge University Press.

Newsome, A.E. (1969) A population study of house mice permanently inhabitating a reedbed in South Australia. *J. Anim. Ecol.* **38**, 361–377.

Seddon, H.R. (1953) *Diseases of domestic animals in Australia*, Parts 1–6. Commonwealth of Australia, Div. Vet. Hyg. Serv. Publ. No. 1–6.

Strecker, R.L. (1962) Population levels. *Pacific Island Rat Ecology* (ed. by T. I. Storer), pp. 74–79. Bull. Bishop Mus. Honolulu 225, 274 pp.

Tamarin, R.H. & Malecha, S.R (1971) The population biology of Hawaiian rodents: demographic parameters. *Ecology*, **52**, 383–394.

Wirtz, W.O. (1972) Population ecology of the Polyesian rat, *Rattus exulans*, on Kure Atoll, Hawaii. *Pacific Science*, **26**, 433–464.

Wolfe, S.L. (1979) Population ecology of the Kulan. *Symposium on the ecology and behaviour of wild and feral equids* (ed. by R. H. Denniston), pp. 205–220. University of Wyoming, Laramie.

Wood, D.H. (1980) The demography of a rabbit population in an arid region of New South Wales, Australia. *J. Anim. Ecol.* **49**, 55–78.

Section III
Biological Mosaics and Tree/Grass Ratios

The essential descriptor of various savanna ecosystems is the relative amount and spacing of woody plants to herbaceous cover, often referred to as the tree/grass ratio. Belsky compares savannas and determinants of their physiognomic structures on a broad scale. She reviews several models of plant community structure and/or function, especially how the woody plant to grass ratio changes across gradients of precipitation and/or nutrients; she argues that the open savanna grasslands of eastern Africa are edaphically-derived, and not as the result of low rainfall or herbivore pressure.

While it is likely that in some higher rainfall areas savannas may originate and/or be maintained by the activities of herbivores, it is clear that in drier areas the introduction of herbivores may lead to grass being replaced by shrubs. Adámoli *et al.* document this in the dry Chaco of Argentina, where overstocking of cattle is the main cause of shifts in the woody plant to grass ratio, with shrubs replacing grass. (See also Section IV on this point.) Adámoli *et al.* also model the dynamic geomorphology of the riverine systems in the Chaco, producing a complex history, structure and composition of the gallery forests which lace through the savanna.

Most of the other papers in this section deal in one way or another with patches of woody plants within savannas. Blackmore *et al.* present a most interesting scenario wherein nutrient-rich patches of *Acacia*-dominated communities within savannas near Nylsvley, South Africa, have their origin not in geology or topography, but in their occupation by Iron Age Tswana tribesmen several hundred years ago and that there must be some positive-feedback mechanism which has allowed the long-term persistence of these patches.

Indeed, patches of woody plants elsewhere seem to persist and even serve as foci for expansion under certain conditions. Archer describes a successional process of physiognomic conversion to increased woodiness of North American subtropical savannas – a process that involves the establishment and outward spread of clusters of woody vegetation, influenced by variations in annual precipitation and degree of grazing pressure. Menaut *et al.* present an excellent, comprehensive, and general simulation model designed to examine tree dynamics using age-specific responses of plants to fire and to competition. In the Ivory Coast savannas they predicted that clumps of trees formed, increased in numbers, and grew in area under all conditions except for the most extreme (e.g. episodic severe burning).

Other papers which deal with this subject are Medina & Silva, Scholes (Section II); Burrows *et al.* (Section IV); and Braithwaite, McKeon *et al.* (Section I).

P.A.W.

12/Development and stability of grass/woody mosaics in a subtropical savanna parkland, Texas, U.S.A.

STEVE ARCHER *Department of Range Science, Texas A&M University, College Station, TX 77843-2126, U.S.A.*

Abstract. The potential natural vegetation of southern Texas and northern Mexico has been classified by plant geographers as savanna. However, many of the present landscapes in this subtropical region are dominated by thorn woodlands. Evidence for replacement of grasslands and savannas by woodlands is based largely on historical accounts, many of which are conflicting. This paper reviews and integrates a series of recent studies addressing the following questions: (1) Have woodlands replaced grasslands or savannas? (2) If there was a physiognomic conversion (a) what successional processes were involved; (b) what time scale would have been required; and (c) what were the causes?

Key words. *Prosopis glandulosa*, succession, disturbance, grazing.

INTRODUCTION

Savannas are characterized by the dual significance of herbaceous and woody plants. Factors regulating the balance between these contrasting plant lifeforms through time include climate, soils, disturbance (e.g. fire, grazing) and their interaction. Changes in one or more of these factors may enable one lifeform to increase and/or the other to decrease. In many systems, increases in woody plant abundandance are accompanied by decreases in herbaceous production and undesirable shifts in composition. Once such shifts in composition and production have been effected, succession to previous states may be slow or may not occur at all. As a result, progressive management strategies that minimize the probability of shifts to increased woody cover should be developed for existing savannas. An understanding of factors regulating the mixture of herbaceous and woody plants in savannas may enable us to identify disturbance and transition thresholds and thereby anticipate undesirable changes in plant community structure and adjust land management practices accordingly. In areas where high levels of woody plant cover already exist, anthropogenic manipulation (fire, herbicides, mechanical treatments, seeding, etc.) may be used to enhance herbaceous productivity. However, effectiveness of such manipulations may depend on their type, timing and sequencing (Scifres *et al.*, 1983), placement on the landscape (Scifres *et al.*, 1988) and subsequent management practices.

Quantitative and historical assessments suggest woody plant dominance has increased substantially in grasslands and savannas during the last 50–300 years in many parts of the world, including Africa (Barnes, 1979; van Vegten, 1983), Australia (Harrington, Wilson & Young, 1984), South America (Schofield & Bucher, 1986; Bucher, 1987), India (Singh & Joshi, 1979) and North America (Buffington & Herbel, 1965; Blackburn & Tueller, 1970; Herbel, Ares & Wright, 1972; Hobbs & Mooney, 1986). Remaining grasslands and savannas may become increasingly susceptible to woody plant encroachment with the anticipated global climatic changes (Emanuel, Shugart & Stevenson, 1985a, b). Successful management of savannas will require an understanding of their stability and resilience at appropriate spatial and temporal scales. To achieve this understanding, answers to two questions are paramount: (1) What kinds, frequencies or intensities of use might cause long-lasting, undesirable changes in composition and productivity, and (2) How much can a savanna change and still recover its original composition (Walker, 1985)?

A transition from grassland or savanna to shrubland or woodland may result (1) if climate and/or disturbance regimes change to (a) enable native woody species to extend their geographic range or (b) increase in stature and density within historical ranges; or (2) if introduced woody species successfully establish and reproduce. Although encroachment of woody plants into grasslands and increased densities of woody plants in savannas have been widely observed, the rates, patterns and dynamics of the process and its causes have seldom been quantified.

The Rio Grande Plains of southern Texas and northern Mexico offer some distinct examples of processes involved in the physiogonomic conversion of grasslands and savannas to woodlands. The potential natural vegetation of this region has been classified as *Prosopis–Acacia–Andropogon–Setaria* savanna (Küchler, 1964). However, much of the present vegetation is subtropical thorn woodland (Blair, 1950). A component of the Tamaulipan Biotic Province (Dice, 1943), the thorny shrubs and small trees in

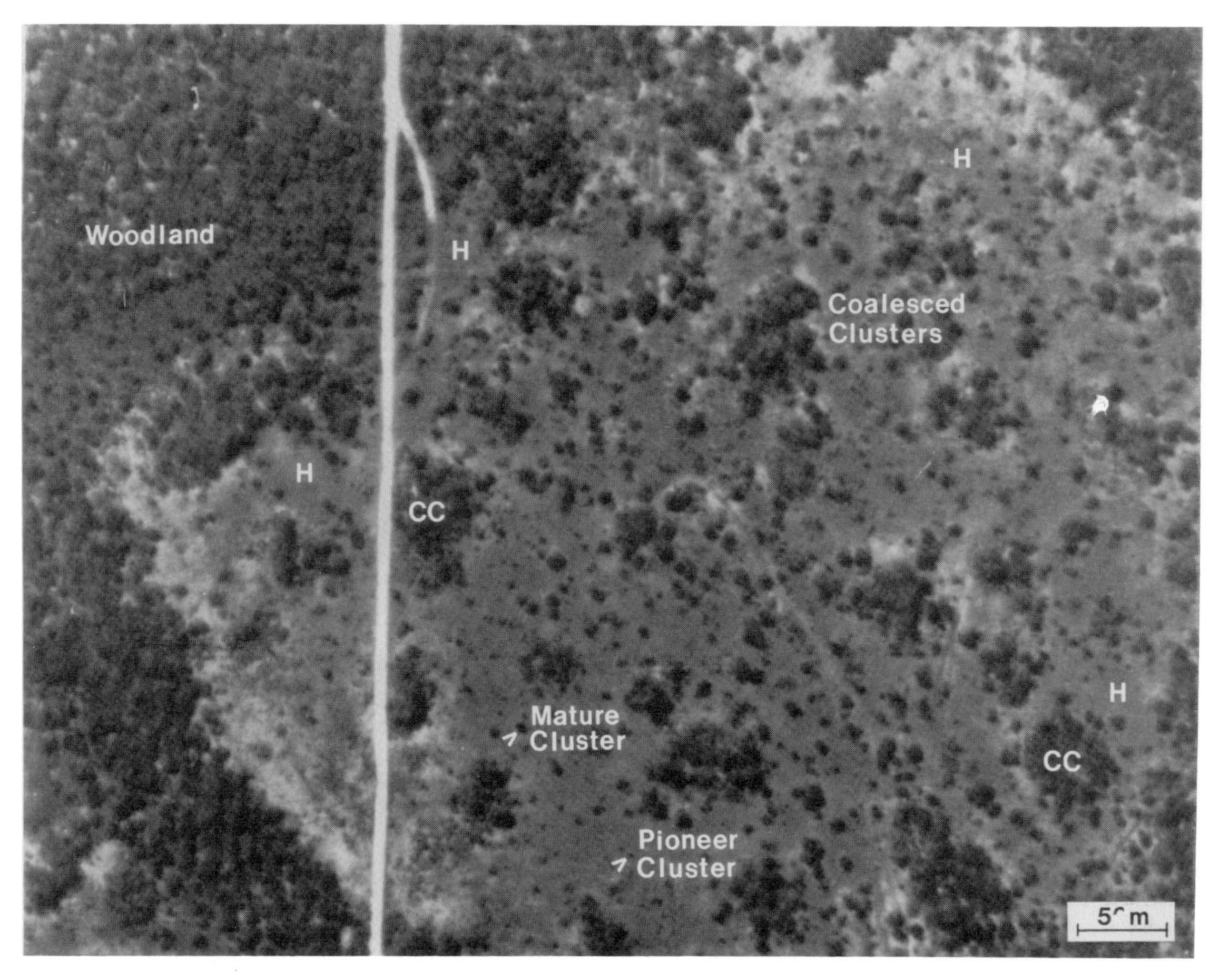

FIG. 1. Aerial view of a two-phase pattern of discrete clusters of woody vegetation embedded in a grass-dominated herbaceous matrix (H) on the La Copita Research Area in southern Texas, U.S.A. The clusters, which were organized about a *Prosopis glandulosa* nucleus (Fig. 2), represented chronosequences (Fig. 3) in which species composition at later stages of development were similar to those of adjacent closed-canopy woodlands (Archer *et al.*, 1988). It is hypothesized the two-phase pattern, which characterized 80% of the reseach area, is moving toward a closed-canopy woodland as new clusters are initiated with existing clusters expand (Table 2) and coalesce (cc = coalesced clusters) (Fig. 6).

this system have counterparts throughout much of the world's tropical and subtropical zones (Brown, 1982). In many instances it is believed these vegetation types have replaced large areas of former grasslands (cf. Schofield & Bucher, 1986). However, the basis for this contention is largely from indirect, historical sources, many of which are conflicting (Malin, 1953). This paper reviews the results of a recent series of studies in southern Texas which have focused on the following questions: (1) Have thorn woodlands replaced grasslands or savannas in recent history, and (2) If there was a physiognomic conversion (a) what were the mechanisms involved; (b) what were the rates and dynamics of the process; and (c) what were the causes?

STUDY SITE

The research reviewed in this paper was conducted on the Texas Agricultural Experiment Station's La Copita Research Area 65 km west of Corpus Christi, Texas (27°40′N; 98°12′W). The site has a history of moderate to heavy cattle grazing since the mid-1800s. Topography is generally level with slopes ≤3%. Elevation ranges from 75 to 90 m. The climate is subtropical with warm winters and hot summers. Mean annual temperature is 22.4°C and the growing season is 289 days. Average annual rainfall is 680 mm with maxima in May and September.

The predominant vegetation of the area was savanna parkland (Fig. 1) consisting of discrete clusters of woody vegetation organized beneath the arborescent legume, honey mesquite (*Prosopis glandulosa* (Torr.) var. *glandulosa*). Common understorey shrubs beneath the *Prosopis* canopy were *Zanthoxylum fagara* (L.), a slightly coriaceous (leathery-textured), broad-leaved evergreen; *Celtis pallida* (Torr.) and *Condalia hookeri* (M. C. Johnst.), each malacophyllous (herbaceous) and deciduous; *Diospyros texana* (Scheele.) a coriaceous, broad-leaved evergreen; *Schaefferia cunefolia* (Gray) and *Ziziphus obtusifolia* (T. & G.), each drought deciduous; and *Berberis trifoliolata* (Moric.),

FIG. 2. A *Prosopis*-mixed shrub cluster at an advanced stage of development (see Fig. 3 for additional details).

a sclerophyllous evergreen (plant nomenclature follows Correll & Johnston, 1979). Vegetation in herbaceous zones was dominated by C_4 grasses such as *Paspalum setaceum* (Michx.), *Bouteloua rigidiseta* (Steud.), *Chloris cucullata* (Bisch.), *Aristida* spp., *Bouteloua trifida* (Thurb.) and *Cenchrus incertus* (M. A. Curtis). Herbaceous dicots included *Evolvulus* spp., *Eupatorium* spp., *Verbesina* spp. and *Zexmania hispida* (H.B.K.). This savanna parkland vegetation complex occurred on fine sandy loam to sandy clay loam soils and occupied 80% of the 1093 ha research area. See Scifres & Koerth (1987) for more complete descriptions of soils, vegetation and climate.

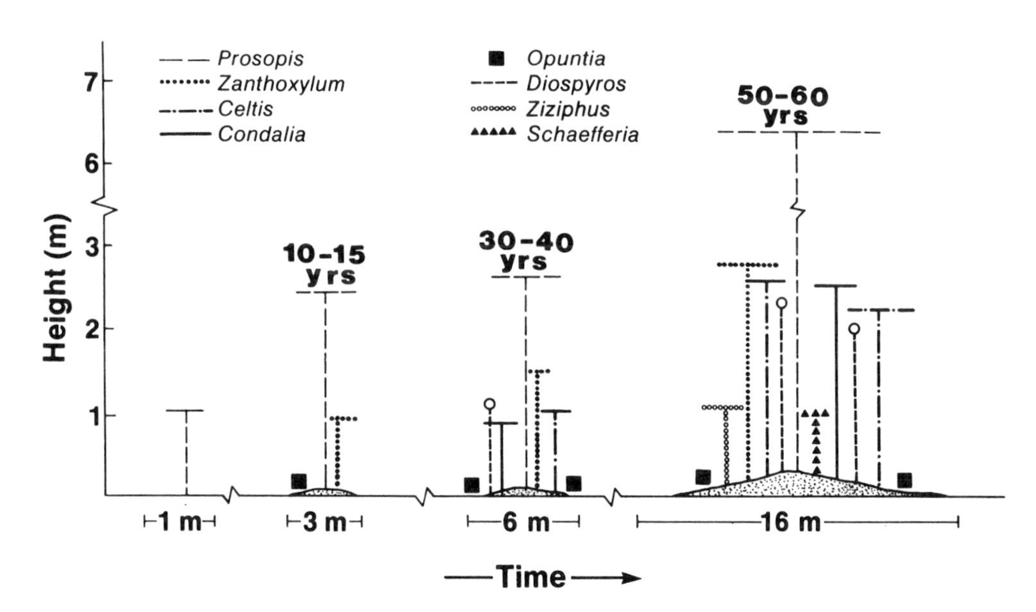

FIG. 3. Development of woody clusters on sandy loam uplands in southern Texas (from Archer *et al*., 1988). It is hypothesized *Prosopis glandulosa* invades herbaceous zones and (1) adds vertical structure which attracts birds disseminating seeds of other woody plants; and (2) modifies soils and microclimate to facilitate germination and/or establishment of the subordinate woody species. The first woody species appear 10–15 years after *Prosopis* establishment; within 60 years clusters average thirteen woody species (Archer, 1989).

HAVE SAVANNAS BEEN CONVERTED INTO WOODLANDS IN RECENT HISTORY?

Inferences from patterns of succession

Sandy loam uplands on the research site were characterized by distinct clusters of woody vegetation embedded in a matrix of C_4 grasses (Fig. 1). Initial investigations focused on quantifying the structure of woody clusters and examining the herbaceous zone for seedlings and saplings of woody plants. Field data (Archer *et al.*, 1988) indicated: (1) although seedlings and samplings of eight woody species were encountered in the herbaceous matrix the great majority, in terms of density and frequency, were of one species, *Prosopis glandulosa*; (2) individual woody clusters ranged in size from 1 to 16 m diameter; (3) there was typically one *Prosopis* plant centrally located in each cluster and it was usually the largest plant, in terms of height, basal diameter and canopy area (Fig. 2); (4) the number of subordinate woody species in clusters, which ranged from one to fifteen, was positively correlated with the size of the *Prosopis* plant ($r^2 = 0.86$); (5) with the exception of *Prosopis*, most species in clusters would be bird-dispersed; and (6) species composition and relative abundance in large clusters was comparable to that of closed-canopy woodlands in neighbouring ephemeral drainages. These data suggest *Prosopis* colonized grass-dominated sites and served as the nucleus of cluster organization, apparently facilitating the ingress and/or establishment of additional woody species otherwise restricted to other habitats (Fig. 3).

As the abundance of woody species beneath *Prosopis* increased, herbaceous production decreased (Fig. 4). If *Prosopis* plants continue to invade and develop in the herbaceous zones, new clusters will form. At the same time, existing clusters will enlarge as new species are added and canopies of established plants develop. The present savanna parkland may thus represent an intermediate stage in the conversion of a grassland to woodland. Closed-canopy woodlands in the region appear to represent sites where this process has already occurred.

Direct assessments of vegetation change

The above scenario was developed from inferences derived from a space-for-time substitution study of successional patterns. Direct assessments of changes in grass-woody composition of these landscapes using aerial photographs and stable carbon isotopes provide additional evidence that woody plants have displaced grasses.

A comparison of aerial photographs from three different landscapes on the study area indicated total woody plant cover increased on the two-phase upland from 13% in 1941 to 36% in 1983 (Archer *et al.*, 1988). During this same time period, mean cluster area increased from 494 to 717 m^2 and coalescence of discrete clusters was evident. If this trend continues, the present savanna parkland may develop into a closed-canopy woodland.

Tissues from C_3 and C_4 plants have distinctive $^{13}C/^{12}C$ ratios (Smith & Epstein, 1971) expressed as $\delta^{13}C$. In addition, $\delta^{13}C$ ratios in soil organic matter generally reflect the relative contribution of C_3 and C_4 plants to site productivity

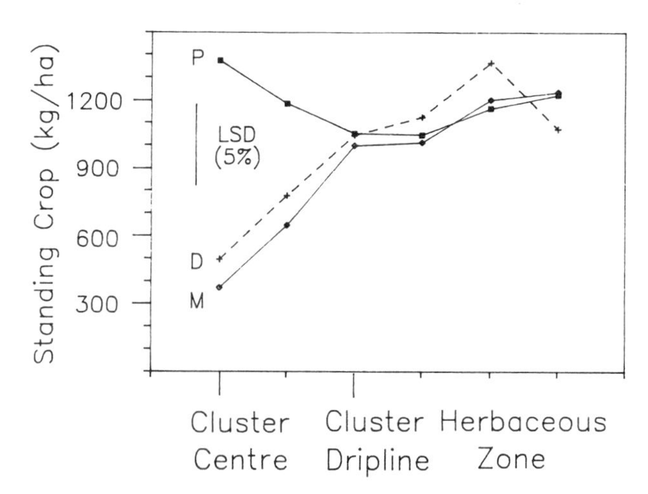

FIG. 4. Herbaceous standing crop (June 1986) along gradients extending from the herbaceous zone into the centre of woody clusters at various stages of development (P = pioneer [e.g. lone Prosopis], D = developing, and M = mature [e.g. Fig. 3]). Although herbaceous production was enhanced following Prosopis establishment, the subsequent ingress of additional woody species adversely affected this parameter. Bare ground increased from an average of 10% in the herbaceous zone to 45% and 68% within developing and mature clusters, respectively (from Scanlan, 1988).

integrated over long periods of time (Troughton, Stout & Rafter, 1974). The $\delta^{13}C$ values indicate the deviation (‰) of the $^{13}C/^{12}C$ ratio of the sample from that of an international standard. The $\delta^{13}C$ of plants and soils associated with the chronosequence in Fig. 3 were determined as described by Dzurec *et al.* (1985). Woody species in clusters possessed the C_3 photosynethetic pathway ($\delta^{13}C$ range = −27 to −32‰), whereas vegetation of grass-dominated zones between clusters was characterized by plants with C_4 pathway ($\delta^{13}C$ range = −13 to −17‰) (Table 1). If shrubs have been a long-term constituent of these landscapes, the $\delta^{13}C$ signature of soils beneath them should reflect this and fall in the −27 to −32‰ range. However, if C_3 shrubs have displaced C_4 grasses: (1) the soil $\delta^{13}C$ value would be greater (less negative) than −27 to −32‰; (2) the degree of departure from the expected ratio would decrease as time of site habitation by shrubs increases; and (3) soil $\delta^{13}C$ values would become less negative with depth along the chronosequence. An analysis of soil organic carbon $\delta^{13}C$ confirmed these predictions.

The organic carbon of soils beneath herbaceous zones was strongly C_4 and reflected the composition of the current vegetation throughout the profile ($\delta^{13}C$ = −14 to −18‰) (Table 1). In contrast, the mean $\delta^{13}C$ ratio in the upper horizon of soils in the centre of clusters at early and late stages of development was −21 and −23‰, respectively. These values were significantly less than the signature of the associated vegetation ($\delta^{13}C$ range = −27 to −32‰) and indicate the presence of C_4-derived carbon on the site. The decrease in $\delta^{13}C$ from −21 to −23‰ appears to reflect the additional input of C_3 carbon associated with the passage of time required for *Prosopis* plants and clusters to develop on the respective sites. Strength of the C_4 signature increased to a depth of 60 cm among soils supporting woody vegetation, converging on the values observed for the herbaceous zones.

TABLE 1. Vegetation attributes and $\delta^{13}C$ values (‰) for vegetation and soil organic carbon along the grassland-to-woodland chronosequence depicted in Fig. 3 (mean ±SE; $n = 3$) (Archer & Tieszen, unpublished). Pioneer clusters consisted of lone *Prosopis* plants. For $\delta^{13}C$ values, analysis of variance indicated vegetation state, depth and the vegetation × depth interaction were significant at $P<0.05$.

	Herbaceous zones	Pioneer clusters	Well-developed clusters
Prosopis height (m)	—	1.5 ± 0.1	5.1 ± 0.6
Prosopis basal diameter (cm)		5.7 ± 1.1	27.0 ± 2.4
Cluster canopy area (m²)	—	2.9 ± 1.2	52.9 ± 11.8
No. of woody species	0 ± 0	1.0 ± 0.0	11.7 ± 1.5
Vegetation $\delta^{13}C$			
Shrub layer	—	−27.2 ± 1.4	−29.6 ± 0.6
Ground layer*	−13.4 ± −0.2[a]	−17.1 ± 2.8[a,b]	−23.0 ± 1.8[b]
Soil $\delta^{13}C$ at depth:			
0–5 cm†	−17.9 ± 0.4[a]	−20.7 ± 0.2[b]	23.1 ± 0.3[c]
15–30 cm	−14.3 ± 0.5[a]	−16.2 ± 1.2[a]	−18.3 ± 0.2[b]
45–60 cm	−13.9 ± 0.6[a]	−14.3 ± 0.6[a]	−17.3 ± 0.3[b]

* Includes litter.
† Means within a row followed by same letter were not significantly different ($P>0.05$).

These data constitute direct evidence that shrubs have displaced grasses on the site. But when did these changes occur? To address this question, information on cluster growth rates obtained from a sequence of aerial photographs spanning a 42 year period was used reconstruct site history.

Rates and dynamics of succession

Changes in total woody plant cover and cluster size were quantified for three sites using aerial photographs taken in 1941, 1960 and 1983 and related to patterns of annual precipitation (Archer *et al.*, 1988). Although there was a net increase in woody cover between 1941 and 1983, the development of woody assemblages were markedly affected by variations in annual precipitation. Between 1941 and 1960, a period characterized by a severe 7-year drought, woody cover decreased on each of three sites examined (Fig. 5). The decline in cover was the combined result of a reduction in the density (mortality) of small clusters (area <5 m²), and a reduction in total canopy area and fragmentation of large clusters. In the subsequent pluvial period (1960–83), growth rates of clusters increased (Table 2) along with the density of small clusters (recruitment), more than compensating for reductions in woody cover incurred between 1941 and 1960. Thus, while there were instances of cyclic replacement and/or fluctuation on these sites between 1941 and 1983, there was an overall succession toward woodland. Assuming (1) the processes operational over the 1941–83 period continue; and (2) edaphic or density-dependent factors will not regulate cluster development or distribution, the time to canopy closure may depend on the frequency and sequencing of drought periods (Fig. 6). The validity of these assumptions remains to be demonstrated.

Estimates of cluster growth rates under different precipitation regimes (Table 2) were used to model growth and determine size/age relationships for *Prosopis* plants and the woody clusters that form beneath them (Archer, 1989). This model indicated establishment of woody species beneath invading *Prosopis* occurred within 10–15 years (Fig. 3). Species richness increased rapidly from 35–45 years, becoming asymptotic at an average of thirteen species per cluster. Estimated age of the largest *Prosopis* plant found in clusters was 172–217 years. However, 90% of the *Prosopis* plants and clusters on the site appear to be <100 years of age and the age-class distribution was that of a young population expanding geometrically (Fig. 7). Model projections concur with numerous historical accounts from the late

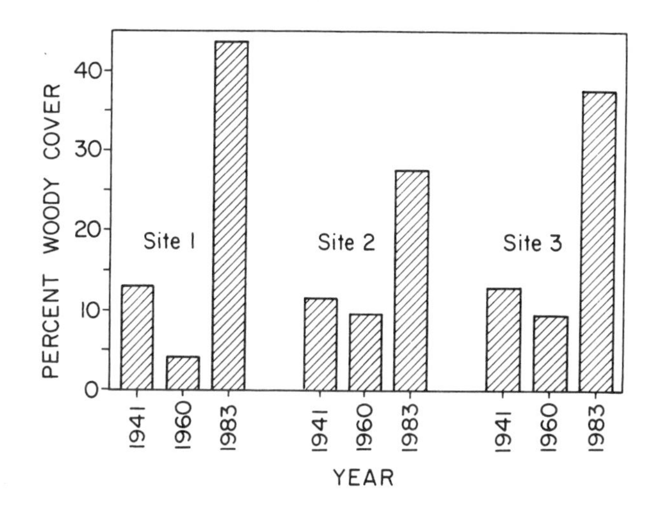

FIG. 5. Changes in total woody plant cover between 1941 and 1983 for three sites on the La Copita Research Area (from Archer et al., 1988). The 1941–1960 period was characterized by a severe 7-year drought in the 1950s, whereas the 1960–83 period received normal to above-normal annual precipitation. Slight decreases in woody cover between 1941 and 1960 were attributed to mortality (disappearance) of small clusters and death of individual plants in large clusters. Increased woody cover between 1960 and 1983 resulted from the initiation of new clusters and rapid growth of existing clusters (Table 2). During the dry 1941–60 period, herbaceous production would have been reduced and dissemination of Prosopis seed by livestock was likely substantial. In the subsequent pluvial period, conditions might thus have favoured Prosopis recruitment.

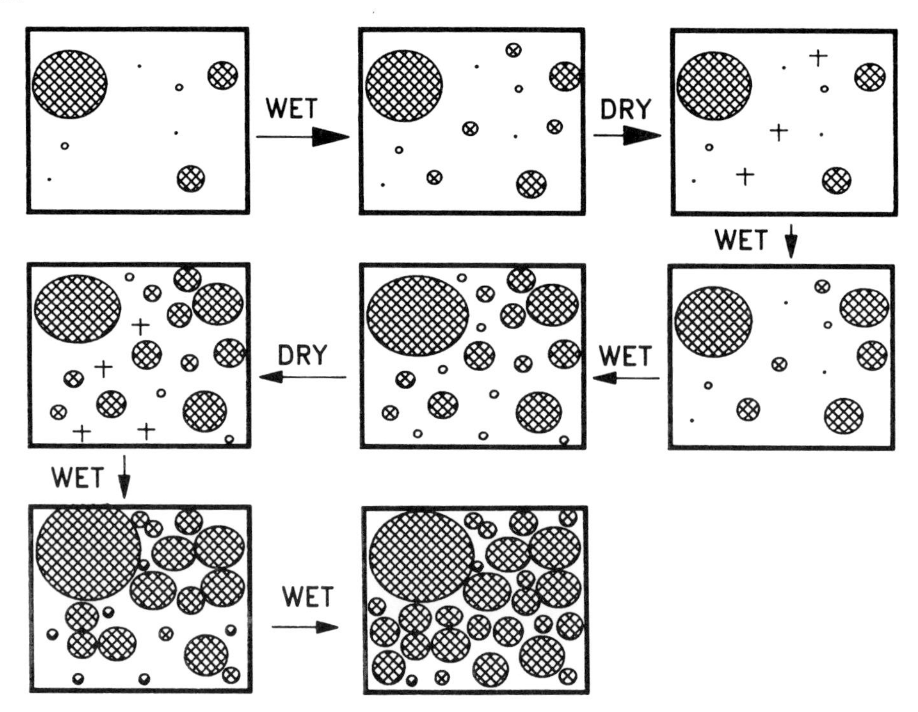

FIG. 6. Hypothesized pattern of succession from savanna to woodland on the La Copita Research Area in southern Texas, U.S.A. Evidence from aerial photographs indicated mortality of small clusters (+) in dry years; however, the appearance of new clusters (•) in normal to above-normal rainfall years more than offset these losses. Although rates of clusters expansion were dynamic, they were generally positive (Table 2). Unless edaphic, pyric or density-dependent processes limit establishment and growth of clusters, the two-phase landscape will develop into a closed-canopy woodland. The rate of transformation may depend upon the frequency and sequencing of dry periods (e.g. Fig. 3).

TABLE 2. Relative growth rates (RGR) of *Prosopis*-mixed shrub clusters in the Rio Grande Plains, Texas, determined by Archer *et al.* (1988) from aerial photographs. 'Dry' denotes growth during the 1941–60 period characterized by a major drought, 'wet' refers to the period of normal to above-normal precipitation from 1960 to 1983. The negative growth among large clusters in the dry period reflects canopy reductions resulting from the loss of individual woody plants.

	RGR ($m^2 m^{-2} y^{-1}$)	
Cluster size (m^2)	Dry (1941–60)	Wet (1960–83)
<100	0.10	0.16
100–400	0.01	0.03
401–1000	0.01	0.02
>1000	–0.08	0.01

1880s which describe these landscapes as savanna or grasslands with 'mottes' (small patches of woods in a prairie land) (Crosswhite, 1980).

CAUSES FOR SUCCESSION FROM SAVANNA TO WOODLAND

The population structure of *Prosopis* (Fig. 7) and woody clusters on the site indicate something may have happened 100 to 200 YBP to destabilize lifeform interactions and shift the balance to favour woody plants over grasses. Data from aerial photographs indicated *Prosopis* plants and clusters >30 years old were relatively persistent landscape features (Archer *et al.*, 1988; Archer, 1989). The age class distribution of *Prosopis* in Fig. 7 may therefore reflect enhanced recruitment rather than decreased mortality in recent history. Because of the pivotal role of *Prosopis* in woody cluster formation and development (Fig. 3), factors regulating its ingress and establishment are of primary importance.

Prosopis possesses numerous attributes which make it a successful invader of grasslands: it produces abundant seeds which are potentially long-lived in the soil; germination and establishment can occur across a broad range of soils, water and light regimes; it is capable of fixing nitrogen early in its life cycle; seedlings can vegetatively regenerate following top removal within 2 weeks of germination; survival of 2- and 3-year-old seedlings following very hot fires can exceed 80% and plants can survive prolonged, severe droughts (Archer *et al.*, 1988, and references therein). In addition, root development in *Prosopis* seedlings is rapid (Table 3), enabling them to effectively partition soil moisture with grasses early in their life cycle (Table 4). *Prosopis* is thus an aggressive plant not easily eliminated from a site once established.

Prosopis has been in North America since the Pliocene (Axelrod, 1937) and wood of *Prosopis* dated 3300 YBP has been recovered in archaeological sites in southern Texas (Hester, 1980). Available evidence indicates its geographic range has changed little in the past 300–500 years (Johnston, 1983). These observations, in conjunction with evidence indicating recent increases in abundance on upland sites, suggest *Prosopis* has been present but largely

TABLE 3. Above ground height and biomass, tap root length and biomass and partial root to total plant biomas ratio of *Prosopis* seedlings excavated on 3 May and 22 August 1985 on the La Copita Research Area in southern Texas, U.S.A. Seedlings were newly emerged (cotyledons only = NE; n = 10) and recently established with at least one true leaf (EST; n = 10) on 3 May and approximately 4 months old (4 mo; n = 7) on 22 August. Each value is mean ±SE. Excavations were terminated at 50 cm (from Brown & Archer, 1990). Nearly 90% of the herbaceous root biomass was within 30 cm of the surface of this site.

	Above ground		Below ground		Partial root to total plant ratio (g/g)
Age	Height (cm)	Biomass (g)	Length (cm)	Biomass (g)	
NE	2.8 ± 0.2	0.11 ± 0.01	5.5 ± 0.0	0.04 ± 0.00	0.27 ± 0.09
EST	7.4 ± 0.8	0.63 ± 0.11	20.7 ± 3.7	0.45 ± 0.12	0.41 ± 0.13
4 mo	8.1 ± 1.1	0.68 ± 0.15	41.3 ± 5.1	0.73 ± 0.14	0.52 ± 0.15

TABLE 4. Best two-factor models, r^2 and significance level (*P<0.05; **P<0.01) for variables regulating stomatal conductance in 1-year-old *Prosopis* seedlings, mature *Prosopis* trees, and *Chloris* tillers. Factors evaluated were % soil moisture (SM) in the upper (<30 cm depth), middle (30–90 cm depth) and lower (>90 cm depth) horizons, vapour pressure deficit (VDP) and soil temperature. No models with more than two factors significantly improved r^2. All multiple r^2 values were significant at P<0.01. VPD was negatively correlated with conductance; other variables reported were positively correlated (from Brown & Archer, 1990).

Plant category	Variables	Partial r^2	P	Multiple r^2
Chloris tillers	SM-Upper	0.64	**	
	SM-Middle	0.21	**	0.85
Prosopis seedlings	SM-Middle	0.47	*	
	SM-Lower	0.24	*	0.71
Mature *Prosopis*	SM-Lower	0.65	**	
	VPD	0.15	*	0.80

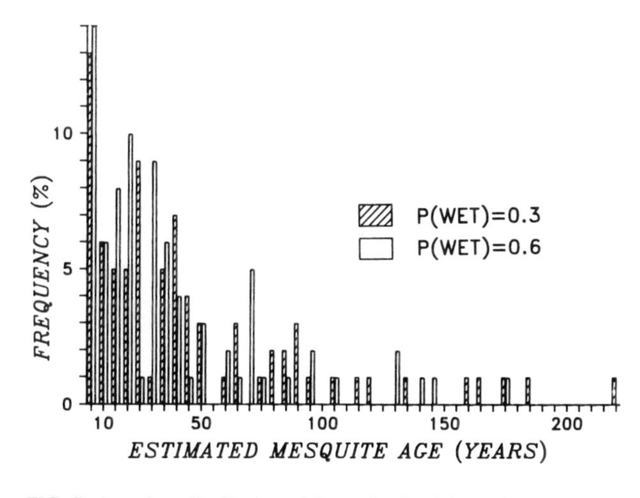

FIG. 7. Age-class distribution of Prosopis glandulosa plants on landscapes at the La Copita Research Area, Texas (from Archer, 1989). Projections are based on a size/age model run under two annual precipitation regimes [P(WET) = 0.3 and 0.6] thought to reasonably represent this part of North America since the mid-1800s. P(WET) refers to the frequency with which growth rates in Table 2 were assigned to time-steps in each run of the model. For the P(WET) = 0.3 run, the 1960–83 growth rates (relatively wet period) were randomly assigned to one-third of the time-steps and the 1941–60 growth rates (relatively dry period) were assigned to the other two-thirds of the time-steps. For P(WET) = 0.6, the proportion of time-steps assigned the 1960–83 and 1941–60 growth rates was two-thirds and one-third, respectively.

confined to other habitats (e.g. drainages, playas and escarpments) throughout the Holocene. What has happened in the past 200 years to enable *Prosopis* to increase its abundance in other habitats? Most hypotheses centre around changes in climate, fire and grazing regimes.

There is evidence suggesting climate in recent history may have shifted to favour woody plants over grasses in portions of North America (Neilson, 1986, 1987). However, Madany & West (1983) documented a case in which a savanna was maintained despite low fire frequency (and possible climatic change) on one site, while succession to woodland coincided with the advent of livestock grazing on a nearby edaphically similar site. Data from these respective studies suggest that although climatic change or variation in recent history may have been necessary, it was not sufficient to have caused a shift from savanna to woodland.

The apparent increase in abundance of woody plants since the 1800s on the La Copita site in southern Texas coincides with the development of the livestock (cattle, sheep and horses) industry in that region (Jackson, 1986). Large numbers and high concentrations of cattle, sheep and horses could have facilitated the spread of *Prosopis* by increasing dispersal into new habitats (Table 5) and enhancing seed germination and seedling establishment. *Prosopis* pods are nutritious and may be heavily utilized by livestock, especially during drought periods when the availabil-

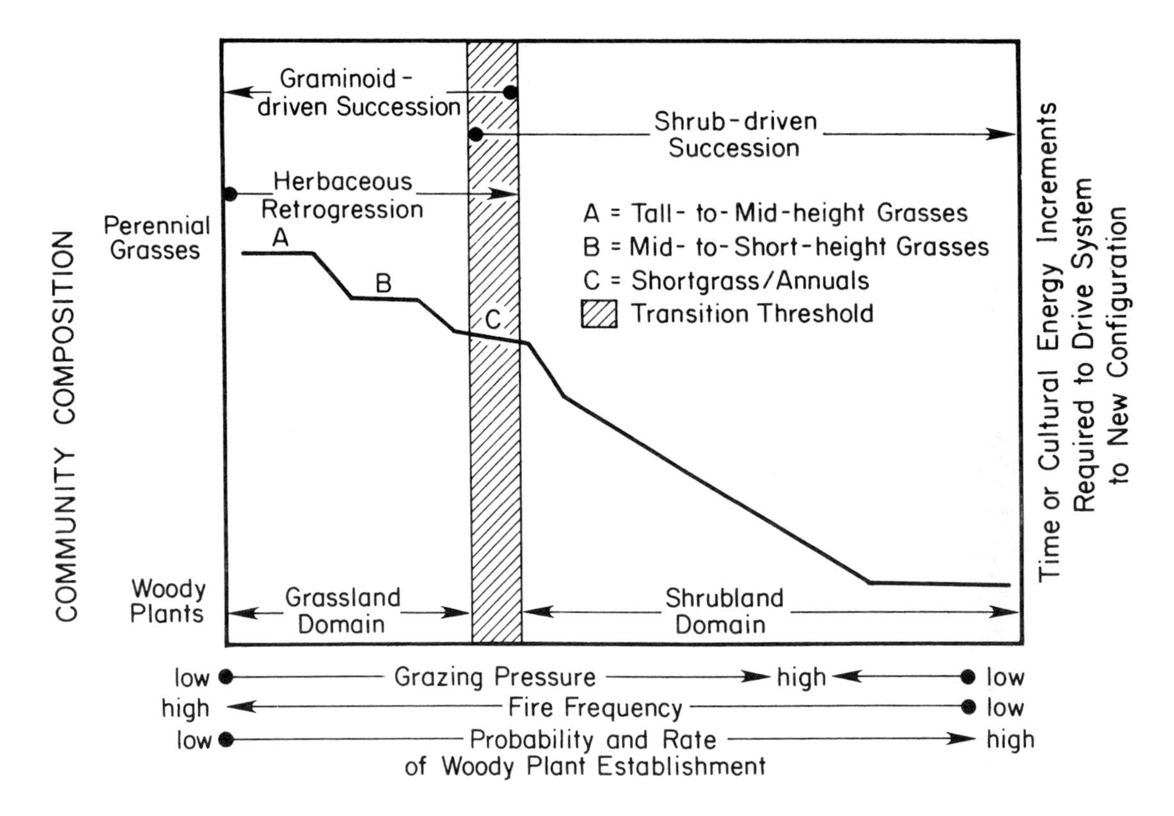

FIG. 8. Conceptual model of changes in community structure as a function of grazing pressure (from Archer, 1989). Within the grassland domain grazing alters herbaceous species composition and productivity, while decreasing fire frequency and intensity. Where seed sources of woody legumes are available, livestock may contribute substantially to dispersal as well as establishment. These factors interact within a variable climate to increase the probability of woody plant establishment. If grazing pressure is reduced prior to some critical threshold, succession toward higher condition grasslands could potentially occur. However, if sufficient numbers of woody plants become established, shrub-driven successional processes begin to move the system toward a new steady-state as a positive feedback situation develops. As woody cover continues to develop (Fig. 6), herbaceous production will decrease (Fig. 4), lowering the carrying capacity for grazers and increasing the grazing pressure in the remaining interstitial zones unless stocking rates are reduced. Once in the shrub- or woodland domain, the soils, seed bank and vegetative regenerative potentials are altered and the site may not revert to grassland or savanna, even following removal of grazing. Anthropogenic manipulation of woody vegetation (e.g. prescribed burning, herbicides, mechanical treatments) can reduce woody cover and enhance herbaceous production. However, without proper grazing management and stategic follow-up treatments, succession back to woodland would be expected.

TABLE 5. Survey results of mean (±SE) *Prosopis* seed and seedling density in areas with and without cattle on an upland site in a Texas savanna parkland, August 1985. Canopy refers to seeds and seedlings encountered beneath adult *Prosopis* plants; open refers to occurrences in herbaceous zones (from Brown & Archer, 1987). The absence of seedlings on the area without cattle suggests native wildlife on the site were relatively poor agents of *Prosopis* dispersal.

	With cattle		Without cattle	
	Canopy	Open	Canopy	Open
Area sampled (%)	8	92	9	91
Seedlings m^{-2}	$12 \pm 2^{a*}$	15 ± 2^{a}	0 ± 0^{b}	0 ± 0^{b}
Seeds m^{-2}	33 ± 7^{a}	11 ± 2^{b}	33 ± 8^{a}	0 ± 0^{c}

* Means within row followed by different letters differed significantly at $P<0.05$.

ity of grass is low. A high percentage of the hard seeds ingested with the pods escape mastication, are scarified in the digestive tract and deposited in a moist, nutrient-rich medium (dung) away from parent plants harbouring host-specific predators (Brown & Archer, 1987, and references therein). Germination of seed dispersed in this manner can be high and establishment is facilitated because herbaceous interference and fire frequency and intensity have been reduced by defoliation (e.g. Brown & Archer, 1989). In the absence of periodic fire, *Prosopis* plants in uplands would develop in stature, providing (1) seed for additional dispersal; (2) vertical structure attractive to avifauna dispersing seeds of other woody species; and (3) a microhabitat potentially conducive to germination and/or establishment of additional woody species (e.g. altered soils, microclimate and decreased herbaceous biomass; Fig. 4). The punctuated development of woody assemblages on these landscapes with a history of heavy grazing by cattle (Fig. 5) may reflect livestock mediation of *Prosopis* seed dispersal, germination, growth and development interacting with the sequencing of dry and wet years.

LANDSCAPE RESISTANCE, RESILIENCE AND MANAGEMENT

Based on inferences from successional processes (Fig. 3), *Prosopis* age-class distributions (Fig. 7) and $\delta^{13}C$ values of soils associated with woody clusters (Table 1) it appears grasslands or open savannas have become savanna parklands in recent history. If the processes observed between 1941 and 1983 continue, savanna parklands (Fig. 1) may become closed-canopy woodlands (Fig. 6) with extremely low herbaceous productivity (Fig. 4). The conceptual summary in Fig. 8 suggests how and why vegetation may be in the process of shifting from one steady-state toward another.

In the absence of edaphic constraints or climatic change, the continued initiation and expansion of clusters might be arrested by grazing management aimed at (1) reducing dispersal of *Prosopis*; (2) enhancing the capacity of grasses to competitively exclude *Prosopis* seedlings; and (3) enabling periodic use of fire to keep new clusters from forming and existing clusters from increasing in areal extent. Among landscape components where seed bank and vegetative regeneration favour post-disturbance woody plant re-establishment, a long-term sequence of vegetation manipulation technologies may be required to drive the system back toward its previous configuration (cf. Scifres *et al.*, 1983).

ACKNOWLEDGMENTS

My thanks to colleagues, research associates and graduate students, past and present, who have contributed to this research. Rob Flinn, Guy McPherson, Tom Thurow and Steve Whisenant reviewed early drafts of the manuscript and made helpful suggestions for improvement. Marcella Smith did her usual fine job of word processing under pressure. The research was conducted in connection with projects H-6717, H-1922 and MS-6131 of the Texas Agricultural Experiment Station (TAES) and USDA/CSRS GRANT 86-CRSR-2-2925. This paper was approved for publication by the Director, TAES as manuscript 24430.

REFERENCES

Archer, S. (1989) Have Southern Texas savannas been converted to woodlands in recent history? *Am. Nat.* **134**, 545–561.

Archer, S., Scifres, C.J., Bassham, C.R. & Maggio, R. (1988) Autogenic succession in a subtropical savanna: conversion of grassland to thorn woodland. *Ecol. Monogr.* **58**, 111–127.

Axelrod, D.L. (1937) A Pliocene flora from the Mt. Eden beds, southern California. *Carnegie Inst. Washington Publ.* **476**, 125–183.

Barnes, D.L. (1979) Cattle ranching in east and southern Africa. *Management of semi-arid ecosystems* (ed. by B. H. Walker), pp. 9–54. Elsevier, Amsterdam.

Blackburn, W.H. & Tueller, P.T. (1970) Pinyon and juniper invasion in black sagebrush communities in east-central Nevada. *Ecology*, **51**, 841–848.

Blair, W.F. (1950) The biotic provinces of Texas. *Texas J. Sci.* **2**, 93–117.

Brown, D.E. (ed.) (1982) Biotic communities of the American southwest-United States and Mexico. *Desert Plants*, **4**, 101–106.

Brown, J.R. & Archer, S. (1987) Woody plant seed dispersal and gap formation in a Northern American subtropical savanna woodland: the role of domestic herbivores. *Vegetatio*, **73**, 73–80.

Brown, J.R. & Archer, S. (1989) Woody plant invasion of grasslands: establishment of honey mesquite (*Prosopis glandulosa* var. *glandulosa*) on sites differing in herbaceous biomass and grazing history. *Oecologia (Berl.)*, **80**, 19–26.

Brown, J.R. & Archer, S. (1990) Water relations of a perennial grass and seedling versus adult woody plants in a subtropical savanna, Texas. *Oikos*, **57**, (in press).

Bucher, E.H. (1987) Herbivory in arid and semi-arid regions of Argentina. *Revista Chilena de Historia Natural*, **608**, 265–273.

Buffington, L.C. & Herbel, C.H. (1965) Vegetational changes on a semidesert grassland range from 1858 to 1963. *Ecol. Monogr.* **35**, 139–164.

Correll, D.S. & Johnston, M.C. (1979) *Manual of the vascular plants of Texas*. University of Dallas, Richardson, Texas.

Crosswhite, F.S. (1980) Dry country plants of the South Texas Plains. *Desert Plants*, **2**, 141–179.

Dice, L.R. (1943) *The biotic provinces of North America*. University of Michigan Press, Ann Arbor.

Dzurec, R.S., Boutton, T.W., Caldwell, M.M. & Smith, B.N. (1985) Carbon isotope ratios of soil organic matter and their use in assessing community composition changes in Curlew Valley, Utah. *Oecologia (Berl.)*, **66**, 17–24.

Emanuel, W.R., Shugart, H.H. & Stevenson, M. (1985a) Climatic change and the broad-scale distribution of terrestrial ecosystem complexes. *Climatic Change*, **7**, 29–43.

Emanuel, W.R., Shugart, H.H. & Stevenson, M. (1985b) Response to comment: climatic change and the broad-scale distribution of terrestrial ecosystem complexes. *Climatic Change*, **7**, 457–460.

Harrington, G.N., Wilson, A.D. & Young, M.D. (1984) *Management of Australia's rangelands*. CSIRO, Melbourne.

Herbel, C.H., Ares, F.N. & Wright, R.A. (1972) Drought effects on a semi-desert grassland. *Ecology*, **53**, 1084–1093.

Hester, T.R. (1980) *Digging into south Texas prehistory*. Corona Publ. Co., San Antonio.

Hobbs, R.J. & Mooney, H.A. (1986) Community changes following shrub invasion of grassland. *Oecologia (Berl.)*, **70**, 508–513.

Jackson, J. (1986) *Los mestenos: Spanish ranching in Texas, 1721–1821*. Texas A&M University Press, College Station, Texas.

Johnston, M.C. (1963) Past and present grasslands of southern Texas and northeastern Mexico. *Ecology*, **44**, 456–466.

Küchler, A.W. (1964) *The potential natural vegetation of the conterminous United States*. American Geographical Society, New York.

Madany, M.H. & West, N.E. (1983) Livestock grazing–fire regime interactions within montane forests of Zion National Park, Utah. *Ecology*, **64**, 661–667.

Malin, J.C. (1953) Soil, animal, and plant relations of the grassland, historically recorded. *Scientific Monthly*, **76**, 207–220.

Neilson, R.P. (1986) High resolution climatic analysis and southwest biogeography. *Science*, **232**, 27–34.

Neilson, R.P. (1987) Biotic regionalization and climatic controls in western North America. *Vegetatio*, **70**, 135–147.

Scanlan, J.C. (1988) Spatial and temporal vegetation patterns in a subtropical *Prosopis* savanna woodland, Texas. Ph.D. dissertation, Department of Range Science, Texas A&M University, College Station.

Scifres, C.J., Hamilton, W.T., Inglis, J.M. & Conner, J.R. (1983) Development of integrated brush management systems (IBMS): decision-making process. *Brush Management Symposium*, pp. 97–104. Texas Tech Press, Lubbock.

Scifres, C.J., Hamilton, W.T., Koerth, B.H., Flinn, R.C. & Crane, R.A. (1988) Bionomics of patterned herbicide application for wildlife habitat enhancement. *J. Range Management*, **41**, 317–321.

Scifres, C.J. & Koerth, B.H. (1987) Climate, vegetation and soils of the La Copita Research Area. *Texas Agric. Exp. Sta. Bull. MP-1626*, p. 28. Texas A&M University, College Station.

Schofield, C.J. & Bucher, E.H. (1986) Industrial contributions to desertification in South America. *TREE*, **1**, 78–80.

Singh, J.S. & Joshi, M.C. (1979) Ecology of the semi-arid regions of India with emphasis on land-use. *Management of semi-arid ecosystems* (ed. by B. H. Walker), pp. 243–273. Elsevier, Amsterdam.

Smith, B.N. & Epstein, S. (1971) Two categories of $^{13}C/^{13}C$ ratios for high plants. *Plant Physiol.* **47**, 380–384.

Troughton, J.H., Stout, J.D. & Rafter, T.A. (1974) Long-term stability of plant communities. *Carnegie Inst. Washington Yearbook*, **73**, 838–845.

van Vegten, J.A. (1983) Thornbush invasion in a savanna ecosystem in eastern Botswana. *Vegetatio*, **56**, 3–7.

Walker, B.H. (1985) Structure and function of savannas: an overview. *Ecology and management of the world's savannas* (ed. by J. C. Tothill and J. J. Mott), pp. 83–91. Australian Academy of Science, Canberra.

13/The origin and extent of nutrient-enriched patches within a nutrient-poor savanna in South Africa

A. C. BLACKMORE, M. T. MENTIS and R. J. SCHOLES *Resource Ecology Group, Department of Botany, University of the Witwatersrand, Wits 2050, Republic of South Africa*

Abstract. There are distinct patches of an *Acacia tortilis* (Forssk.) Hayne-dominated community within the *Burkea africana* Hook.-dominated savanna at Nylsvely, South Africa. Both occur on sandy soils derived from the same parent material.

The levels of Ca, Mg, K and PO_4 in the soil under the *Acacia* communities are 10–100 times higher than those under the *Burkea* community. The levels of Na are approximately equal.

A previously postulated geomorphic origin for the patches is rejected on the grounds of the landscape position they ocupy, and the absence of a significant textural difference.

An anthropogenic origin is supported by the presence of Iron Age artefacts within the *Acacia* alone. Order-of-magnitude calcualtions indicate that the activities within an Iron Age settlement would be sufficient to account for the nutrient accumulations observed.

Key words. Anthropogenic, dystrophic, eutrophic, Iron Age, savanna, South Africa.

INTRODUCTION

The savanna biome study site at Nylsvley, South Africa (24°40′S, 28° 40′E) is located on sandy soils about 1 m deep, derived from sandstones of the Waterberg series (South African Geological Survey, 1978). The Nylsvley areas is of low relief, undulating between 1080 m (Nyl River floodplain) and 1140 m (low hill) altitude. The climate has been summarized by Anon. (1975) and Huntley & Moris (1978, 1982). Approximately 90% of the mean annual rainfall of 620 mm occurs during the period October–March. The mean daily maximum summer temperature is 28.4°C and the mean daily maximum winter temperature is 22.3°C. Relative humidity is low, leading to an annual potential evapotranspiration greatly exceeding annual rainfall. The vegetation is a broad-leafed savanna, dominated by *Burkea africana* in the tree layer and *Eragrostis pallens* Hack. in the grass layer. About 15% of the study site is occupied by a fine-leafed savanna community, dominated by *Acacia tortilis* with *Cenchrus ciliaris* L. and *Aristida congesta* Roam & Schult in the grass layer (Fig. 1). *Acacia nilotica* (L.) Willd. ex Del, *Dichrostachys cinerea* (L.) Wight & Arn and *Sclerocarya birrea* (A. Rich) Hochst. are common trees in the *Acacia* savanna.

The *Acacia* communities occur in discrete patches 50–500 m in diameter, and is associated with soils with markedly higher nutrient status than the surrounding soils (Harmse, 1977). This phenomenon is not limited to the Nylsvley area. Localized nutrient-rich soils and concomitant vegetation have been documented in eastern Botswana (Denbow, 1979) and elsewhere in the northern, eastern and central regions of South Africa (Evers, 1975; Maggs, 1976; Mason, 1968, 1969). These sites have associated with traces of Iron Age settlements, but other nutrient enrichment processes may be responsible, such as the action of grazing animals, termites or erosion. Since the distinction between nutrient-rich and nutrient-poor savannas is fundamental in Africa (Scholes, this issue), a situation where they occur in close proximity on soils derived from the same parent material deserves close examination.

The distinctness of the patches and the observation that they are ecologically different from the surrounding *Burkea* community led to speculation regarding their origin. Tinley (1981) suggested that pockets of mudstone on the hill had weathered to yield soils with a slightly finer texture and higher nutrient status downslope. He based his argument on the location of the sties, which are mostly around the base of a low hill.

Archaeological investigations by Fordyce (1980) revealed Iron Age artefacts (pottery, bones, slag, charcoal, furnaces and hut floors) in all known *Acacia* patches at Nylsvley including the one studied here, but none in the surrounding *Burkea* vegetation. On the basis of the style of pottery and ^{14}C dating of charcoal the sites were estimated to have been

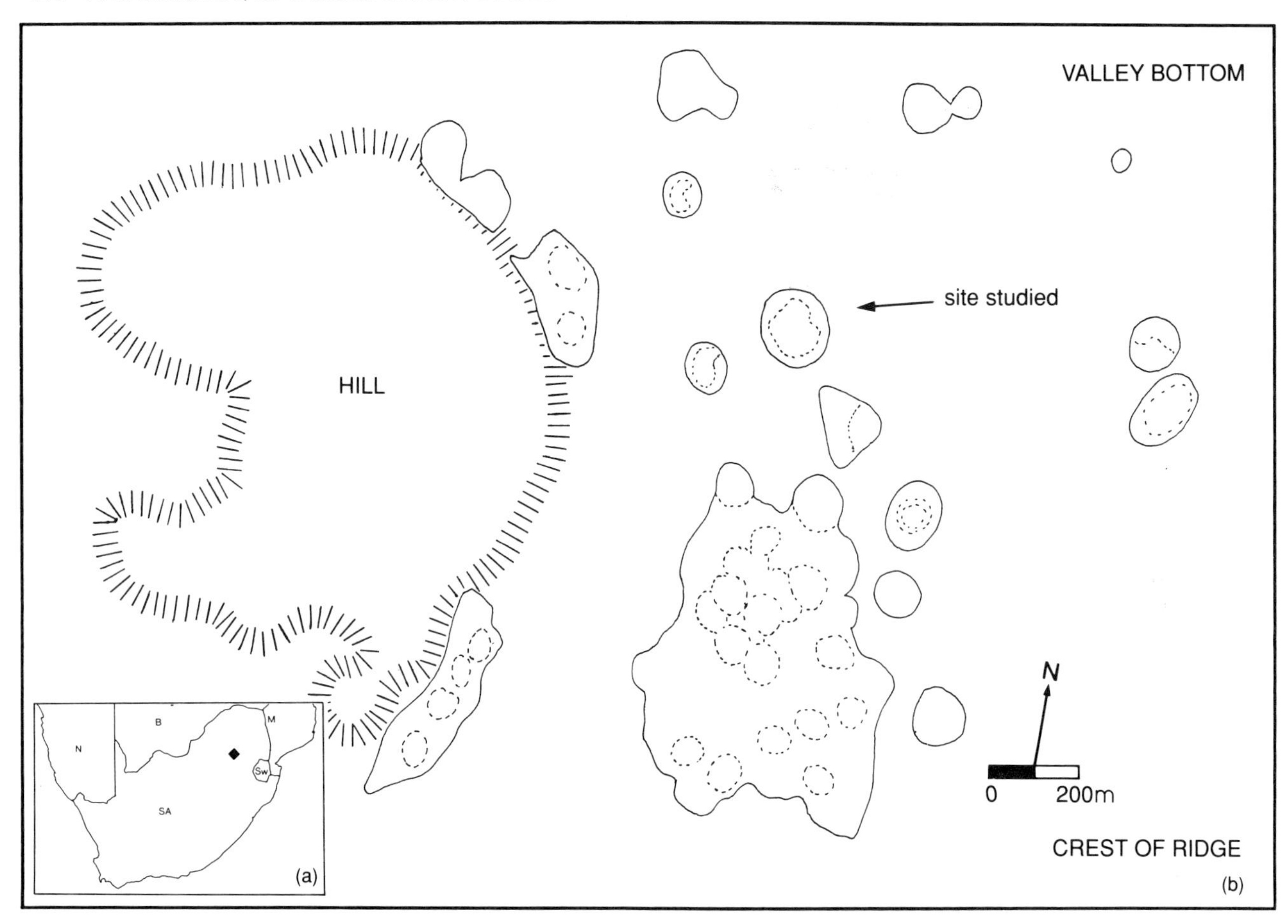

FIG. 1. (a) The geographical position of Nylsvley (◆) within South Africa (SA) (N = Namibia, B = Botswana, M = Mozambique, SW = Swaziland). (b) The relative position of fertile patches (—) and dense *Acacia* clumps (– – –) within a *Burkea* savanna.

occupied around A.D. 1300 (see Evers, 1975; Fordyce, 1980). No recent artefacts have been found. It was therefore suggested that the nutrient enrichment resulted from a relatively short period of human occupation of the sites many years ago. Typical life-span of a Tswana Iron Age village is said to range from 10 to 50 years. Proponents of the geomorphic origin argued that human settlement was drawn to the already enriched sites for purposes of agriculture.

The aims of this study were to (i) define the soil physical and chemical changes that occur across the transition from *Burkea*- to *Acacia*-dominated communities, and relate them to changes in vegetation; (ii) calculate the total nutrient accumulation within one representative site; and (iii) test the hypotheses that human activities alone were sufficient to result in the observed nutrient accumulations.

METHODS

Aerial photographs (scaled at 1:40,000) of the study area were examined for discrete *A. tortilis–C. ciliaris* communities. A roughly circular, isolated settlement stand consisting of a distinct *Acacia* centre within a well-defined clearing was selected approximately 800 m to the east of the hill (Fig. 1). A 230 m transect was laid out at right angles to the slope passing through the centre of the *Acacia* patch and overlapping 50 m into the *Burkea* vegetation (Fig. 2).

Location, basal circumference and height of the trees and shrubs within 5 m on either side of the transect were recorded. The total woody biomass was calculated using equations given in Rutherford (1979). Herbaceous density was determined in contiguous 0.5×0.5 m quadrats along the transect and the data accumulated for each 10 m interval. Standing crop of the herbaceous layer was recorded at consecutive 10 m intervals by clipping and oven drying the biomass of a 1×1 m quadrat. Nutrient pools of Mg, K and Na within both the woody and herbaceous components were calculated from tables given in Frost (1985).

Soil cores were taken to an average of 0.8 m (soil depth varied between 0.65 and 1.05 m) at 10 m intervals along the transect. A (0.0–0.35 m) and B (0.35–0.8 m) horizons were separated, dried and analysed for Ca, Mg, K, Na, P and pH (FSSA, 1980), organic carbon using the wet digestion procedure (Nelson & Sommers, 1975) and texture using the pipette method (Black, 1965).

RESULTS

The discreteness of the enriched patches is clear from Fig. 3. The ten-fold increase in Ca, Mg, K and PO_4 in the A horizon over a horizontal distance of 10 m is reflected by both the woody vegetation (Fig. 3a) and the herbaceous

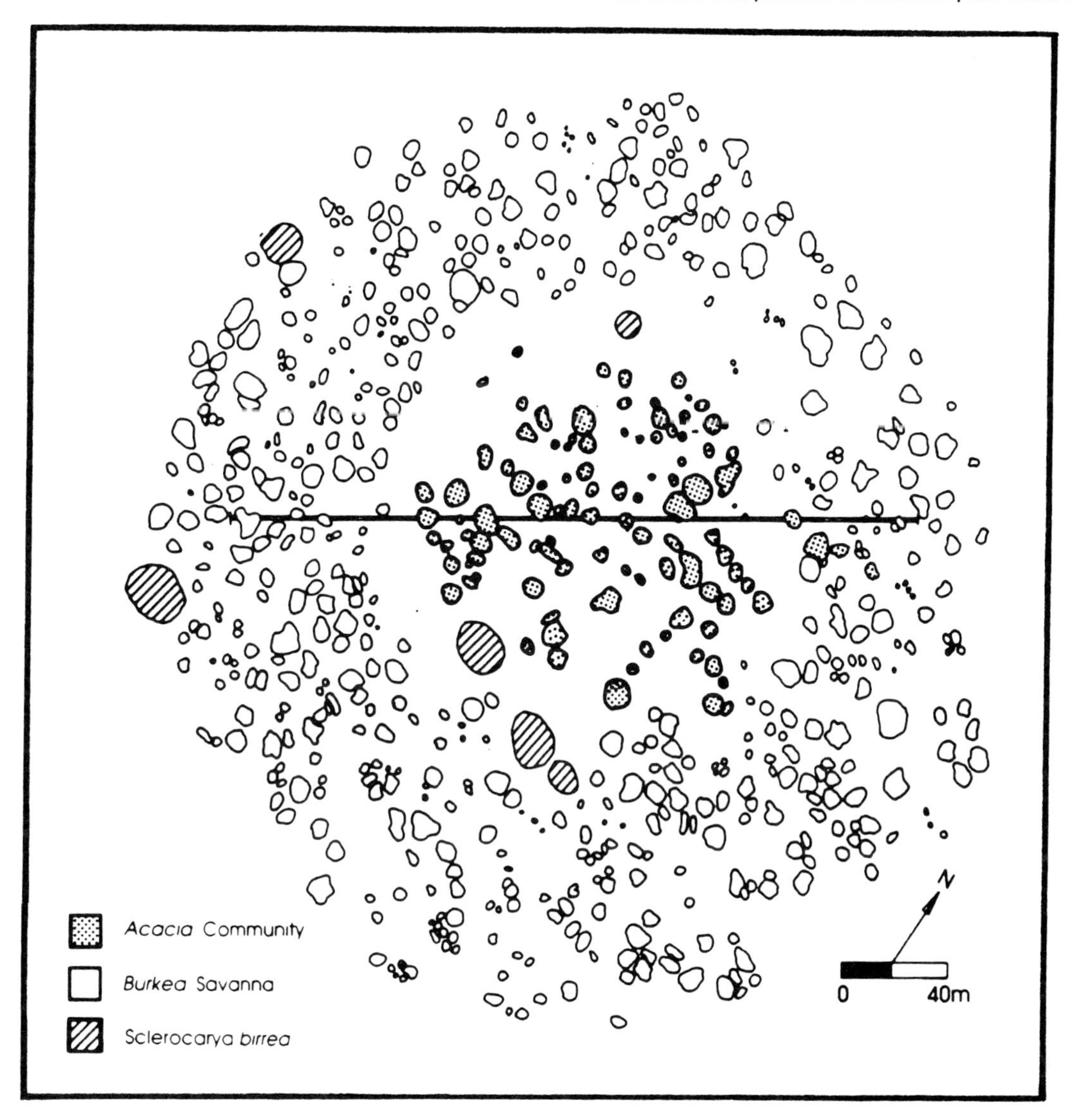

FIG. 2. An areal view of the study site indicating position of tree canopies and layout of the sampling transect.

vegetation (Fig. 3b), although the woody plants tend to be absent from the periphery of the patches. Towards the centre of the patch, levels of Ca, Mg, K and PO_4 (Fig. 3c) were up to 100 times those in the *Burkea* area, and were associated with clumps of *C. ciliaris*, a known indicator of phosphorus-rich soils (Denbow, 1979). The high concentration of basic cations was associated with an increase in pH and a five-fold increase in organic carbon in the A horizon. There was a negligible change in the nutrient status of the B horizon. Sodium increased by a small amount in the A horizon only. There was a small (2%) increase in the silt plus clay fraction within the boundaries of the *Acacia* patch.

Both the woody and herbaceous vegetation reflected the changes in soil chemistry accurately. There was very little overlap in the species composition of the two communities. *Panicum maximum* Jacq. occurred in both, but its distribution was primarily related to the position of tree canopies. The contribution of non-grass species to the herbaceous layer was much higher in the *Acacia* than the *Burkea* area. The *Burkea* woody community consists of greater than ten species while the *Acacia* community is completely dominated by *A. tortilis* and *A. nilotica*, with *Grewia flava* DC. as a shrub. The transition between communities is remarkably sharp (occurring over a horizontal distance of 5 m), and faithfully reflects the onset of the nutrient enhancement.

The *Acacia* trees are only found about 20 m in from the boundary of the site (as indicated by the soil and herbaceous layer changes), leading to a distinct treeless periphery of the patches.

By summation of the nutrients in all horizons down to bedrock and integrating over the extent of the *Acacia* patch, it was possible to calculate the total nutrient accumulation within this particular study site (Table 1). For this purpose the site was assumed to be circular and radially symmetrical. Relative to the *Burkea* soils there was approximately 11.4 times as much Ca, 9.3 times more PO_4, and 2.7 times more Mg and K. Sodium levels were, however, approximately 0.8 times less.

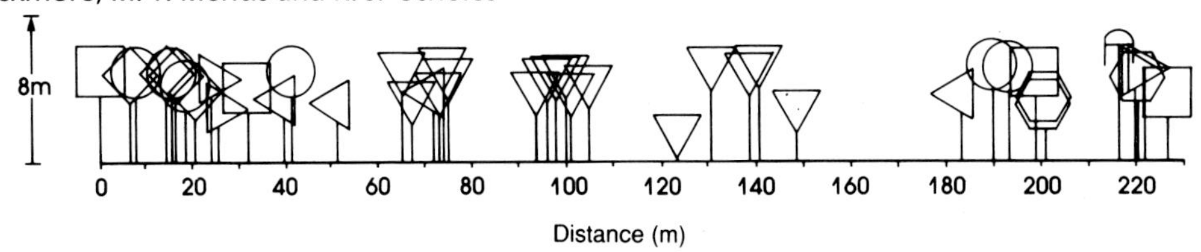

FIG. 3(a). Position of woody plants with a basal circumference greater than 0.4 m along the transect through the *Acacia* site (▽, *Acacia tortilis/nilotica*; ⌓, *Boscia albitrunca* (Burch.) Gilg & Benedict; □, *Burkea africana* ▷, *Dichrostachys cinerea*; ◁, *Dombeya rotundifolia* (Hochst.) Planchon; ◇, *Lannea discolor* (Sonder) Engl.; ⬡, *Securidaca longipedunculata* Fresen; ○, *Terminalia sericea* Burch. ex DC.

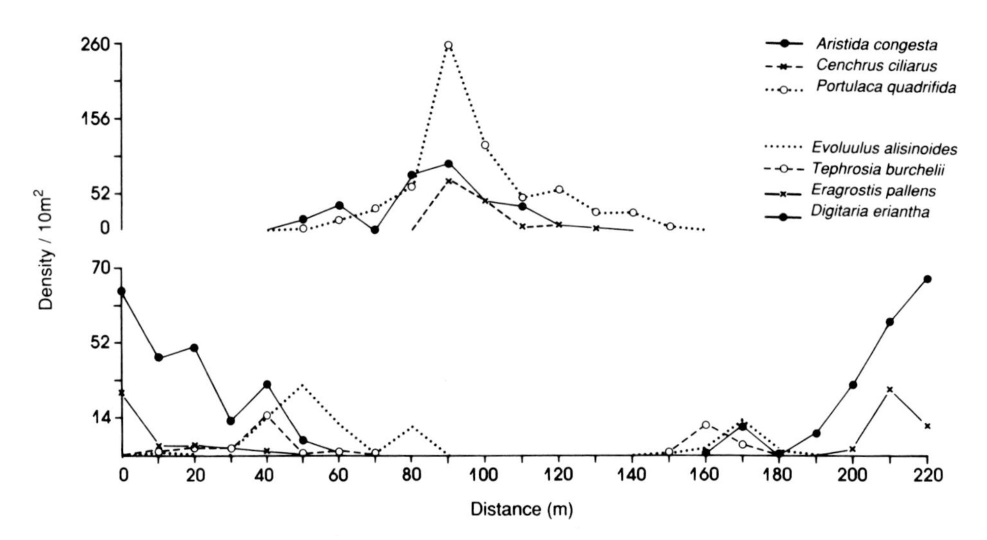

FIG. 3(b). Density patterns of selected herbaceous species.

TABLE 1. Total nutrient concentrations (kg ha^{-1}) in an *Acacia* community and a *Burkea* community. The approximate extent of the *Acacia* patch studied was 1.131 ha. Soil nutrients refer to total amounts in the case of organic carbon and phosphorus, and extractable (1 M ammonium bicarbonate) quantities in the case of Ca, Mg, K and Na. Plant nutrients are total quantities.

	Mg	P	K	Ca	Na
Acacia community					
Woody	2.9	2.3	9.6	—	—
Herbaceous	0.5	0.5	2.4	—	—
Soil	1612.6	2030.1	1768.2	14284.3	76.2
Total	1616.0	2033.1	1780.2	14284.3	76.2
Burkea community					
Woody	8.0	6.4	10.1	—	—
Herbaceous	0.5	0.5	2.4	—	—
Soil	582.0	212.2	639.4	1257.4	99.8
Total	590.6	219.2	651.9	1257.4	99.8

DISCUSSION

If the origin of the patches was geomorphic, an increase of the fine soil fractions would be expected. The observed increase was negligible at the threshold of resolution of the method used. It could readily be explained as a result of clay imported for the construction of huts and furnances. Futhermore, although most of the sites are around the base of the hill (which has no large mudstone deposits), a few are several kilometres away from, and above, any putative source of nutrient rich material. On the basis of this evidence the geomorphic origin of the patches is refuted. The tendency of the sites to be in the vicinity of the hill, which is the only significant landscape feature for many kilometres, can be explained in terms of the preference of Tswana people for such locations, which afford protection from the elements and raiding parties (Mason, 1969).

The absence of nutrient enhancement from the subsoil horizons, and the circular shape of the patch both indicate that leaching of the nutrients out of the site is slow. There is

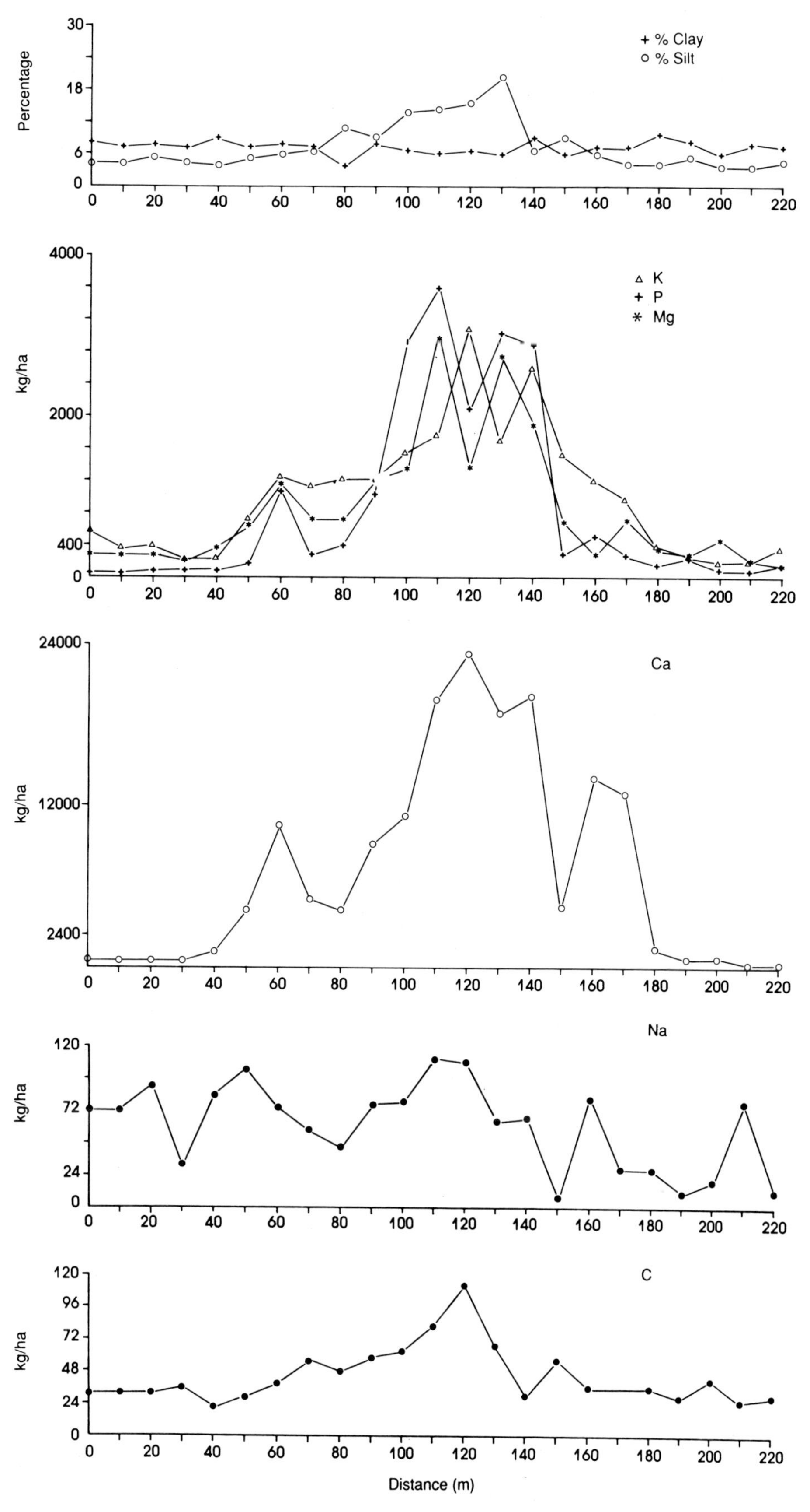

FIG. 3(c). Concentration of nutrients within the soil profile.

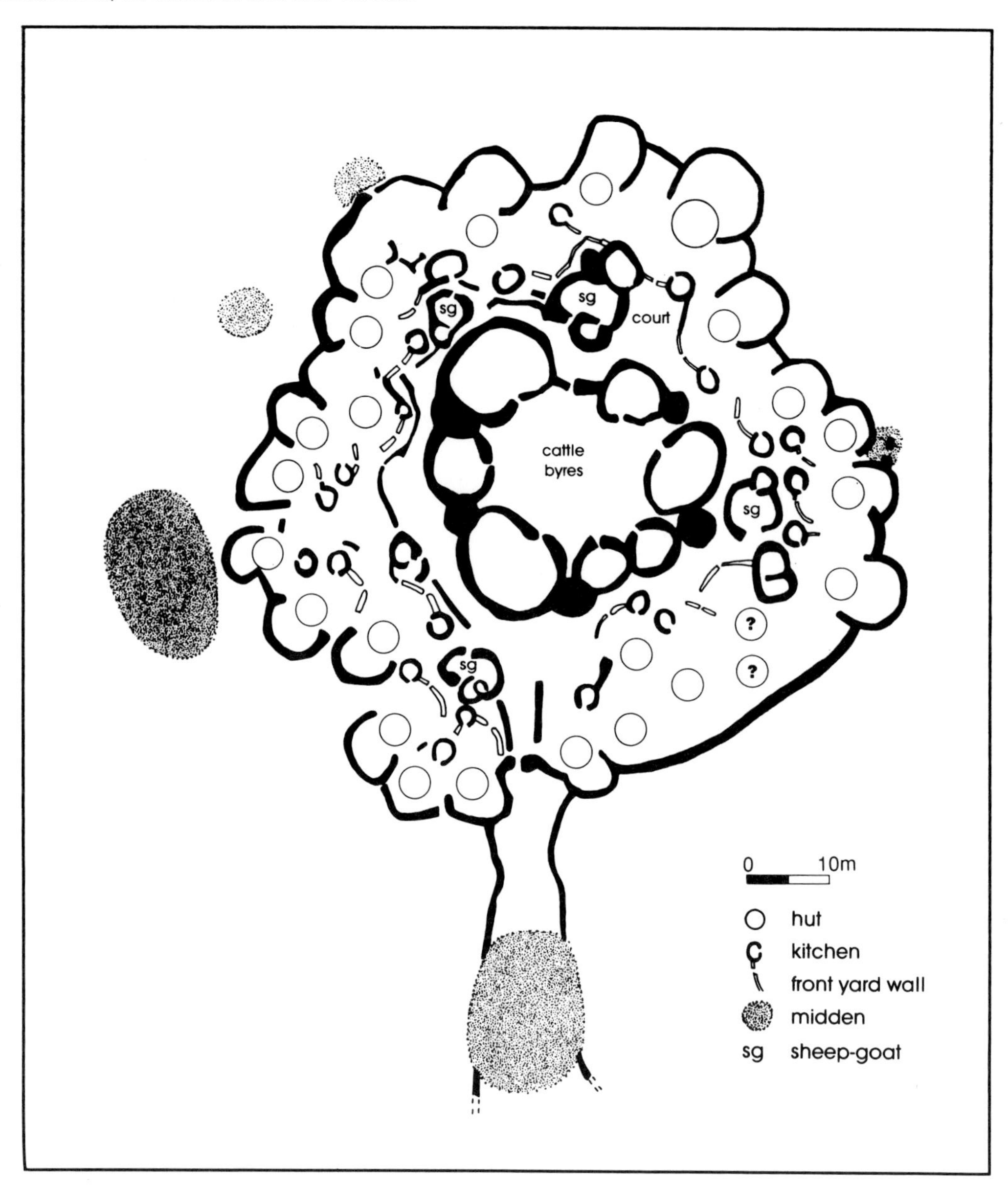

FIG. 4. Diagrammatic representation of a typical Iron Age Tswana settlement. Reproduced from Huffman (1986).

no evidence of downslope migration of the nutrients despite evidence that significant sub-surface water flow occurs. The water-holding capacity of the soil in the *Burkea* areas is 9% (volumetric), while that in the *Acacia* site is about 30%, due to the increase in organic carbon. The water-holding capacity of the profile is frequently exceeded, given the average rainfall of 620 mm during the October–April season. Therefore some leaching would be expected. The absence of a sodium enhancement may reflect a leaching loss, since sodium is not an essential nutrient for most plants, and is therefore not immobilized in live and dead plant tissues.

Meredith (1987) showed that leaching losses did occur from the A horizons of both *Acacia* and *Burkea* soils, albeit at a low rate unless the soils were disturbed. It therefore seems necessary to postulate some biological feedback mechanism to reinforce the nutrient accumulations against long-term nutrient loss. The effect of fires, which occur with an approximately 5-year frequency, would also be expected to dissipate the nutrient accumulations over the long period since occupation. The roots of *Burkea africana* are known to extend for many metres beyond the canopy (Rutherford, 1982), and could therefore in theory penetrate nearly to the middle of the patch studied. This would also constitute a nutrient depletion mechanism.

The foraging range of the ungulate herbivores within the study area is larger than the scale of the patches, which could also lead to an evening-out of nutrient distribution in the landscape. However, they have been observed to spend more time in the *Acacia* patches than would be expected from the area of the patches. If this preference exceeds the higher grass productivity in the *Acacia* patches, mammalian herbivory could represent the

TABLE 2. Predicted minimum import of nutrients (kg ha^{-1}) by a Tswana settlement of thirty-five inhabitatns for a period of 15 years. For assumptions, see text.

	Mg	P	K
Cattle	4698	1620	3670
Wood	457	369	1555
Total	5155	1989	5225

inward nutrient flux needed to balance attrition by other mechanisms.

The higher carbon and nitrogen content of the soil within the patches could have two origins: they may be relics of dung and charcoal from the occupation period; or they may be secondary consequences of the high soil fertility status, which would be expected to lead to higher primary production. The dominant trees in both communities are legumes, but neither has been demonstrated to support symbiotic nitrogen fixing bacteria, despite extensive investigation (Zietsman, Grobbelaar & van Rooyen, 1988). The litter nitrogen content is broadly similar in both sites, but the leaves of the dominant trees of the *Burkea* vegetation contain large amounts of secondary compounds, which may account for their slow decomposition rate (Bezuidenhout, 1980). A thick layer of tree litter is a conspicuous feature of the *Burkea* site which is absent from the *Acacia* site. Why a slow decomposition rate should lead to a low soil carbon content is not known.

The resemblance in size, shape, and nutrient distribution between the sites and known Tswana settlement sites is striking (compare Figs. 2 and 4).

By making certain assumptions it is possible to calculate, in broad terms, the probable nutrient accumulation which would have resulted from a period of occupation by Tswana Iron Age people. The principal activities which would have led to nutrient imports are (i) the practice of allowing cattle to graze extensively during the day, but enclosing them within the village at night for protection against predators, and (ii) the collection of firewood. The lower bounds of nutrient accumulation indicated in Table 2 were calculated by assuming the period of occupation of a single site to be from 15 to 30 years, and the number of inhabitants to be between 35 and 115 (Huffman, 1986). Wood use per capita is approximately 10 kg day^{-1} (M. Gandar, Institute of Natural Resources, University of Natal, Pietermaritzburg, South Africa, personal communication), and the number of cattle per family unit to be approximately between 1 and 8 (T. N. Huffman, Department of Archeology, University of the Witwatersrand, Johannesburg, South Africa, personal communication). The defaecation rate of cattle is approximately 10 kg $head^{-1}$ day^{-1} (Church, 1980). The nutrient content of dead wood was obtained from Frost (1985), and that of dung was obtained by analysis of fresh specimens from cattle grazing on similar vegetation.

The calculated potential imports exceed the observed accumulations in all cases. This represents a strong test of the anthropogenic origin hypothesis, which cannot be rejected on these grounds.

This study was based on the intensive investigation of a single *Acacia* patch. It is believed that the trends and quantities observed here are broadly representative of other *Acacia* patches within the vicinity, given their similarities in vegetation and the limited amount of soil data available from other sites. The period of occupation of a single site is believed, on ethnographic and archaeological evidence, to have been short, but the total duration of occupation in the complex of sites may have extended to several centuries.

CONCLUSIONS

Above-ground vegetation, especially the herbaceous layer, is a reliable indicator of the position of nutrient accumulations within the Nylsvley study site. The origin of the accumulations could not have been geomorphic, and the most likely origin is occupation, for a brief period several hundred years ago, by Tswana tribesmen. The long-term persistence of the sites would appear to require some feedback mechanism to balance the loss of nutrients by erosion, leaching, herbivory, fire and rooting. Establishment of the mechanism of persistance is a key objective to ongoing research at the site.

ACKNOWLEDGMENTS

This research was performed with financial and logistic support from the Foundation for Research Development of the Council for Scientific and Industrial Research. Our special thanks are due to Professor A. Bailey for his helpful hints and comments on the manuscript.

REFERENCES

Anonymous (1975) A description of the Savanna Ecosystem Project, Nylsvley, South Africa. *South African National Scientific Programs Report 1*, pp. 24. CSIR, Pretoria.

Bedzuidenhout, J.J. (1980) Ekologiese studie van die ontbinding van bogrondse plantreste in die Nylsvley savanne-ekosisteem. D.Sc. thesis, University of Pretoria.

Black, C.A. (1965) *Methods of soil analysis, I&II*, pp. 1572. American Society of Agronomy, Inc., Wisconsin.

Church, D.C. (1980) Excretion and digestion. *Digestive physiology and nutrition of ruminants* (ed. by D. C. Church), pp. 115–131. O&B Books Inc., Corvallis.

Denbow, J.R. (1979) *Cenchrus ciliaris*. An ecological indicator of Iron Age middens using aerial photography in eastern Bostwana. *S. Afr. J. Sci.* **75**, 405–408.

Evers, T.M. (1975) Recent Iron Age research in the Eastern Transvaal, South Africa. *S. Afr. archeol. Bull.* **30**, 71–83.

Fertiliser Society of South Afrcia (FSSA) (1980) *Soil analysis*, 4th edn. Pretoria.

Fordyce, B. (1980) The prehistory of Nylsvley. *Progress Report to the National Programme for Environmental Science*, pp. 12. Typescript. CSIR, Pretoria.

Frost, P.G.H. (1985) Organic matter and nutrient dynamics in a broad leafed savanna. *Ecology and management of the worlds savannas* (ed. by J. C. Tothill and J. J. Mott), pp. 200–206. Australian Academy of Science, Canberra.

Harmse, H.J. von M. (1977) Grondsoorte van die Nylsvley-natuurreservaat. *South African National Scientific Programmes Report 15*, pp. 64. CSIR, Pretoria.

Huffman, T.N. (1979) African origins. *S. Afr. J. Sci.* **75**, 233–237.

Huffman, T.N. (1986) *Iron Age settlement of patterns and the origins of class distinction in southern Africa*, pp. 219–388. Advances of World Archaeology.

Huntely, B.J. & Morris, J.W. (1978) Savanna Ecosystem Project: Phase I summary and Phase II progress. *South African National Scientific Programmes Report 29*, pp. 52. CSIR, Pretoria.

Huntley, B.J. & Morris, J.W. (1982) Structure of the Nylsvley Savanna. *Ecology of tropical savannas* (ed. by B. J. Huntley and B. H. Walker), pp. 433–455. Springer, Berlin.

Maggs, T. (1976) Iron Age communities of the southern highveld. *Occasional Publication* No. 2. Natal Museum, Pietermaritzsburg.

Mason, R. (1968) Transvaal and Natal Iron Age settlement revealed by aerial photography. *African Studies*, pp. 167–180.

Mason, R. (1969) *Prehistory of the Transvaal*, p. 498. Witwaterssrand University Press.

Meredith, F. (1987) The effect of soil and vegetation disturbance on the movement of selected nutrients in savanna soils. M.Sc. thesis, University of the Witwatersrand, Johannesburg.

Nelson, D.W. & Sommers, L.E. (1975) A rapid and accurate procedure for estimation of organic carbon in soils. *Proc. Ind. Acad. Sci.* **84**, 456–462.

Rutherford, M.C. (1979) Above ground subdivisions in woody species. *South African National Scientific Programmes Report 36*, pp. 33. CSIR, Pretoria.

Rutherford, M.C. (1982) Woody plant biomass distribution in *Burkea africana* savannas. *Ecology of tropical savannas* (ed. by B. J. Huntley and B. H. Walker), pp. 120–141. Ecological Studies 42. Springer, Berlin.

South African Geological Survey (1978) 1:250,000 Geological Series Map 2428 Nylstroom. Government Printer, Pretoria, South Africa.

Tinley, K.L. (1981) Salient landscape features of the Nylsvley area and it regional environs. *Report to the National Programme for Environmental Sciences*, pp. 4, Typescript. CSIR, Pretoria.

Zietsman, P.C., Grobbelaar, N. & van Rooyen, N. (1988) Soil nitrogenase activity of the Nylsvley Nature Reserve. *S. Afr. J. Bot.* **54**, 21–27.

14/Tree community dynamics in a humid savanna of the Côte-d'Ivoire: modelling the effects of fire and competition with grass and neighbours

J. C. MENAUT, J. GIGNOUX, C. PRADO and J. CLOBERT *Laboratoire d'Ecologie (URA CNRS 258), Ecole Normale Supérieure, 46 rue d'Ulm, 75230 Paris Cédex 05, France*

Abstract. Humid savannas are often made of woody groves and grassy patches in which a few woody individuals develop. A simulation model has been built to explore (1) the role of dispersal and individual growth in community structure; (2) the rôle of local-neighbourhood competition on seedling and adult survival; (3) the interaction between fire and vegetation structure.

To study local interactions and neighbourhood relationships, computations were performed at the individual level and space is explicitly taken into account. Competition has been treated as a whole on the basis of above-ground relationships between individuals, and has a relatively weak effect. Competition for water in the upper horizons of the soil should be more efficient at limiting clump development. The average fire regime cannot prevent these savannas from being invaded by trees. Only a combination of strong competition between individuals and episodic fierce burning should regulate tree density in the long term.

Key words. Modelling, savannas, dynamics, competition, fire.

INTRODUCTION

The co-occurrence of trees and grasses is a major characteristic of savannas. The continuous grass layer is dominated by woody individuals at variable densities and promotes the recurrence of fire (Menaut, 1983). Savanna-like formations are reported to have existed even before the Eocene (van der Hammen, 1983), and their persistence raises several questions (Walker & Noy-Meir, 1982; De Angelis *et al.*, 1985): (1) Which properties maintain the overall pattern? (2) What is the role of fire (intensity and regime)? (3) Which life-history traits of the individuals have the greatest impact at population level and subsequently on community dynamics?

The structure and dynamics of savanna vegetation is generally expressed as the balance between trees and grasses, and more easily by tree density and distribution. While tree density has been related to soil water availability (Walter, 1971) or to disturbance phenomena (Walker & Noy-Meir, 1982), no general hypotheses address tree distribution patterns. In humid savannas, most trees and shrubs are aggregated in more or less loose clumps, giving room to open patches of strong grasses in which some scattered woody individuals develop. Apart from the case of waterlogged savannas in which trees find refuge on old termite-mounds, no evidence yet shows that tree clumps are bound to particular soil conditions.

It has been demonstrated that fire controls tree density in savannas, especially in those of the Guinea zone where the high grass biomass production induces severe burnings (Dereix & N'Guessan, 1976; Brookman-Amissah *et al.*, 1980; Olla-Adams & Adegbola, 1982). Fire is then often considered as a stabilizing element, preventing savanna from evolving towards a woodland (Gillon, 1983). Most savanna fires are surface fires, burning through the grass layer, and controlling seedling establishment (Frost & Robertson, 1987). Seedling survival is favoured under tree clumps where the weaker grass layer cannot sustain intense fire. Once established, seedlings and young saplings still experience fire. Each fire burns the above-ground shoots and bring the individuals back to a previous stage, until they escape fire, when tall enough, or die (Trollope, 1984). By its effects on individuals, fire affects vegetation structure at the plot level.

Savanna species possess a set of vital attributes (Noble & Slatyer, 1980) which determines the nature of vegetation structure and dynamics. We hypothesize that constraints such as fire may change the distribution parameters (size of the individuals and density of the community) but do not affect the nature of the structure (type of distribution).

We have developed a simulation model to explore the causal mechanisms of vegetation structure in savanna communities. The principal aim is to analyse how a spatially explicit formulation can account for the role of fire in savannas and to what extent the individual performance of trees is responsible for the observed spatial pattern. The model follows a Gleasonian approach, integrating the biology and fate of all individuals throughout their life cycle. As space is an explicit factor, species performance is linked to the effect of local environmental conditions on different life-stages of plants within the community (Harper, 1977; Leps & Kindlmann, 1987). We assume that the spatial pattern of savannas depends on the species dispersal ability, and the capacity of individuals to survive at seedling stage. The final outcome of these processes (the actual spatial pattern) is linked to the local environmental conditions of germination sites and to the neighbouring competition interactions.

Our objectives are to assess: (1) the rôle of dispersal and individual growth in community structure; (2) the rôle of local-neighbourhood competition on seedling and adult survival; (3) the interaction between fire and vegetation structure.

THE DATA BASE

Site characteristics

Field data were collected on the Lamto Research Station, located in the Côte-d'Ivoire (West Africa: 5°02′W; 6°13′N; site description in Menaut & César, 1979), at the edge of the rain forest domain (annual rainfall *c.* 1300 mm). Most of the area is covered by open tree/shrub savannas. Each year, in the heart of the dry season, they experience severe burning because of the high grass production. If fire is a process which maintains savannas, this should be more easily identifiable than in drier savannas, not submitted to forest pressure.

The vegetation is composed of three strata: a continuous grass layer (except under dense woody groves) up to 2 m high, a tree/shrub stratum (2–8 m) and an independent stratum of tall palm-trees. Four species account for 90% of the woody community. In their first stage of development, they behave as hemicryptophytes, producing annual suckers whose size depends on root age. After a few years of periodic reduction by fire, they decline and die if no shoot is able to grow above 2 m high and escape fire, from which stage trees behave as phanerophytes. Some individuals grow in the open grassy patches, but most of them are distributed in more or less loose clumps. Less frequent tree species appear as taller isolated individuals, dominating or not a shrub cluster.

SAMPLING PROCEDURE

Seven plots (50×50 m), with different tree density, had been established in 1969. All the individuals, from seedling to dead tree, were mapped on a plan and measured in basal diameter, height and crown surface. Two plots protected from fire since 1964 and 1969 were similarly treated. The same sampling design was kept throughout the study. Data provided information on spatial structure, size-class distribution, reproductive ability, growth and mortality of the woody individual over time (Menaut, 1971; Menaut & César, 1979). Details on the statistical analysis of the data and on the calculations are found in Gignoux (1988). A series of results is used as input parameters, another kept for validation.

MODEL DESCRIPTION

Main features

The model is based on three life-history stages (seedling, juvenile and reproductive adult) and is characterized by the explicit treatment of the spatial structure of tree stands. Computations are thus performed at the individual level to study local interactions and neighbourhood relationships (Prado, 1988). Trees, located by their coordinates on the plane, are distributed in a continuous space with wrap around margins to avoid edge effects, as described in Van Tongeren & Prentice (1986) and Crawley & May (1987).

Emphasis is given to the tree layer. Grasses are treated as an heterogeneous, tree- and fire-dependent pattern of constraints, which affects tree recruitment. The grass cover, spatial pattern and biomass, is a function of tree pattern and density, decreasing down to total disappearance under the clumps. Depending on rainfall, the grass cover can be irregularly burnt (intensity and pattern of fire). Given the complexity of treating all possible spatial data in the model, this variability has not been taken into account in this first version of the model. An approach to the phenomenon is made by attributing a variance to the mean value of tree seedling survival in the grass layer; survival values were then randomly computed within that variance.

Species survival depends on the time–space constraints that affect the performance of individuals throughout their life cycle. Survival constraints are represented by two factors: fire and neighbourhood competition functions. Resources are not explicitly treated. Species are characterized by seven parameters: reproductive output (expressed as seedlings), dispersal, reproductive life-time, life-span (age-specific survival processes) and individual's growth rate, maximum diameter and height. In the present version of the model, only one 'synthetic' species is treated. In order to explore the nature of the dynamics, it was not felt necessary, as a first step, to take into account the relatively small differences in the behaviour of the major species constituting the community. Nevertheless, the present formalism enables us to implement a multi-species system. In the discussion, we introduce another type of species, far less abundant, fire-sensitive, but with a longer life-span and reaching a taller height. Fig. 1 represents the structure and Fig. 2 one iteration of the model.

VARIABLES OF THE MODEL

The inputs of the model, running on a yearly basis, are:

1. An initial distribution (spatial pattern and size classes) over a plot of 50×50 m. It is based either on an actual field situation (for validation against field data) or using a β-law giving a size-class distribution for a given density, a built-in algorithm enabling to generate simultaneously two kinds of spatial pattern (aggregated and random).

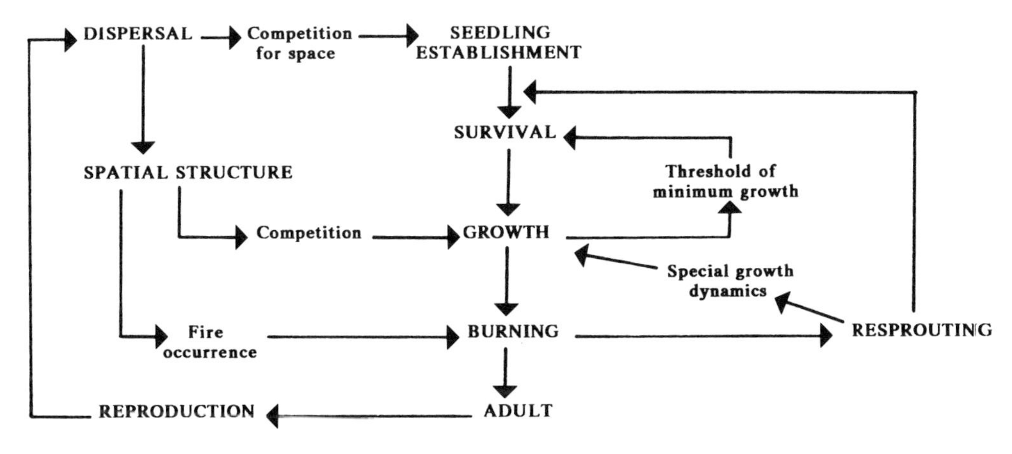

FIG. 1. Structure of the model.

2. Four life-history attributes: age specific survival rate, reproductive life-time, seed production and dispersal.

3. Allometric relationships between diameter, height and crown surface.

4. Growth rules of seedlings, suckers and mature trees.

5. Height of the grass cover and the function of biomass decrease in relation to tree density. Grass biomass is not introduced as such, but through its effects on tree growth and survival.

6. Fire intensity, introduced as a function of grass biomass, and fire regime: basically once a year, with the possibility to test different regimes (once every two, three ... years and total protection).

The outputs of the model are:

1. A map of the plot with the position of each individual, giving density and type of pattern.

2. The dimensions and age of each individual, enabling calculation of the size and age class structure of the community.

MODELLED PROCESSES

Plant growth

As in most individual tree-based models, a basic equation is used to increment the size of each tree. We have followed the approach developed by Botkin, Janak & Wallis (1972), in which tree growth is expressed by the annual volume increment as a function of diameter (D) and height (H). Botkin's equation gives the potential growth and reads:

$$\frac{d[D^2H]}{dt} = G \cdot D^2 \cdot \frac{1 - D \cdot H}{D_{max} \cdot H_{max}} \tag{1}$$

where G = growth rate, D = diameter at breast height, H = height, D_{max} = maximum diameter at breast height, and H_{max} = maximum height.

Height and diameter are related by allometric functions. For instance, Ker & Smith (1955) relate height (H) to diameter at breast height (D) by the equation:

$$H = b_1 + b_2 \cdot D - b_3 \cdot D^2 \tag{2}$$

parameter b_1 being breast height. Assuming that a tree reaches its maximum height with its maximum diameter ($dH/dD = 0$ and $H = H_{max}$ when $D = D_{max}$), it is possible to obtain the solutions for the parameters b_2 and b_3 as functions of maximum diameter (D_{max}) and maximum height (H_{max}) with $b_2 = 2(H_{max} - b_1)D_{max}$ and $b_3 = (H_{max} - b_1)/D^2_{max}$ (Botkin *et al.*, 1972).

In Lamto savannas, field data show that basal diameter is a better predictor of height than diameter at breast height is (Table 1). Consequently, for $b_1 = 0$, equation (2) now reads:

$$H = 2 \cdot \frac{H_{max}}{D_{max}} \cdot D - \frac{H_{max}}{D_{max}} \cdot D^2 \tag{3}$$

TABLE 1. Data on the regressions height on diameter, according to the equation: $H = b_1 + b_2 \cdot D - b_3 \cdot D^2$ (significance: 5%).

	Parameters value	R^2	D_{max} (cm)	H_{max} (cm)
Initial equation	$b_1 = 51 \pm 6$ $b_2 = 48.6 \pm 1.4$ $b_3 = 0.97 \pm 0.05$	0.77	25	660
Modified equation with $b_1 = 0$	$b_1 = 0$ $b_2 = 58.5 \pm 0.8$ $b_3 = -1.26 \pm 0.04$	0.93	23	679

Deriving equation (3) and combining with the growth function (eq. 1), we obtained the height and basal diameter formulations used in the model:

$$\frac{dH}{dt} = \frac{(D_{max} - D) \cdot H_{max} \cdot G \cdot D(1 - D \cdot H/D_{max} \cdot H_{max})}{H \cdot D_{max}^2 + 2 \cdot D \cdot H_{max} \cdot (D_{max} - D)} \tag{4}$$

$$\frac{dD}{dt} = \frac{D_{max}^2 \cdot G \cdot D(1 - D \cdot H/D_{max} \cdot H_{max})}{2 \cdot H \cdot D_{max}^2 + 2 \cdot D \cdot H_{max} \cdot (D_{max} - D)} \tag{5}$$

The surface of the horizontal projection of a given tree crown needs to be calculated in order to estimate the distance beyond which no competition occurs between trees

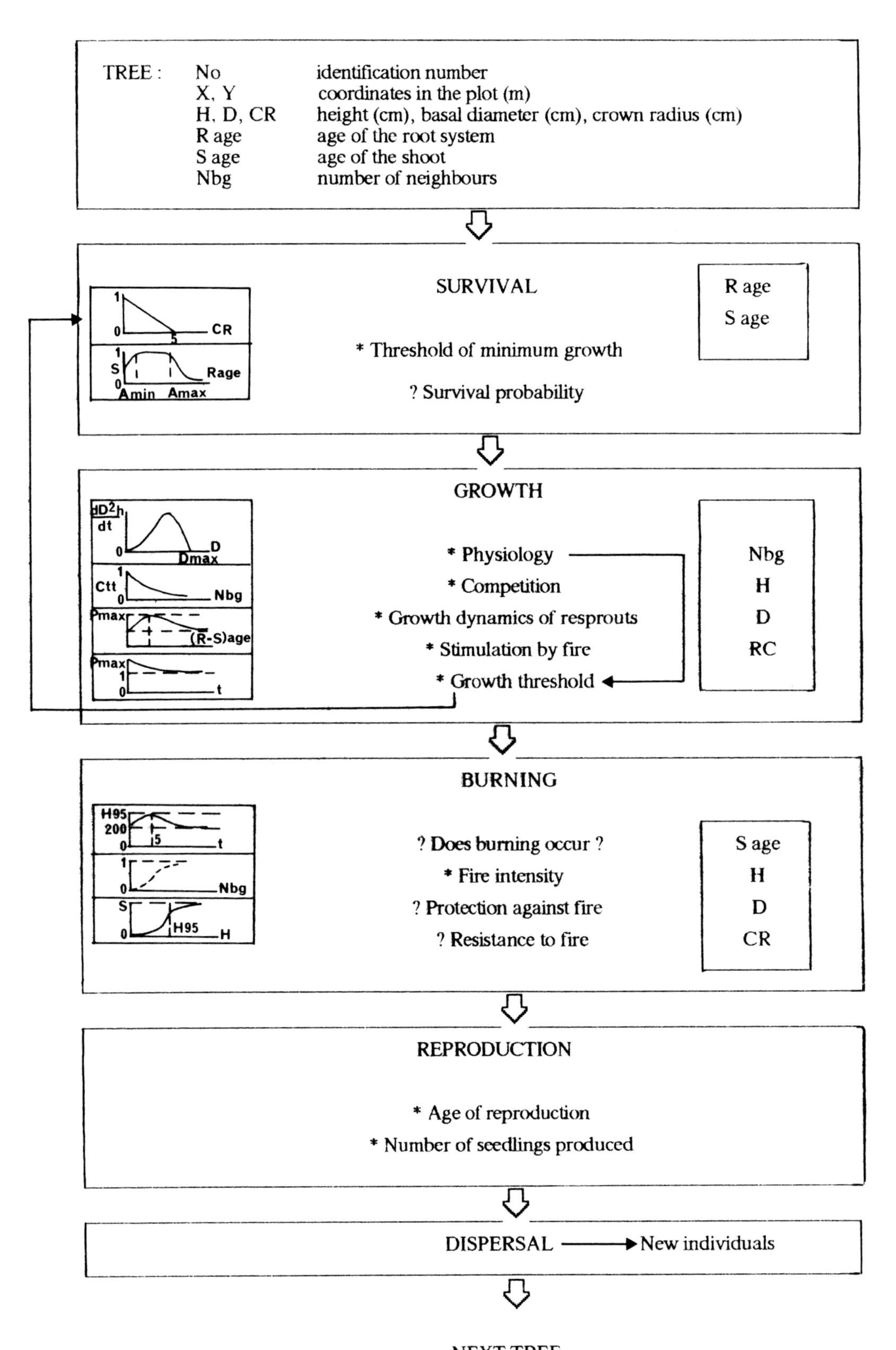

FIG. 2. One iteration of the model (* deterministic steps, ? stochastic steps).

(root competition is not taken into account) and the areas under which most seedlings fall and survive.

Crown surface (C_s) is treated as a function of tree height (H), based on field data. This function approximates crown surface to a circle and takes the form of:

$$\log C_s = -1.3 + 2 \cdot \log H (R^2 = 0.82)$$

For burnt individuals, the growth of the annual suckers is enhanced with the increasing size of the root system. We thus define an enhancement parameter (P_{max}) expressing the maximum difference in growth rate between a genet and a sucker the year following burning. In chaparral vegetation, Keeley & Keeley (1981) have recorded differences in size between seedlings and 1-year suckers ranging from 1.6 to 5.8, depending on the species and the age of the resprouting individual. Assuming that whenever an individual is burned, shoot age (S_{age}) is set back to zero while roots maintain the genet's real age (R_{age}), the actual enhancement is then a function of the difference ($R_{age}-S_{age}$). The integrative parameter R_s traducing the effect of fire on resprouting can be expressed as (Fig. 3):

$$R_s = 1 + a \cdot P_{max} \cdot (R_{age} - S_{age}) \cdot e^{-b(R_{age} - S_{age})} \qquad (6)$$

The constants a and b are chosen in such a way that the parameter be maximum for a given difference between R_{age} and S_{age} (e.g. 10 years on Fig. 3). Finally, we consider that, beyond a certain difference in the development of roots and shoots, the photosynthetic assimilation by the aerial parts cannot sustain the maintenance of the root system. In Lamto, it has been shown that, except for a particular species developing a greater number of suckers each year, phanerophytes cannot be maintained in their hemicryptophytic stage beyond 5–10 years, depending on species and individuals (César & Menaut, 1974). In the model, after a given $R_{age}-S_{age}$, R_s is affected in such a way that the individual dies within the next 5 years.

Using the growth function (eq. 1) to which we add the effect of resprouting (eq. 6), we compute the instantaneous potential growth for each tree as follows:

$$\frac{d[D^2H]}{dt} = G \cdot D^2 \frac{1 - D \cdot H}{D_{max} \cdot H_{max}} \cdot R_s \qquad (7)$$

with R_s equal to 1 when a shoot reaches 2 m high, whatever $R_{age}-S_{age}$ might be.

Competition

Competition is treated at the level of each individual's neighbourhood. We include two kinds of competitive interactions: tree–grass competition and tree-to-tree competition. In both cases, this effect is asymmetrical, being driven by a size-hierarchy in favour of the tallest individual (one-sided competition). The competition factor modifies growth equations, taking values that range between a variable lower bound greater than zero, and 1.

Tree–grass interactions are handled by means of a competition factor which reflects the feedback effects from trees on the herbaceous cover (shading) on the one hand, and those of grasses on tree growth on the other. The herbaceous layer acts only on trees that are smaller than 2 m high, according to a function with takes the form (Fig. 4):

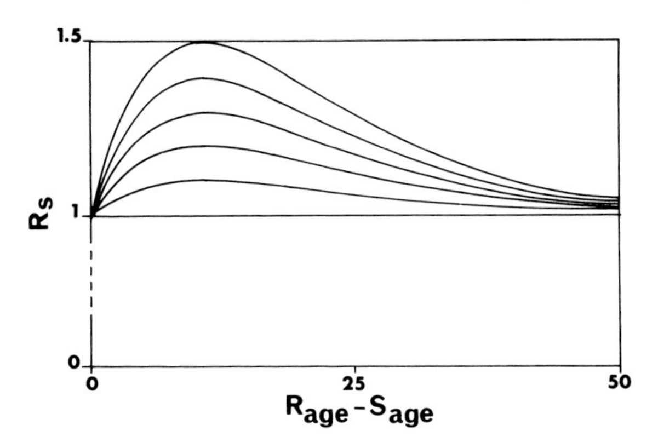

FIG. 3 Growth enhancement factor of suckers (R_s) as a function of the difference in age between below-ground and above-grounds parts ($R_{age}-S_{age}$).

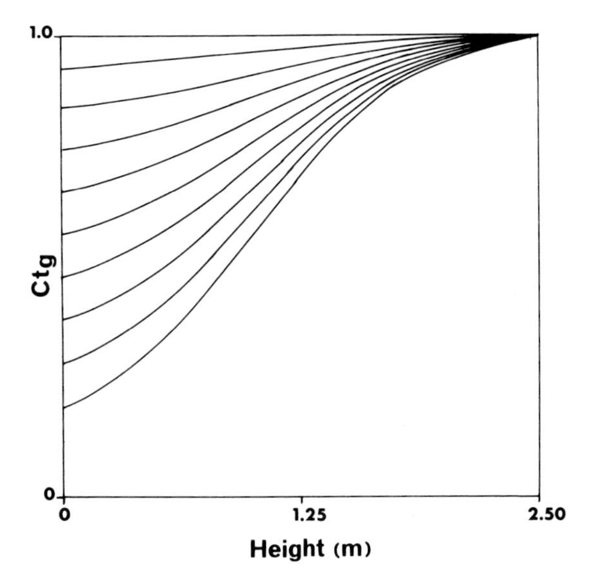

FIG. 4. Effect of competition by grasses on trees (C_{tg}) in relation to tree height.

$$C_{tg} = C_g + \frac{1 - C_g}{1 + b \cdot e^{(-c \cdot H/100)}} \qquad (8)$$

where C_g = factor of competition by grass, H = tree height (cm), b and c = tuning parameters chosen in such a way that C_{tg} = 0.99 for a tree reaching 2 m high.

The competitive influence of grasses on small trees depends on grass biomass, which falls in relation to the number of contiguous or overlapping trees overtopping the grass layer (Obot, 1988). We thus introduce in the model a linear function, such that six neighbours reduce the herbaceous cover to the point where it has a null effect on tree growth. Using equation (8), for a given lower bound (intensity of competition by an open grass layer on a small tree), a family of six curves can be obtained in relation to the number of neighbouring trees. The equation synthesizing tree–grass competition now reads:

$$C_{tg} = C_g + \frac{1 - C_g}{1 + b \cdot e^{(-c \cdot H/100)}} \cdot \frac{N_{bg}}{6} \qquad (9)$$

where N_{bg} is the number of neighbouring trees.

Tree-to-tree competition is here modelled as a spatial relationship (Harper, 1977; Pacala & Silander, 1985; Silander & Pacala, 1985). Neighbourhood can be defined according to several criteria, such as density, distance to individuals or available area in relation to the position and nature of neighbours (Mithen, Harper & Weiner, 1984). Here, we consider that individuals are neighbours whenever their crowns overlap, and that reduction on individual growth occurs as a function of the number of taller neighbours. Such a criterion has the major advantage of a good biological interpretation (shading effects: Begon, 1984). Using field data, we adjust individual-tree volume to the number of neighbours. A negative exponential function is obtained, approximating the effect of competition on growth. This coefficient takes the form of:

$$C_{tt} = e^{-\alpha \cdot N_{bg}} \qquad (10)$$

where α is the adjustment parameter and N_{bg} the number of neighbours (Fig. 5).

The instantaneous actual growth of a tree is thus a function of its potential growth (eq. 7), limited by competition with grasses (eq. 9) and neighbours (eq. 10). It reads:

$$\frac{d[D^2H]}{dt} = G \cdot D^2 \frac{1 - D \cdot H}{D_{max} \cdot H_{max}} \cdot R_s \cdot C_{tg} \cdot C_{tt} \qquad (11)$$

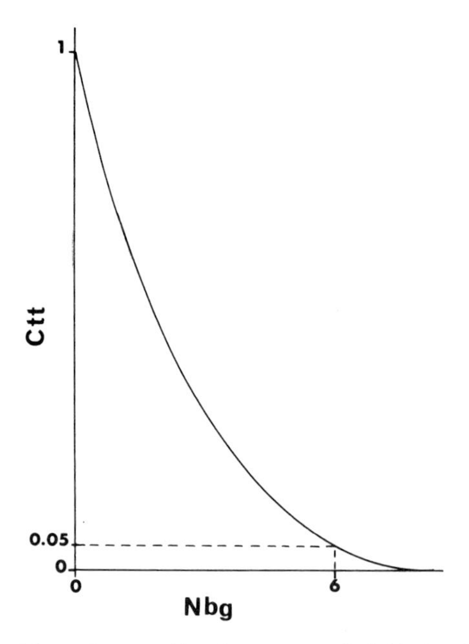

FIG. 5. Effect of competition between tree individuals (C_{tt}) as a function of the number of neighbours (N_{bg}).

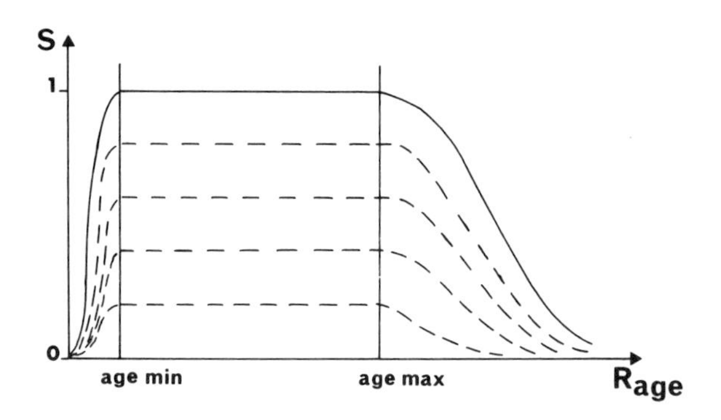

FIG. 6. Tree survival (S) as a function of the age of the root system (R_{age}).

Survival

In this model, survival is dependent on age (Platt, Evans & Rathbun, 1988), competition with neighbours and fire. Analysing field data (Gignoux, 1988), we obtained the size-, age- and neighbour-related mortality percentages, which enabled us to develop an algorithm of tree survival. The resulting curve (Fig. 6) is characterized by three parts: (i) a low survival section ($0 < R_{age} < age_{min}$) that affects the seedling life-stage: mortality related to small size and low resistance to microclimatic conditions and invertebrate consumption; (ii) a plateau ($age_{min} < R_{age} < age_{max}$) where adult survival is maximum; (iii) a smooth fall ($R_{age} < age_{max}$) representing senescence.

Survival in relation to age and competition with neighbours

Survival depends on the age of the root system of the individual (genet's age = R_{age}), and survival probability (S) is computed as follows:

$$\text{IF } R_{age} \begin{cases} \leq age_{min} : S = S_{min} + c_1 \cdot R_{age} \cdot e^{-(R_{age}/age_{min})} \\ > age_{min} \text{ and } R_{age} \leq age_{max} : S = 0.999 \\ > age_{max} : S = c_2 \cdot R_{age} \cdot e^{-(1/3) \cdot (R_{age}/age_{max})3} \end{cases}$$

where age_{min} = minimum age to reach maturity (beginning of the plateau), when survival is maximum (S = 0.999); age_{max} = maximum age before senescence; S_{min} = survival probability of new individuals, fixed at 0.2 according to field data and Rutherford (1981); $c_1 = (0.999 - P_{min})/age_{min}$, and $c_2 = e^{0.33}/age_{max}$.

To include the effect of competition, we introduce a competition-survival factor [S_c, $(0 \leq S_c \leq 1)$] that reduces survival probability (S). To obtain S_c, we calculate a threshold of growth (G_{min}), which on the basis of the instantaneous potential growth (eq. 7), represents the minimum photosynthetic production necessary to ensure plant maintainance ($P_m = 0.95$):

$$G_{min} = G \cdot D^2[(1 - D \cdot H)/(D_{max} \cdot H_{max})] \cdot R_s \cdot P_m)$$

We then compare G_{min} to the actual instantaneous growth (eq. 11). If the actual growth is lower than G_{min} is, S_c is diminished according to a negative linear equation. Its effect is such that if, during five consecutive years G_{min} is lower than the actual growth is, S_c takes a value of zero (broken lines on Fig. 6).

Survival in relation to fire

Survival in fires is based on the following rules: (1) fire only affects the aerial part of the tree; (2) almost all savanna trees can resprout; (3) fire intensity depends on litter biomass and water content (climatic conditions); (4) a threshold height determines if a tree will survive fire or not.

If a tree is reached by fire, its survival probability depends on its height (H) as compared to a threshold (H_{95}) above which the individual has a probability of 95% to survive fire. If H does not reach H_{95}, the aerial parts are burned: shoot age (S_{age}) is set back to zero while roots maintain the genet's actual age (R_{age}) (Fig. 7). For a tree without neighbours and experiencing fire each year, H_{95} is set at 2 m. Of course, H_{95} is a function of fire intensity, i.e. of grass biomass. As previously shown, grass biomass decreases with tree density ($G = G - G \cdot N_{bg}/6$) and so does H_{95}. On the other hand, in protected areas, total phytomass increases with litter accumulation, inducing a greater H_{95}. After a few years (2–5 years in Lamto according to local conditions; Mentis & Tainton, 1984), the phytomass decreases due to reduced production and litter decay. Fig. 8 shows how H_{95} changes with fire frequency and intensity (i.e. grass biomass). A stochastic procedure is added to take into account the variability due to the hydric state of the phytomass.

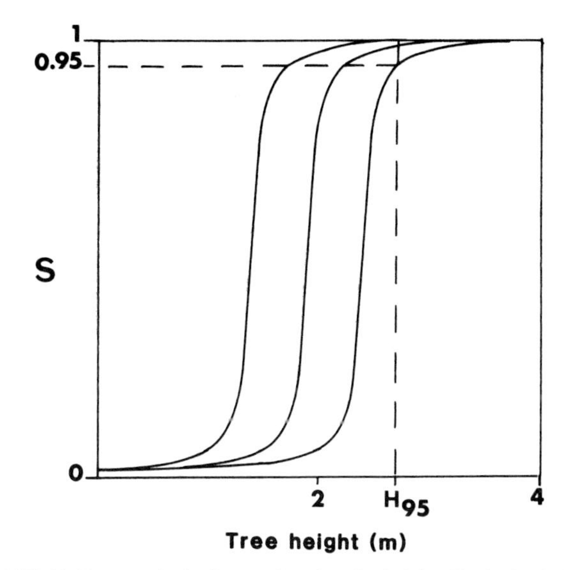

FIG. 7. Tree survival (S) as a function the height; H_{95} is the height above which a tree has a probability of 0.95 to escape fire.

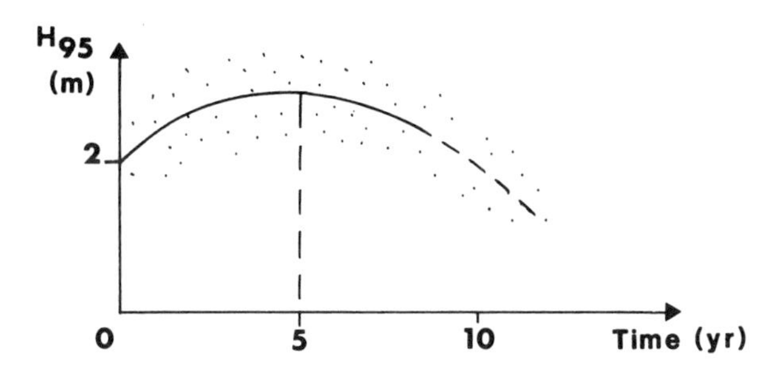

FIG. 8. H_{95} (height above which a tree has a probability of 0.95 to escape fire) as a function of the duration of protection against fire (i.e. grass biomass, hence fire intensity).

Reproduction

According to Harper (1977), the theoretical curve of seed production increases from the first age of reproduction up to a plateau from which seed production decreases with senescence. In the absence of accurate field data, we assume for simplification that reproduction starts and ends with the age_{min} and age_{max} previously defined for survival. If reproducers exist, the rain of propagules that they produce should maintain a certain quantity of established seedlings (Botkin *et al.*, 1972). Seed bank dynamics is then not included in the model, and seedlings are the only reproductive parameter taken into account.

Douglas & Werk (1986) have demonstrated the existence of a relationship between vegetative biomass and seed production. The actual seedling production of an individual is thus a function of its size and age, which allows to take competition and fire effects into account. The equation reads:

$$S_r = S_p \cdot (C_s/C_{smax}) \cdot (1 - S_{age}/age_{max}) \qquad (12)$$

where, for a given individual, S_r = number of seedlings produced; S_p = potential seedling production; C_s = actual size (expressed in crown diameter); C_{smax} = potential size (expressed in crown diameter); S_{age} = age of the above ground parts, and age_{max} = maximum reproductive age.

Dispersal and recruitment

Both processes are rather simply treated. Each reproductive adult acts as a dispersal centre. Seeds are poorly dispersed by wind and mostly fall under tree canopy, either directly or after consumption by birds. The dispersal algorithm allocates a fraction of viable seedlings below the tree crown (80%) and a fraction outside of it (20%). Both fractions are randomly distributed in their respective area of fall. As a first approach, the actual space available on the ground for seed germination (distribution of tree stumps and grass tufts) is not taken into account by the model. As explained above, each dispersed individual is an established seedling, and, as in most gap models (Shugart, 1984), trees are established at the size of small saplings (0.25 m). From this moment onwards, each recruited individual is followed throughout its life cycle, survival being dependent on the environmental sieves previously prescribed.

RESULTS

Fig. 9 examplifies the type of results provided by the model. The spatial evolution can be paralleled to the structural change in age- and size-classes.

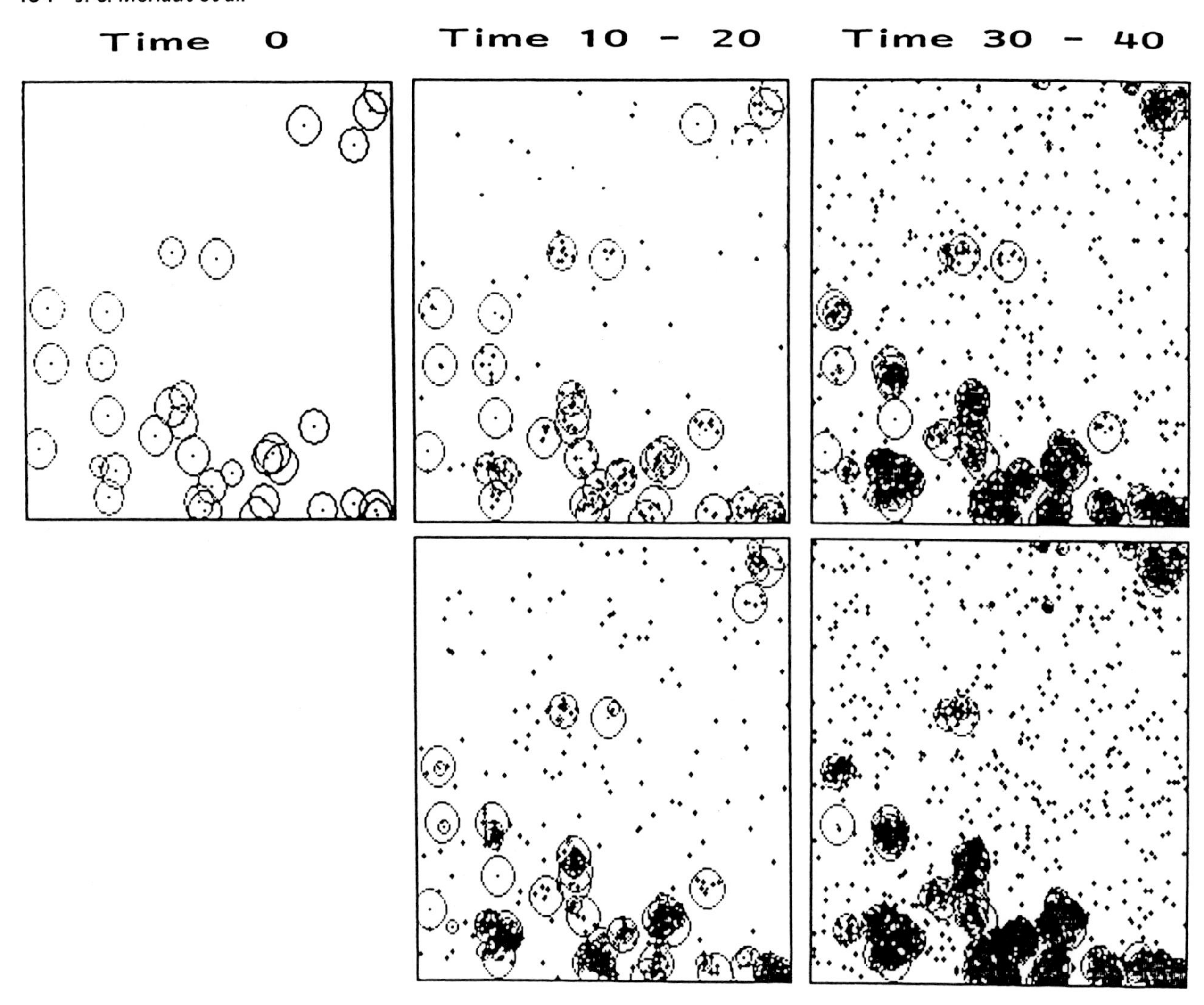

FIG. 9. Simulation of the evolution during 40 years of the spatial pattern of the woody community.

The sensitivity analysis has been difficult to perform. The model is slow to run, due to the high number of parameters and variables taken into account and to the way space is treated. A faster way will now consist of keeping the same approach to analyse the fate of a clump, and of using a more simple and synthetic one for the whole plot. The sensitivity analysis is all the more important to achieve as some fundamental processes, such as suckers versus genets growth rates, reproductive life-time and number of seedlings generated according to size and age, are only roughly estimated from field data or taken from the literature from other vegetation types. We therefore analysed the effects of changes in only some parameters, such as initial situation, pattern of fire in the grass layer, competition, recruitment and growth of suckers. Some preliminary results appear.

Spatial variability

Aggregration seems an unavoidable process. An aggregated initial situation is always reinforced. A random one is always disrupted when the introduced random variability in fire spreading and effects leads several neighbouring saplings to escape fire and initiate a clump. Spatial variability in fire spread (expressed in the model as a randomly distributed proportion of the area covered by grasses) also controls the appearance of isolated individuals: the higher the proportion, the greater is their number.

The number of clumps appearing after a hundred years remains relatively stable for a given initial situation, and varies with changes in the initial structure. The location of new clumps is highly predictable as a function of the initial situation (freq. = $-0.11 + 0.36\ N_{bg}$; $\alpha = 5\%$; $R^2 = 0.80$), as is the size and structure of these clumps. A better parameterization of the process should in the future enable one to predict the evolution of the spatial structure in the long term.

Temporal variability

If the location of the appearance of clumps is predictable, the moment of appearance is highly variable. Analysing the rates of increment of the tree population in a series of simulations, the trajectories over time are significantly different.

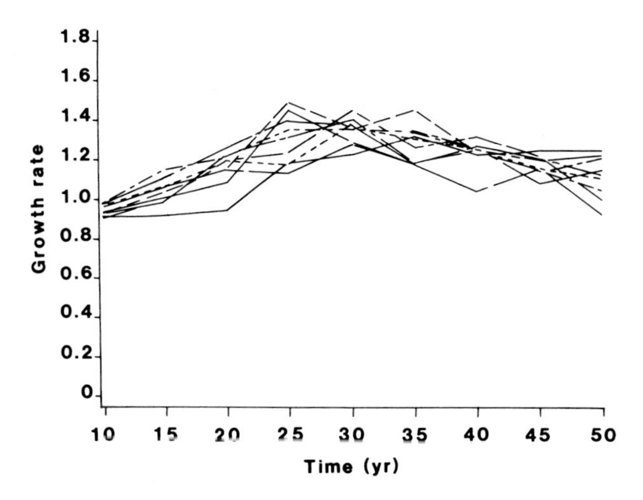

FIG. 10. Simulated evolution in tree density with time.

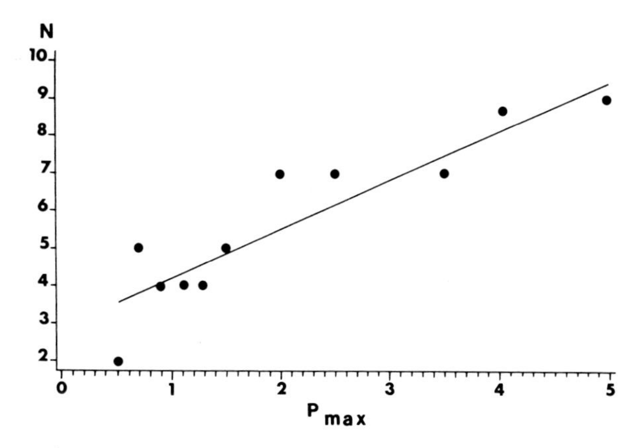

FIG. 11. Number of clumps (N) as a function of the growth enhancement factor (P_{max}).

A tendency to peak after 30 years appears, starting from the initial situation depicted in Fig. 9, but cannot yet be confirmed (Fig. 10). Such a 30 years oscillation has not been confirmed by simulations done in another initialization.

Sensitivity to competition

Competition was expected to control the density of the community. As a matter of fact, the first analyses show that a threefold increase in the competition factor from 0.3 to 0.9) only reduces density by 50%. Further tests must be done, but it seems that competition only has a drastic effect at the highest values.

Sensitivity to recruitment

Age_{min} expresses the recruitment capabilities of the woody population. In the model, the same age_{min} is used to define (1) the age at which individuals become mature and can produce seeds, and (2) the age before which they have the lowest survival probability (young stages and sensitivity to fire). Further research should provide data to determine an age-specific behaviour for each process.

Analysing the rôle of varying age_{min} on the rate of increment of the population revealed the existence of a threshold, between 10 and 20 years, separating a dynamics of extinction from a dynamics of explosion of the population. Reproduction, as influenced by age_{min}, has a weak effect on community dynamics has compared to the one it has on the survival of young individuals.

Sensitivity to suckers regrowth

Any increase in the growth enhancement factor (P_{max}) results in an increase in the number of clumps (NC) appearing in the plot ($NC = 2.92 + 1.30 R_{max}$; $\alpha = 5\%$; $r^2 = 0.80$) (Fig. 11). However, beyond a high value of P_{max}, clumps rapidly coalesce and the relationship is no more valid.

CONCLUSION

At this stage, the model raises many problems and solves few of them. No stable situation can be reached, whatever may be the changes in the value of the input parameters. Only extreme values may lead to extinction of the woody community. Otherwise clumps multiply and extend. In the long term, the plot should be totally invaded by trees and shrubs. Changes in the parameters only slow down or speed up the process.

Thus is posed the problem of savanna dynamics, and of the adequacy of the model to simulate it: Are savannas submitted to 'different types of concurrent multidirectional succession, occurring over different time scales, and confounded by equally important non-directional changes of other kinds' (Walker, 1981)? Or are they in a permanent progression towards woodland (Archer, Scifres & Bassham, 1988)?

A common opinion is that savannas are 'stabilized' in the long term by recurrent disturbances, among which fire would play a major role (Sanford, Obot & Wari, 1982). In this case, what would limit clump development? According to the model, fire alone does not seem to have this ability. However, there remains the possibility that, under extreme conditions (high winds, high air temperatures, low relative humidities), surface fires develop into crown fires and destroy the clump (Trollope, 1984). Crown fires spreading from a tree to another have never been reported in Lamto savannas. Another hypothesis, under test, is that competition with another type of woody species would result in periodic clump reduction. In Lamto savannas, a few long-living tree species grow above the average woody cover (<8 m high) and may reach 12–15 m (César & Menaut, 1974). They are always found as isolated individuals within or outside clumps. Young individuals always grow in sites protected from fire, mostly in the heart of clumps from which the grass cover, hence fire, is excluded. The scenario would be that these species only establish in an existing clump, and dominate it. Competition would then seriously reduce the growth of the understorey (more than inter-shrub competition does) and increase its mortality rate. Facilitated by the fact that the shrubby individuals at the origin of the clump are within the same age-range, the gap thus opened would result in the invasion by grasses, the return of fire and destruction of seedlings. The combined effects of competition and fire would then exclude the understorey and account for the occurrence of these tall isolated trees. Anyhow, the treatment of competition is certainly a shortcoming of the model. Overall competition factors were assigned on above-ground relationships between plants (seedlings in the grass layer and crowns overlap for mature trees). A particular attention should be given to root competition for water in the upper horizons of the soil (Scholes, 1987). Such competition between seedlings and grasses on the other hand, and especially between seedlings and mature trees on the other, should severely limit seedling establishment and clump extension.

The question of savanna 'stabilization' is clearly a time-scale problem. Some authors claim that vegetation never becomes static, but changes in a 'wave-like' or cyclical pattern over long phases (Botkin & Sobel, 1975). Is this the case for Lamto savannas? In 20 years, tree density in annually burnt plots has increased by *c.* 30% (Dauget & Menaut, unpublished data); which would confirm the results of the model. Extreme events or episodic co-occurrent disturbances should then be responsible for the maintenance of savannas in the very long term (Frost *et al.*, 1985; Walker & Graetz, 1989). However, because of the life-span of savanna trees (from 50 to over 200 years), and because any disturbance will likely affect them for several generations, many centuries are required before any real pattern can be seen (Sanford, 1982).

Though not yet able to generate a fully 'realistic' situation, the model helps in exploring the influence of varying key biological processes and disturbances on savanna structure and dynamics, and suggesting lines for further research are provided. Climatic changes are liable to affect plant reproductive ability directly and indirectly through changes in disturbances (e.g. fire intensity and regime). The resulting vegetation would then be more responsible for changes in energy and nutrient transfers than the immediate effect of climatic changes on primary production and decomposition.

REFERENCES

Archer, S., Scifres, C. & Bassham, C.R. (1988) Autogenic succession in a subtropical savanna: conversion of grassland to thorn woodland. *Ecol. Monogr.* **58**, 111–127.

Begon, M. (1984) Density and individual fitness: asymmetric competition. *Evolutionary ecology* (ed. by B. Shorrocks), pp. 175–194. Blackwell Scientific Publications, Oxford.

Botkin, D.B., Janak, J.F. & Wallis, J.R. (1972) Some ecological consequences of a computer model of forest growth. *J. Ecol.* **60**, 849–872.

Botkin, D.B. & Sobel, M.J. (1975) Stability in time-varying ecosystems. *Amer. Nat.* **109**, 625–646.

Brookman-Amissah, J., *et al.* (1980) A re-assessment of a fire protection experiment in north-eastern Ghana savanna. *J. appl. Ecol.* **17**, 85–99.

César, J. & Menaut, J.C. (1974) Le peuplement végétal. *Bull. Liasison Chercheurs de Lamto*, Special issue no. 2, 1–161.

Crawley, M.J. & May, R.M. (1987) Population dynamics and plant community structure: competition between annuals and perennials. *J. theor. Biol.* **125**, 475–489.

De Angelis, D.L., *et al.* (1985). Ecological modelling and disturbance evaluation. *Ecol. Model.* **29**, 399–419.

Dereix, C. & N'Guessan, A. (1976) *Etude de l'action des feux de brousse sur la végétation: les parcelles feux de Kokondekro, résultats après 40 ans de traitement*, p. 31. Centre Technique Forestier Tropical, Bouaké.

Douglas, A.S. & Werk, K.S. (1986) Size-dependent effects in the analysis of reproductive effort in plants. *Am. Nat.* **127**, 667–680.

Frost, P.G., *et al.* (1985) Responses of savannas to stress and disturbance. *Biol. Intern.*, Special issue 10, 1–82.

Frost, P.G. & Robertson, F. (1987) The ecological effects of fire in savannas. *Determinants of tropical savannas* (ed. by B. H. Walker), pp. 93–140. IRL Press, Oxford.

Gignoux, J. (1988) *Modélisation de la dynamique d'une population ligneuse. Application à l'étude d'une savane africaine*, p.83. DEA INAPG-Univ. Paris XI.

Gillon, D. (1983) The fire problem in tropical savannas. *Tropical savannas* (ed. by F. Bourlière), pp. 617–641. Elsevier, Amsterdam.

Harper, J. (1977) *Population biology of plants*, p. 892. Academic Press, London.

Keeley, J.E. & Keeley, S.C. (1981) Post-fire regeneration of southern California chaparral. *Am. J. Bot.* **68**, 524–530.

Ker, J.W. & Smith, J.H.G. (1955) Advantages of the parabolic expression of height–diameter relationships. *For. Chron.* **31**, 235–246.

Leps, J. & Kindelmann, P. (1987) Models of the development of spatial patterns of an even-aged plant population over time. *Ecol. Model.* **39**, 45–57.

Menaut, J.C. (1971) Etude de quelques peuplements ligneux d'une savane guinéenne de Côte-d'Ivoire, p. 153. Doctoral dissertation, Univ. Paris VI.

Menaut, J.C. (1983) The vegetation of African savannas. *Tropical savannas* (ed. by F. Bourlière), pp. 109–149. Elsevier, Amsterdam.

Menaut, J.C. & César, J. (1979) Structure and primary productivity of Lamto savannas, Ivory Coast. *Ecology*, **60**, 1197–1210.

Mentis, M.T. & Tainton, N.M. (1984) The effect of fire on forage production and quality. *Ecological effects of fire in South African ecosystems* (ed. by P. de V. Booysen and N. M. Tainton), pp. 245–254. Springer, Berlin.

Mithen, R., Harper, J.L. & Weiner, J. (1984) Growth and mortality of individual plants as a function of 'available area'. *Oecologia*, **62**, 57–60.

Noble, I. & Slatyer, R.O. (1980) The use of vital attributes to predict successional changes in plant communities subject to recurrent disturbances. *Vegetatio*, **43**, 5–21.

Obot, E.A. (1988) Estimating the optimum tree density for maximum herbaceous production in the Guinea savanna of Nigeria. *J. Arid Environ.* **14**, 267–273.

Olla-Adams, B.A. & Adegbola, P.O. (1982) Effects of burning treatments on structure, herbaceous standing crop and litter accumulation of derived savanna in the Olokemeji Forest reserve. *Nigerian savanna* (ed. by W.W. Sanford, H.M. Yefusu and J.S.O. Ayeni), pp. 151–159. Kainji Lake Research Institute, New Bussa.

Pacala, S.W. & Silander, J.A. (1985) Neighborhood models of plant population dynamics. 1. Single-species models of annuals. *Am. Nat.* **125**, 385–411.

Platt, W.J., Evans, G.W. & Rathbun, S.L. (1988) The population dynamics of a long-lived conifer (*Pinus palustris*). *Am. Nat.* **131**, 491–535.

Prado, C. (1988) Un modèle de succession végétale: rôle des traits biologiques des espèces et des contraintes spatiales, p. 212. Doctoral dissertation, Univ. Paris VI.

Rutherford, M.C. (1981) Survival, regeneration and leaf biomass changes in woody plants following spring burns in *Burkea africana–Ochna pulchra* savanna. *Bothalia*, **13**, 531–552.

Sanford, W.W. (1982) Savanna: a general review. *Nigerian savanna* (ed. by W. W. Sanford, H. M. Yefusu and J. S. O. Ayeni), pp. 3–22. Kainji Lake Research Institute, New Bussa.

Sanford, W.W., Obot, E.A. & Wari, M. (1982) Savanna vegetational succession. *Nigerian savanna* (ed. by W. W. Sanford, H. M. Yefusu and J. S. O. Ayeni), pp. 418–432. Kainji Lake Research Institute, New Bussa.

Scholes, R.J. (1987) Response of three semi-arid savannas on contrasting soils to the removal of the woody component, p. 297. Doctoral dissertation. University of Witwatersrand, Johannesburg.

Shugart, H.H. (1984) *A theory of forest dynamics. The ecological implications of forest succession models*, p. 278. Springer, New York.

Silander, J.A. & Pacala, S.W. (1985) Neighborhood predictors of plant performance. *Oecologia*, **66**, 256–263.

Trollope, W.S.W. (1984) Fire in savanna. *Ecological effects of fire in South African ecosystems* (ed. by P. de V. Booysen and N. M. Tainton), pp. 149–176. Springer, Berlin.

Van der Hammen, T. (1983) The palaeoecology and palaeogeography of savannas. *Tropical savannas* (ed. by F. Bourlière), pp. 19–35. Elsevier, Amsterdam.

Van Tongeren, O. & Prentice, I.C. (1986) A spatial simulation on model for vegetation dynamics. *Vegetatio*, **65**, 163–173.

Walker, B.H. (1981) Is succession a viable concept in African savanna ecosystems? *Forest succession* (ed. by D. C. West, H. H. Shugart and D. B. Botkin), pp. 431–447. Springer, New York.

Walker, B.H. & Noy-Meir, I. (1982) Aspects of the stability and resilience of savanna ecosystems. *Ecology of tropical savannas* (ed. by B. J. Huntley and B. H. Walker), pp. 556–590. Springer, Berlin.

Walker, B.H. & Graetz, R.D. (eds.) (1989) *Effects of atmospheric and climate change on terrestrial ecosystems*, p. 61. International Geosphere-Biosphere Programme report no. 5, Stockholm.

Walter, H. (1971) *Ecology of tropical and subtropical vegetation*. Oliver & Boyd, Edinburgh.

15/Tree/grass ratios in East African savannas: a comparison of existing models

A. JOY BELSKY *The Cornell Plantations, Cornell University, Ithaca, New York 14853, U.S.A.*

Abstract. Tropical savannas are commonly described as biomes with continuous grass strata and discontinuous tree or shrub strata. A number of simple models have been developed that allow tropical savannas to be compared to one another. Older models were based on two or three variables such as rainfall, temperature, and soil texture, but newer ones either include a greater number of environmental variables or utilize biologically relevant factors such as the moisture and nutrients available to plants during their growing season. In this paper, the effectiveness of these models in describing the structure and function of the savannas of East Africa is examined.

Key words. Tropical savannas, East Africa, conceptual models, edaphic grasslands, woodlands.

INTRODUCTION

Plant communities or biomes can be compared by contrasting their biotic and abiotic environments or by examining their relative positions within simplified models. Although comparisons of environmental characteristics are instructive, differences in floristic, climatic and edaphic detail may obscure similarities in form and function. As a result, ecologists often use conceptual models to identify key similarities and differences among vegetation types.

An initial attempt to compare the savannas of East Africa to other tropical savannas using realistic existing models exposed a variety of conflicting models and hypotheses, each emphasizing different aspects of savanna ecology (cf. Huntley & Walker, 1982; Bourliere, 1983; Tothill & Mott, 1985; Cole, 1986; Frost *et al.*, 1986; Walker, 1987). All models agree with the core definition that tropical savannas are characterized by a continuous grass stratum and a discontinuous tree or shrub stratum. Beyond that, the models emphasize different aspects of the structure and function of tropical savannas and the different factors limiting their distributions. The goal of this paper is to examine the validity of these models in relation to East African savannas and to determine which models, if any, realistically describe this biome.

GLOBAL MODELS

Whittaker's model

Among the early models were those by Whittaker (1975) and Holdridge (1947), who described the environmental conditions defining and limiting all terrestrial ecosystems, including tropical savannas. In Whittaker's model, tropical savannas are characterized by low to moderate annual rainfall (500–1300 mm) and by high mean annual temperatures (18–30°C) (Fig. 1). They are bounded by woodlands and forests at higher rainfall regimes, by thorn-scrublands and deserts at lower rainfall regimes, and by temperate grasslands and steppes at lower temperatures. Although the savanna biome within this model is limited by rainfall and temperature, savanna physiognomy (i.e. the ratio of woody plant to grass cover) is dependent on edaphic characteristics and recurrent fire. Deshmukh (1986) recently expanded Whittaker's model to include herbivory as a factor altering tree/grass ratios in tropical savannas.

East African savannas fit Whittaker's model (Fig. 1) exceptionally well. Different locations within the Serengeti ecosystem in northern Tanzania, for example, have mean annual rainfalls of 400–1200 mm (Norton-Griffiths, Herlocker & Pennycuick, 1975) and mean annual temperatures of 15–21°C (Schmidt, 1975; Jager, 1982), placing them within, to slightly below, the rainfall and temperature ranges in Whittaker's model. The few other exceptions include savannas in the Kenyan highlands having mean annual temperatures of less than 18°C (Griffiths, 1969), parts of Tsavo and Amboseli National Parks in southern Kenya that receive an average of 300–500 m rainfall (Western, 1983; Wijngaarden, 1985), and savannas of Turkana District in northern Kenya that receive an average of 150–600 mm annual rainfall (Coughenour *et al.*, 1985).

The close fit between East African savannas and Whittaker's model can be attributed to his allowing tree/grass ratios to fluctuate temporally within the climatic borders of savannas. Many of the areas of East Africa that still support relatively natural biotic systems are characterized by non-equilibrium plant communities in which the vegetation

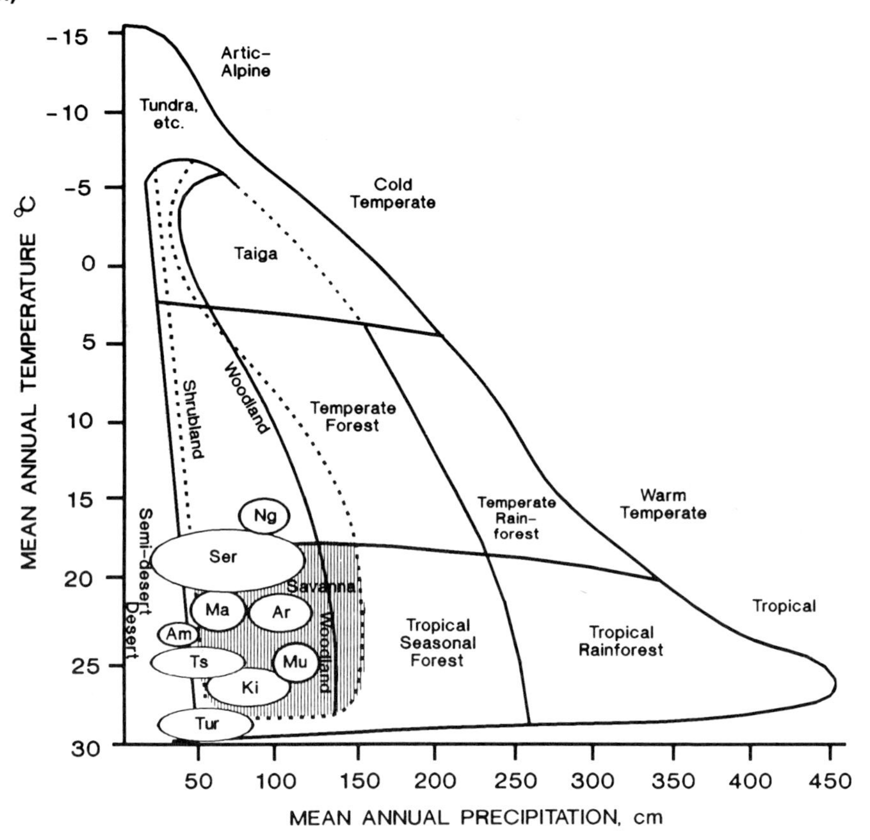

FIG. 1. Relationships of major terrestrial biomes to temperature and rainfall gradients (modified from Whittaker, 1975). The tree/grass ratios within tropical savannas (the shaded area) are determined by edaphic characteristics, fire and large herbivores, especially elephants. The ovals represent the relative positions of Amboseli (Am) and Tsavo (Ts) National Parks and Turkana (Tur) District in Kenya; Arusha (Ar), Manyara (Ma), Ngorongoro (Ng) and Serengeti (Ser) National Parks in Tanzania; and Murchison Falls (Mu) and Kidepo (Ki) National Parks in Uganda. (References are in text.)

cycles between open savanna and woodland (or thorn-scrubland). In areas of high annual rainfall (800–1200 mm) such as the northern, western and central parts of the Serengeti ecosystem in Tanzania, Murchison Falls and Kidepo Valley National Parks in Uganda, and Akagera National Park in Rwanda, woodlands are converted to open savannas during periods of high elephant density or high fire frequency, but they revert back to woodlands when fire frequencies and the densities of elephant and other browsers are reduced (Buechner & Dawkins, 1961; Laws, 1970; Spinage & Guinness, 1971; Harrington & Ross, 1974; Norton-Griffiths, 1979; Pellew, 1983; Smart, Hatton & Spence, 1985; Dublin, 1986). In the Serengeti, this cycling between woodlands and grasslands is accompanied by vegetational changes that have received considerable attention and for which many of the causes are known or can be inferred (Belsky, 1989). In areas of lower rainfall such as in Tsavo National Park, grass productivity and standing dead biomass is low, and fires less frequent. In these areas, natural bush and thorn scrublands are converted to open savannas predominantly by elephants (Thorbahn, 1984; Wijngaarden, 1985).

Holdridge's model

Holdridge devised a triangular model (1947) based on three climate-related parameters – rainfall, temperature and evapotranspiration. In this model, tropical savannas are bounded by dry forest at higher rainfall (>1000 mm), by thorn-forests at lower rainfall (<500 mm), and by thorn steppe and temperate savannas at lower temperatures (<18°C). Although Holdridge increased the number of environmental parameters from two to three, his model is less informative than Whittaker's since the possibility of temporal cycling between grassland and woodland is not included.

REGIONAL MODELS

Simple abiotic models – the Walter/Walker model

Walter (1971) identified rainfall as the major factor governing savanna physiognomy in African savannas. According to his interpretation, in low rainfall areas (<250 mm annually) only grass species, which are shallow rooted, received sufficient moisture to grow, and grasslands dominate. Where rainfall is higher (250–500 mm annually), trees and grasses coexist since rainfall is sufficient for water to percolate to lower soil horizons, allowing the more deeply rooted trees to survive periods of drought. In areas of high rainfall (>500 mm), woodlands dominate since soil moisture is sufficient to support a closed canopy, which shades out the grasses.

Walter's description of savanna function was developed

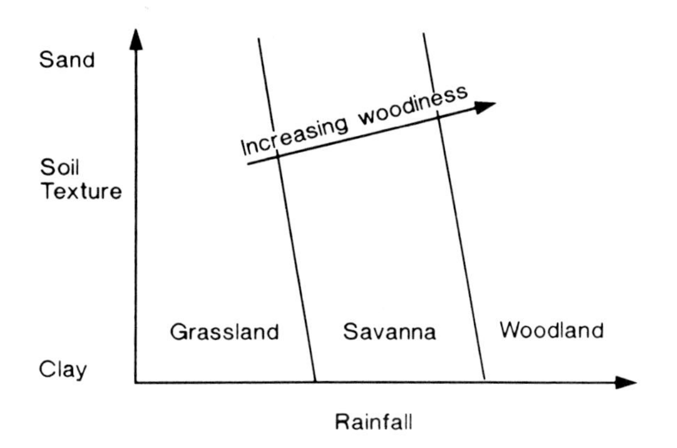

FIG. 2. The Walter/Walker model of tropical savannas (reproduced from Walker & Noy-Meir, 1982). (Permission to reproduce this figure was granted by B. H. Walker.)

and expanded into a series of models by Walker *et al.* (1981) and Walker & Noy-Meir (1982). Their simplest model (Walker & Noy-Meir, 1982) is based on two environmental gradients, rainfall and soil texture (Fig. 2). It predicts an increase in tree density along a gradient of increasing rainfall (but see Walker (1985) for a modified version of this model).

Walter's original model, which was developed for savannas in S.W. Africa, and subsequent models by Walker and his colleagues (1981, 1982) are not valid for East African savannas where communities at the low end of the rainfall gradient are dominated by thorn-scrublands, not by grasslands. In addition, tree/grass ratios in East Africa are determined by fire, soil and herbivory, not by rainfall. Most regions of East Africa that have the proper climatic and edaphic characteristics can, and often do, support a temporal succession of different communities – from grassland to woodland and back again (Belsky, 1989). This capability of supporting a range of tree densities has been demonstrated by vegetational trends monitored within fire- and mammal-proof enclosures. When grassland plots in the central and northern sections of the Serengeti ecosystem were protected from herbivores and fire for more than 2 years, they developed dense thickets of tree seedlings and sprouts, with more than 1000 stems per hectare (Belsky, 1984). Similarly, when open savannas in Tsavo National Park were protected from elephants and fire by deep trenches, they developed into dense thorn-scrublands (Oweyegha-Afunaduula, 1984; Wijngaarden, 1985).

If tree and bush thickets in low rainfall areas are left undisturbed long enough for natural thinning of the trees and bushes to occur, they may eventually develop into open savannas or grasslands. Walter (1971) described the reversion of an *Acacia mellifera* (Vahl.) Benth. subsp. *detinens* (Burch.) Brenan thicket in S.W. Africa to an open savanna after 50 years of protection from grazing. It is also possible that in the absence of fire and elephants the number and density of trees in mature communities would increase along a gradient of increasing rainfall. However, existing pristine vegetation suggests that thorn-scrublands and thickets, not grasslands, would still dominate the lower end of the rainfall gradient. Large sections of Tsavo National Park, for example, are currently covered by *Acacia–Commiphora* scrub, even though rainfall (250–350 mm; Wijngaarden, 1985) and elephant numbers (Ottichilo, 1986) are low.

Natural grasslands do occur in East Africa, but only in areas of unusual edaphic conditions. The Serengeti Plains in the southeastern Serengeti ecosystem consists of treeless grasslands, even when protected from herbivory and fire (Belsky, 1986a, b). This absence of trees has been attributed both to low rainfall (Belsky, 1983, 1987) and to herbivory by large mammals (Bell, 1982; McNaughton, 1983). Low rainfall has been implicated since these grasslands are at the low end of the strong rainfall gradient (400–1200 mm; Norton-Griffiths *et al.*, 1975) that appears to dominate the Serengeti ecosystem. However, 400–700 mm annual rainfall is adequate for tree growth elsewhere in East Africa. In fact, trees grow within the Serengeti Plains on granitic outcrops, called kopjes or inselbergs, where soils are deeper and less alkaline than in the surrounding plains (Anderson & Talbot, 1965; de Wit, 1978). The absence of trees in the plains, therefore, is probably not due to low rainfall.

Bell (1982) assumed that the lack of trees in the Serengeti Plains was due to intense herbivory and McNaughton (1983) concluded that the dominant environmental gradient controlling species composition and community physiognomy in the Serengeti ecosystem was grazing intensity. Attributing the lack of trees in the Serengeti Plains to herbivory, however, is refuted by the fact that trees do not grow in these plains, even when protected from herbivores (Belsky, 1986a, b).

The absence of trees in the Serengeti Plains is due primarily to its unique soils, which are shallow (30–100 cm deep) and highly alkaline (pH 8.0–10.0). The soils are derived from sodic, carbonitic ash originating from nearby volcanoes and are underlain by shallow calcium carbonate (calcrete) layers (de Wit, 1978). It is not clear which factor is more important in limiting tree growth. In the driest part of the Serengeti Plains near Olduvai George (≈400 mm annual rainfall), seedlings of *Acacia tortilis* (Forsk.) Hayne occasionally occur in the grasslands, but they do not grow taller than 50 cm except where soils are deeper (on dunes) or have been modified by high concentrations of cattle dung inside abandoned stock enclosures (Belsky & Amundson, 1986).

The dominant gradient influencing the vegetation of the Serengeti, from short grasslands in the south to dense woodlands in the north, is most properly labelled a 'complex' gradient (Whittaker, 1975) since it consists of a combination of: increasing rainfall, decreasing grazing, and decreasing soil salinity (soil pH declines from 10.0 to 6.0). These three factors are not independent since salts are more strongly leached out of soil by the higher rainfall occurring in the north (Anderson & Talbot, 1965; de Wit, 1978), and grazing patterns are directly related to grass productivity, which increases with increasing rainfall. All three factors affect vegetational composition and distribution independently (McNaughton, 1983; Belsky, 1983, 1986b), but in the Serengeti Plains it is the high salinity and shallow soils that exclude trees, not herbivores or rainfall.

The importance of saline, sodic conditions in creating

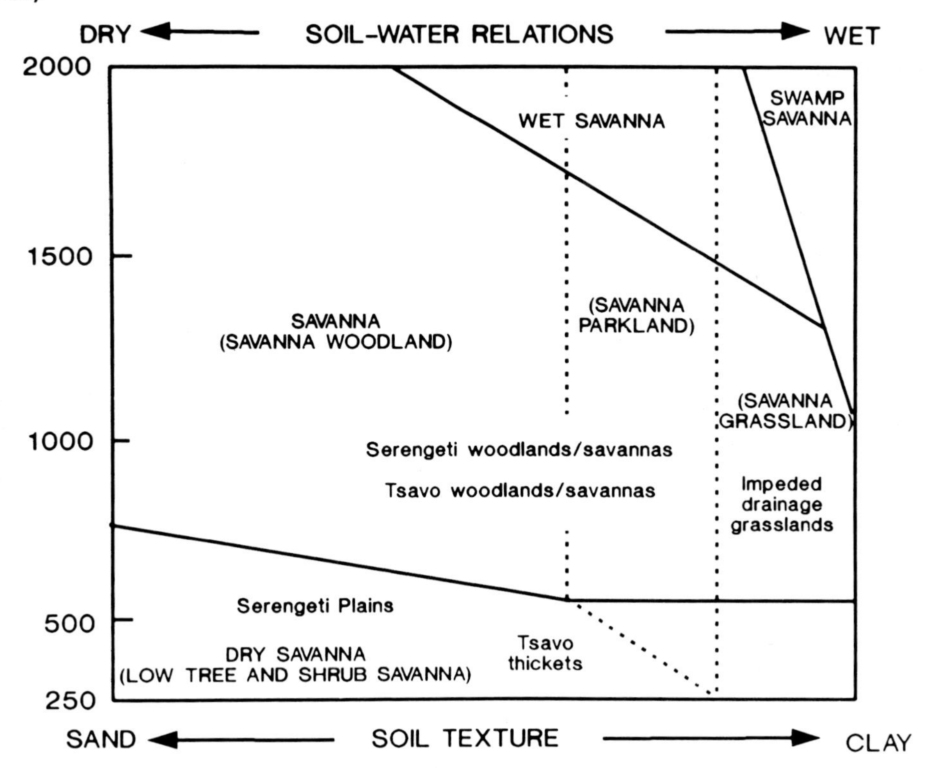

FIG. 3. The Johnson/Tothill model of tropical savannas (modified from Johnson & Tothill, 1985), including the relative locations of East African savannas within the model.

pure or natural grasslands (called edaphic grasslands by Vesey-Fitzgerald, 1973) is confirmed in other parts of East Africa. Trees are absent from saline-sodic soils in Ngorongoro and Manyara National Parks in Tanzania (Anderson & Herlocker, 1973; Prins, 1987) and Amboseli National Park in Kenya (Western, 1983). In Amboseli, neither trees nor shrubs grow inside grazer-proof exclosures erected on sodic soils, but they do occur inside nearby exclosures constructed on more neutral soils (personal observations).

The natural grasslands of East Africa are also strongly associated with poorly drained soils. Trees and shrubs are often excluded from clay-rich black-cotton soils at the bottoms of soil catenas due to poor water penetration, and from river flood-plains due to seasonally anaerobic conditions (Anderson & Talbot, 1985; Anderson & Herlocker, 1973; Vesey-Fitzgerald, 1973, 1974; Pratt & Gwynne, 1979; de Wit, 1978; Prins, 1987; Jager, 1982).

Simple abiotic models – the Johnson/Tothill model

Johnson & Tothill (1985) published a model of savannas (Fig. 3) that is similar to the Walker/Noy-Meir (1982) model in that it is also based on rainfall and soil texture. However, in their model, grasslands occur under the wettest conditions (600–2000 mm annual rainfall and high soil clay concentrations), not under the driest conditions. At the dry end of their rainfall/soil texture gradient (≈375 mm rainfall and sandy soil), savanna-woodlands and parklands are replaced by low tree and shrub savannas, not by grasslands. East African savannas fit comfortably within this model since natural grasslands often occur on the wettest soils. In Fig. 3 the placement of the grasslands of the Serengeti Plains on the dry end of the gradient is due to their unique soil characteristics (see above).

Walker's multi-factor model

Taking a less reductionistic approach to the description of tropical savannas, Walker (1987) developed a model that incorporated all important environmental factors influencing the vegetation structure of tropical savannas. This model shows the complexity of ways that environmental factors interact. A modified example of this model (Fig. 4) illustrates that the vegetation of East African savannas is most strongly influenced (the large arrows) by rainfall, which influences species composition through its effects on soil moisture and chemistry, by soil chemistry, which effects tree and grass species composition (Anderson & Talbot, 1965; Herlocker, 1976; de Wit, 1978; Jager, 1982; Wijngaarden, 1985), by fire and elephants, which reduce tree cover and prevent its regeneration (Laws, 1970; Buechner & Dawkins, 1961; Norton-Griffiths, 1979; Pellew, 1983; Dublin, 1986), and by grazing ungulates, which affect grassland species composition (Lock, 1972; Edroma, 1981; McNaughton, 1983; Belsky, 1984, 1986a, b, 1987).

Functional classification model

Savanna ecologists have recently developed a new functional model of tropical savannas (Frost *et al.*, 1986; Medina, 1987; Goldstein *et al.*, 1988). This model is based

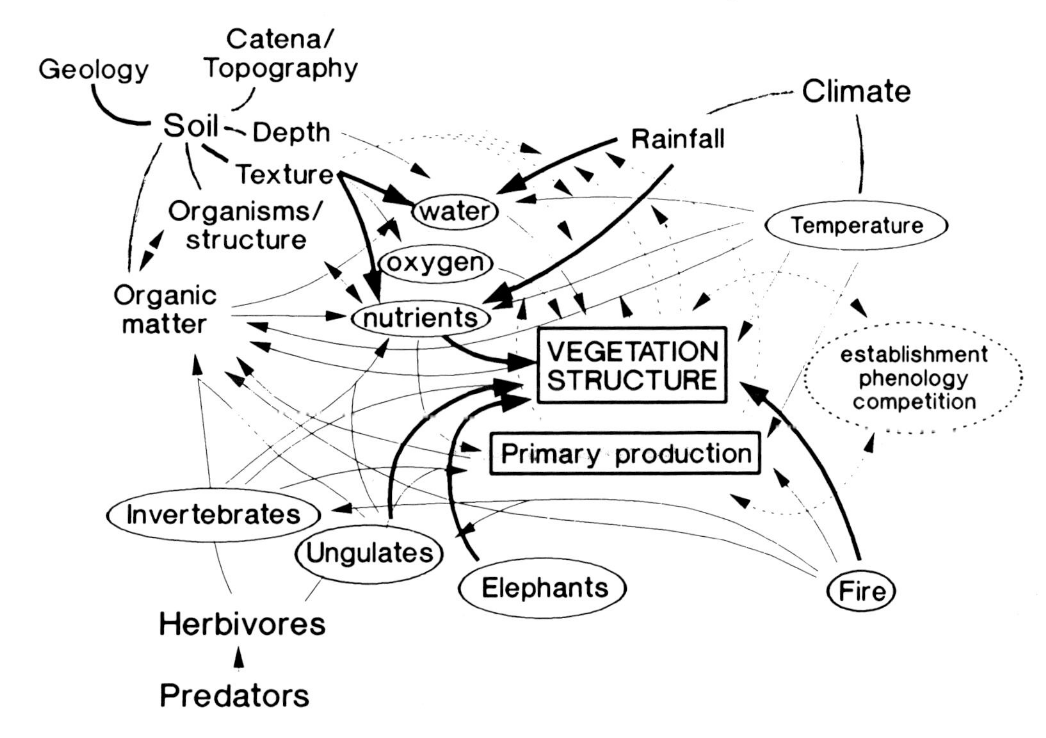

FIG. 4. Walker's multi-factor model of the determinants of vegetation structure in East African savannas (modified from Walker, 1987). The large arrows indicate the major environmental determinants of vegetation structure.

on two variables – plant available moisture (PAM), which integrates rainfall, water infiltration, evapotranspiration, soil texture, and hydrologic regime into a single measure of the soil moisture available to plants, and available nutrients (AN), which is a measure of the nutrients available to plants during their period of growth (Fig. 5). This model is novel since it substitutes biologically meaningful measures (PAM and AN) for purely physical variables.

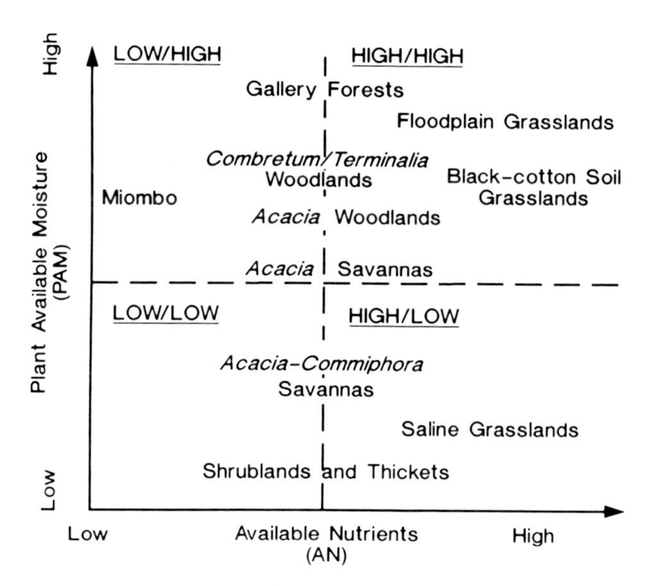

FIG. 5. Functional classification model of tropical savannas along gradients of available moisture and nutrients, and the relative positions of East African savannas within the model (modified from Frost *et al.*, 1986).

Since the axes of the functional model have not yet been well defined, the placement of savannas or community types within the model can only be approximate. East African grasslands were placed high on the AN axis, supporting Bell's (1982) observation that plant biomass in savanna communities decreases with increasing soil nutrients. It does not confirm his theory that the decrease is due to increased herbivory (Bell, 1982). Instead, it is probable that the low plant biomass (i.e. a low density of trees) in East African savannas is due to edaphic characteristics such as high soil salinity and anaerobic conditions that are often positively correlated with high soil-nutrient content.

CONCLUSION

Of the several models available for the comparison of tropical savannas around the world, those by Whittaker (1975) (Fig. 1) and Johnson & Tothill (1985) (Fig. 3) have the most immediate utility. Mean annual rainfall and temperature and descriptions of soil texture are already available in the literature or are easily collected. And they are heuristically acceptable. The functional classification model (Frost *et al.*, 1986; Medina, 1987) allows for the most accurate comparison of savanna function, but until the axes of this model are associated with quantifiable variables, the model provides little improvement over earlier models.

ACKNOWLEDGMENTS

Travel to Darwin, Australia, and research for this paper were supported by the U.S. National Science Foundation (BSR–8516982).

REFERENCES

Anderson, G.D. & Herlocker, D.J. (1973) Soil factors affecting the distribution of the vegetation types and their utilization by wild animals in Ngorongoro Crater, Tanzania. *J. Ecol.* **61**, 627–651.

Anderson, G.D. & Talbot, L.M. (1965) Soil factors affecting the distribution of the grassland types and their utilization by wild animals on the Serengeti Plains, Tanganyika. *J. Ecol.* **53**, 33–56.

Bell, R.H.V. (1982) The effect of soil nutrient availability on community structure in African ecosystems. *Ecology of tropical savannas* (ed. by B. J. Huntley and B. H. Walker), pp. 193–216. Springer, Berlin.

Belsky, A.J. (1983) Small-scale pattern in four grassland communities in the Serengeti National Park, Tanzania. *Vegetatio*, **55**, 141–151.

Belsky, A.J. (1984) The role of small browsing mammals in preventing woodland regeneration in the Serengeti National Park, Tanzania. *Afr. J. Ecol.* **22**, 271–279.

Belsky, A.J. (1986a) Population and community processes in a mosaic grassland in the Serengeti, Tanzania. *J. Ecol.* **74**, 841–856.

Belsky, A.J. (1986b) Revegetation of artificial disturbances in grasslands in the Serengeti National Park, Tanzania. II. Five years of sucessional change. *J. Ecol.* **75**, 937–951.

Belsky, A.J. (1987) Revegetation of natural and human-caused disturbances in the Serengeti National Park, Tanzania. *Vegetatio*, **70**, 51–59.

Belsky, A.J. (1989) Landscape patterns in a semi-arid ecosystem in East Africa. *J. Arid Envir.* **17**, 265–270.

Belsky, A.J. & Amundson, R.G. (1986) Sixty years of successional history behind a moving sand dune near Olduvai Gorge, Tanzania. *Biotropica*, **18**, 231–235.

Bourliere, F. (1983) *Tropical savannas*. Ecosystems of the World No. 13, p. 730. Elsevier, New York.

Buechner, H.K. & Dawkins, H.C. (1961) Vegetation change induced by elephants and fire in Murchison Falls National Park, Uganda. *Ecology*, **42**, 752–766.

Cole, M.M. (1986) *The savannas: biogeography and geobotany*, p. 738. Academic Press, New York.

Coughenour, M.B., Ellis, J.E., Swift, D.M. Coppock, D.L., Galvin, K., McCabe, J.T. & Hart, T.C. (1985) Energy extraction and use in a nomadic pastoral ecosystem. *Science*, **230**, 619–625.

Deshmukh, I. (1986) *Ecology and tropical biology*, p. 387. Blackwell Scientific Publications, Oxford.

Dublin, H. (1986) Decline of the Mara woodlands: the role of fire and elephants. Ph.D. thesis, University of British Columbia, Vancouver.

Edroma, E.L. (1981) The role of grazing in maintaining high species-composition in *Imperata* grassland in Rwenzori National Park, Uganda. *Afr. J. Ecol.* **19**, 215–233.

Frost, P., Medina, E., Menaut, J.-C., Solbrig, O., Swift, M. & Walker, B. (1986) *Responses of savannas to stress and disturbance*, p. 82. Biology Intern. Special Issue No. 10. Intern. Union of Biol. Sci., Paris.

Goldstein, G., Menaut, J.-C., Noble, I. & Walker, B.H. (1988) Exploratory research. *Research procedure and experimental design for savanna ecology and management* (ed. by B. H. Walker and J.-C. Menaut), pp. 13–20. Publication No. 1, Responses of Savannas to Stress and Disturbance.

Griffiths, J.F. (1969) Climate. *East Africa: its peoples and resources* (ed. by W. T. W. Morgan), pp. 107–117. Oxford University Press, New York.

Harrington, G.N. & Ross, I.C. (1974) The savanna ecology of Kidepo Valley National Park: I. The effects of burning and browsing on the vegetation. *E. Afr. Wildl. J.* **12**, 93–105.

Herlocker, D. (1976) *Woody vegetation of the Serengeti National Park*, p. 31. Texas A & M University Press, College Station, Texas.

Holdridge, L.R. (1947) Determination of world plant formations from simple climatic data. *Science*, **105**, 367–68.

Huntley, B.J. & Walker, B.H. (1982) *Ecology of tropical savannas*, p. 669. Springer, Berlin.

Jager, T. (1982) *Soils of the Serengeti Woodlands, Tanzania*, p. 239, Agric. Res. Rep. (Versl. landb. Onderz), 912. Pudoc, Wageningen.

Johnson, R.W. & Tothill, J.C. (1985) Definitions and broad geographic outline of savanna lands. *Ecology and management of the world's savannas* (ed. by J. C. Tothill and J. J. Mott), pp. 1–13. Australian Academy of Science, Canberra.

Laws, R.M. (1970) Elephants as agents of habitat and landscape change in East Africa. *Oikos*, **21**, 1–15.

Lock, J.M. (1972) The effects of hippopotamus grazing on grasslands. *J. Ecol.* **60**, 445–468.

McNaughton, J.S. (1983) Serengeti grassland ecology: the role of composite environmental factors and contingency in community organization. *Ecol. Monogr.* **53**, 291–320.

Medina, E. (1987) Nutrients: requirements, conservation and cycles in the herbaceous layer. *Determinants of savannas* (ed. by B. Walker), pp. 30–65. IUBS Monograph Series No. 3, IRL Press, Oxford.

Norton-Griffiths, M. (1979) The influence of grazing, browsing, and fire on the vegetation dynamics of the Serengeti. *Serengeti: dynamics of an ecosystem* (ed. by A. R. E. Sinclair and M. Norton-Griffiths), pp. 310–352. University of Chicago Press.

Norton-Griffiths, M., Herlocker, D. & Pennycuick, L. (1975) The patterns of rainfall in the Serengeti ecosystem, Tanzania. *E. Afr. Wildl. J.* **13**, 347–374.

Ottichilo, W.K. (1986) Age structure of elephants in Tsavo National Park, Kenya. *Afr. J. Ecol.* **24**, 69–75.

Oweyegha-Afunaduula, F.C. (1984) Vegetation changes in Tsavo National Park (East), Kenya. M.Sc. thesis, University of Nairobi, Nairobi.

Pellew, R.A.P. (1983) The impacts of elephants, giraffe and fire upon the *Acacia tortilis* woodlands of the Serengeti. *Afr. J. Ecol.* **21**, 41–74.

Pratt, D.J. & Gwynne, M.D. (1977) *Rangeland management and ecology in East Africa*, p. 310. Hodder and Stoughton, London.

Prins, H.H.T. (1987) *The buffalo of Manyara*, p. 283. Computekst, Groningen.

Schmidt, W. (1975) Plant communities in permanent plots of the Serengeti Plains. *Vegetatio*, **30**, 133–145.

Smart, N.O.E., Hatton, J.C. & Spence, D.H.N. (1985) The effect of long-term exclusion of large herbivores on vegetation of Murchison Falls National Park, Uganda. *Biol. Cons.* **33**, 229–245.

Spinage, C.A. & Guinness, F.E. (1971) Tree survival in the absence of elephants in the Akagera National Park, Rwanda. *J. appl. Ecol.* **8**, 723–728.

Thorbahn, P.F. (1984) Brier elephant and the brier patch. *Natural History*, **93**, 70–78.

Tothill, J.C. & Mott, J.J. (1985) *Ecology and management of the world's savannas*, p. 384. Australian Academy of Science, Canberra.

Vesey-Fitzgerald, D. (1973) *East African grasslands*, p. 95. East African Publishing House, Nairobi.

Vesey-Fitzgerald, D. (1974) Utilization of the grazing resources by buffaloes in the Arusha National Park, Tanzania. *E. Afr. Wildl. J.* **12**, 107–134.

Walker, B.H. (1985) Structure and function of savannas: an overview. *Ecology and management of the world's savannas* (ed. by J. C. Tothill and J. J. Mott), pp. 83–91. Australian Academy of Science, Canberra.

Walker, B.H. (1987) A general model of savanna structure and function. *Determinants of savannas* (ed. by B. Walker), pp. 1–12. IUBS Monograph Series No. 3, IRL Press, Oxford.

Walker, B.H., Ludwig, D., Holling, C.S. & Peterman, R.M. (1981) Stability of semi-arid savanna grazing systems. *J. Ecol.* **69**, 473–498.

Walker, B.H. & Noy-Meir, I. (1982) Aspects of the stability and resilience of savanna ecosystems. *Ecology of tropical savannas* (ed. by B. J. Huntley and B. H. Walker), pp. 556–590. Springer, Berlin.

Walter, H. (1971) *Ecology of tropical and subtropical vegetation*, p. 539. Oliver and Boyd, Edinburgh.

Western, D. (1983) *A wildlife guide and a natural history of Amboseli*, p. 75. General Printers, Nairobi.

Whittaker, R.H. (1975) *Communities and ecosystems*, 2nd edn, p. 385. Macmillan, New York.

Wijngaarden, W. van (1985) *Elephants–trees–grass–grazers*, p. 159, ITC Publ. No. 4, Enschede, The Netherlands.

Wit, H.A. de (1978) Soils and grassland types of the Serengeti Plains (Tanzania), p. 298. Ph.D. thesis, Agricultural University, Wageningen.

16/Stress and disturbance: vegetation dynamics in the dry Chaco region of Argentina

JORGE ADÁMOLI, ETHEL SENNHAUSER, JOSÉ M. ACERO and ALEJANDRO RESCIA *GESER (Grupo de Estudios Sobre Ecología Regional), Facultad de Ciencias Exactas y Naturales, Universidad de Buenos Aires, Argentina*

Abstract. This paper analyses two processes acting on vegetation dynamics in the dry Chaco region of Argentina. The major one is the response of herbaceous/woody species to overgrazing. Extensive cattle breeding reached a peak shortly before the 1940s, when animal production in the region became saturated. Thus over-exploitation reached a crisis during this decade and there followed a sharp decrease in the number of cattle and the efficiency of the production system. The number of *puestos* (stations), on the contrary, remained fairly constant. Thus overgrazing has led to degradation of the natural systems. Today the most intensely degraded areas form a 25–50 ha fringe around each *puesto* headquarters and represent less than 1% of the total area studied. Here, marked changes in soil physiochemical properties can be seen together with the elimination of trees and shrubs. Away from this fringe, degradation basically involves the herbaceous cover, changing the original landscape of forests and savannas into one of forests and shrub patches. There are not only structural alterations but also changes in the system's dynamics due to modifications in the relationships between its components. The sequence of changes seems to exhibit resilient behaviour showing hysteresis.

The second process is the dynamics of gallery forest resulting from intensive river-bed migrations which characterize the region. A main process model relates the largest and most frequent floods to the complex structure, high diversity and Amazonian lineage of the floristic composition of the forests. It also attributes the null water supply of abandoned river-beds to Chaco lineage forests, adjusted to rainfall seasonality. However, owing to the high morphological instability of the region these courses may become active or inactive throughout time by means of digression or filling in, which leads to well-developed forests growing on ancient river-beds and, conversely, dry forests growing along the margin of permanent rivers. A model is presented which, by analysing river bank community dynamics in terms of the geomorphological instability (constant river-bed migration, river-bed filling, bank formation and disintegration), shows how the structure and composition of these forests would be determined by past floods rather than by the present one.

Key words. Overgrazing, riverine, savanna vegetation dynamics, disturbance, gallery forests, stability, Chaco, Argentina.

INTRODUCTION

The Chaco region is a large sedimentary plain of about 1,000,000 km^2, extending north and south of the Tropic of Capricorn over northern Argentina, western Paraguay, eastern Bolivia and part of southeastern Brazil. Its eastern boundary, running along the Paraguay and Parana rivers, where the yearly rainfall amounts to 1200 mm, is characterized by the prevalance of hyperseasonal savannas (Sarmiento, 1984) alternating with gallery forests. In the central part and mainly the west the climate is drier (500–700 mm annual rainfall) and markedly seasonal, the dry season being about 7 months long. The original vegetation consisted of a mosaic formed by xerophytic forests, gallery forests, and soil-determined or fire-generated savannas (Adámoli *et al.*, 1972).

A regional analysis of the spatial heterogeneity reveals two macrounits: Humid Chaco and Dry Chaco. At the subregional level, fluvial activity causes recurrent patterns which contrast with those present in neighbouring subregions. At the local level, in turn, the lack of relief (generally in terms of centimetres) and virtual lithological uniformity allow topography and soil variations to generate pronounced ecological differences, manifested through the presence of different plant formations (Morello & Adámoli, 1974).

In addition to these static aspects of the spatial heterogeneity, found at different scales, there are dynamic processes at work, such as the relationships between woody and herbaceous species in the savanna formations (Walker & Noy Meir, 1982; Walter, 1987; Sarmiento, Goldstein & Meinzer, 1985; Goldstein & Sarmiento, 1987) and forest dynamics (Shugart, 1984).

This paper discusses the two major phenomena of vegetation dynamics of the Chaco region: changes in herbaceous/woody species caused by overgrazing, and gallery forest dynamics resulting from intensive river-bed migration processes.

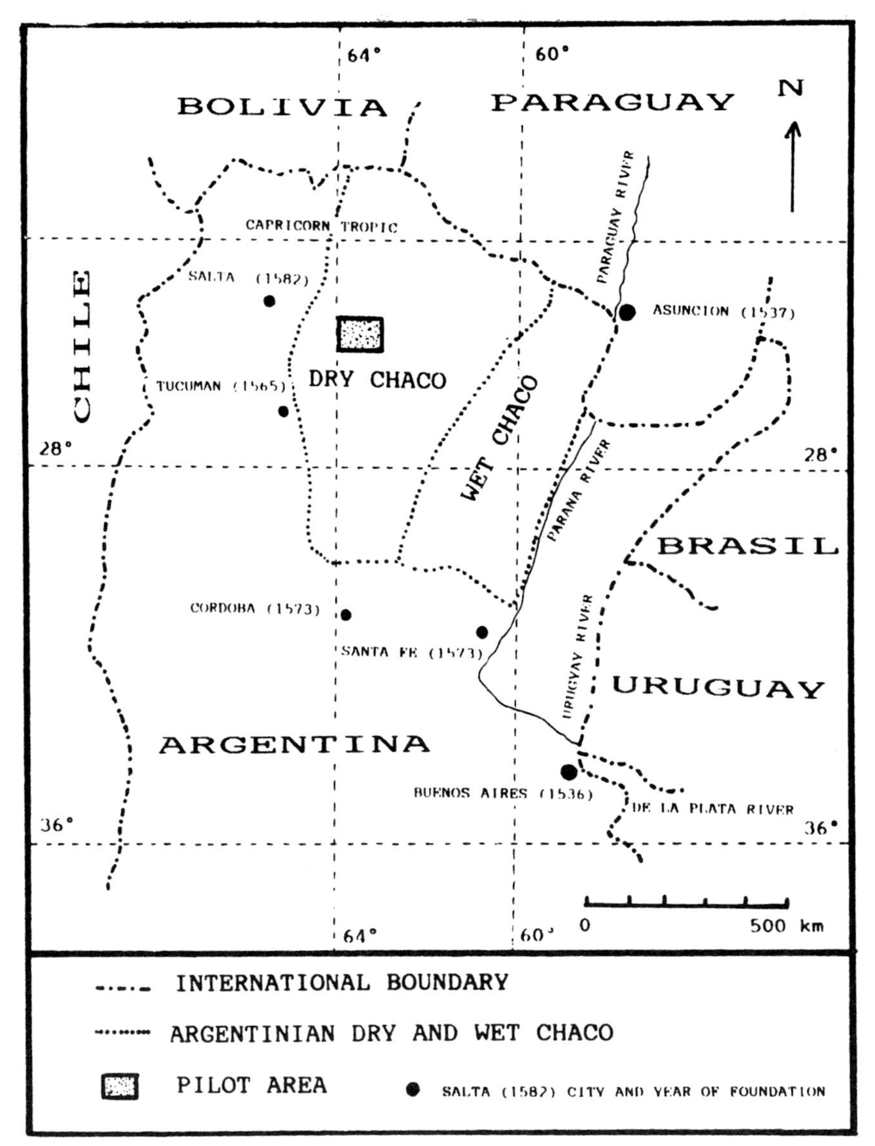

FIG. 1. The Chaco region.

Both processes have regional scope and they generate heterogeneity in space and time. They constitute a source for experimental analysis and development of the concepts of stress, disturbance, stability and resilience, as defined in the document on the Responses of Savannas to Stress and Disturbance Programme (Frost *et al.*, 1986).

PASTURE DEGRADATION AND ALTERATION OF THE HERBACEOUS/WOODY BALANCE IN DRY CHACO

The oldest cities in Argentina were founded by the Spaniards more than 400 years ago, in the periphery of the then forbidding Chaco region (Fig. 1). However, permanent occupation of the area by cattle breeders and timber producers started during the first decades of this century. Morello & Saravia Toledo (1959) assign the arrival of the railroad a decisive role in the settling and consolidation of this inner frontier. Cattle production is based on the exploitation of 5000 ha units with no internal subdivisions or fences marking the perimeters. The station is made up of one or two houses, a well and corrals; this station is called a *puesto*.

The occupation process

The occupation of the Chaco region by cattle caused changes in the environment and amount of pasture. At first there was a sustained increase in the number of cattle, accompanied by an increase in the number of *puestos*, until the 1940s (Figs. 2a and 2b). During most of the 1930s a pluriannual dry period reigned over the Chaco region (Galmarini & Raffo del Campo, 1964) and had a synergistic influence on other events (e.g. Frost *et al.*, 1986).

The early stage of cattle increase was due to an expansion of area. Cattle production was based on low-level technology, hence was limited by the strongly seasonal nature of the rainfall (summer). Shortly before the beginning of the 1940s the production capacity of the region (5 ha/head;

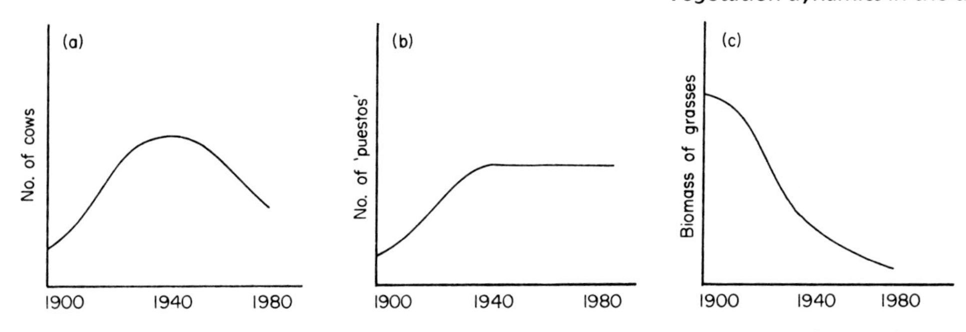

FIG. 2. Change in the number of cows (a), 'puestos' (b) and biomass of grasses (c) in the Chaco region.

0.2 head/ha) had already been saturated. Thc increased grazing pressure led to a gradual decrease in pasture biomass towards a virtual collapse in the 1940s continuing to the present (Fig. 2c). This reduction in pasture caused a sharp decrease in cattle numbers, mostly in the efficiency of production (Fig. 2a), a trend not followed by the number of *puestos* which remained fairly constant (Fig. 2b). Once established, breeders seldom abandon their pastures.

The area under the influence of each *puesto* (about 5000 ha) remains unchanged, but because of lower capacity carry only about 100 heads per *puesto*.

Number of *puestos*, pasture and soil degradation

Hypothesis II in the Responses of Savannas to Stress and Disturbance programme (Frost *et al.*, 1986) states that 'The reversibility of change in plant species composition is inversely related to the degree of change in soil physicochemical properties' (Isichei, 1988). We carried out a research programme on 560,000 ha in the Province of Salta, an area of extensive cattle breeding in Dry Chaco (Fig. 1). In this area 140 *puestos* were singled out and information was gathered, including the date when they were set up, which in most cases fell within the 1920–30 period. Each *puesto*, on average, was approximately 5000 ha. Using aerial photographs taken in 1960 and satellite images from 1975 and 1985, it was possible to obtain three reference points. During those 25 years (1960–85) there were no significant changes in the number of *puestos* (six were set up and four disappeared as a consequence of the expansion of the agricultural frontier (Adámoli *et al.*, 1989).

It is easy to detect *puestos* in aerial photographs and satellite images because they are surrounded by a fringe showing degradation of the soil and virtual elimination of grass and shrubs, due to the concentration of large and small animals (cows and goats) around the water holes and corrals (Fig. 3). This fringe was measured in the field as well as on the photographs and images, being between 500 and 700 m in diameter and between 25 and 50 ha in area. These values are similar to those described by Morello & Saravia Toledo (1959) for measurements in the 1950s.

Our interpretation of these data contradicts many descriptions of the environmental degradation of the Chaco, which is generally described as an expanding process without control. According to our data, soil degradation has stabilized around values reached in the 1950s. Less than 1% of the total area studied has shown intense degradation in soil physical and chemical properties.

In contrast with the intense soil degradation (localized only around *puestos*) almost all the region shows pasture degradation resulting in the virtual elimination of grasses. The ecological consequences of this process as well as the possibilities of reversion must be carefully examined. Recovering degraded pastures is feasible in the short run (Table 1), as demonstrated by experimentation (Saravia Toledo, 1984). This is in contrast to recovery from the intensive soil degradation around the *puestos*, which is very difficult to reverse under similar conditions.

Herbaceous/woody dynamics

Processes which led to pasture degradation also destroyed the herbaceous/woody balance. The original landscape consisted of forest patches alternating with extensive savannas generated by fire or soil properties. This was replaced by communities characterized by trees (partially depleted through timber production) (Fig. 4), thick and continuous shrubs (1.5–3 m high), and very low coverage of grasses (Fig. 4). (In the most degraded areas flowering specimens can only be found under the protection of cactaceous plants or thorn shrubs.)

The dynamics between grazing pressure and pasture biomass is illustrated in Fig. 6. From an initial state 'A' corresponding to the beginning of cattle in Dry Chaco (low grazing pressure and high pasture biomass) an increase in grazing pressure can cause a gradual decrease in biomass until states 'B' or 'C' are reached, where yields are still

TABLE 1. Comparison of pastures with and without shrubs removed. (From Saravia Toledo, 1984.)

	Pasture with shrubs removed	Pasture with shrubs not removed
Grass coverage (%)		
In sandy soils	52.5	1.8
In silty soils	31.9	0.0
Gramineous species (no.)		
In sandy soils	8	3
In silty soils	4	0
Herbaceous vegetation production (kg ha^{-1})	2870	345
Carrying capacity (ha per head of cattle)	4	20

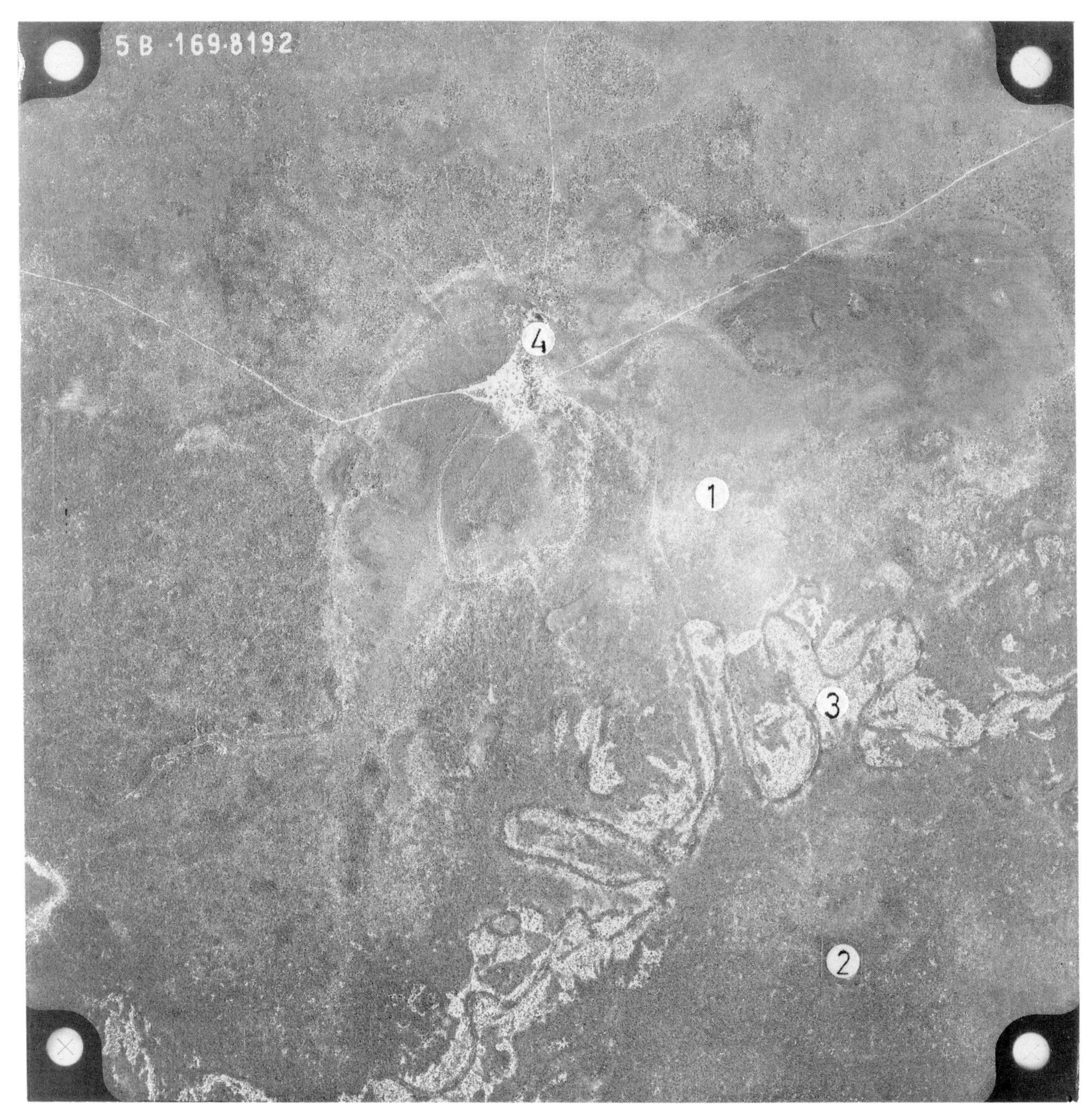

FIG. 3. Vegetation patterns of dry Chaco region: 1, savanna; 2, xerophytic forest; 3, area of intense soil degradation due to overgrazing; 4, 'puesto', note the fringe around it showing degradation of the soil and elimination of grass and shrubs. (Scale 1:35,000; photo: IGM, 1966.)

reasonable and the ecosystem's structural characteristics have not been basically altered. A reduction in grazing pressure allows a return from 'C' to 'B' or 'A'.

Although this trend is represented in Fig. 6 as a straight line, it in fact follows different paths and modalities. The decrease in herbaceous biomass resulting from the increase in grazing pressure, and its subsequent recovery, have delayed responses due to the action of regulating mechanisms characterizing the states defined by 'A', 'B' and 'C'. The path followed suggests hysteresis in the sense of Rabinovich (1981) or Walker (1987). Representing this relationship as a straight line shows the trend but not the complexity of the actual path.

From point 'C' an increase in grazing pressure should push the system towards point 'D'. If in addition to overgrazing, drought is added, the response ceases to be gradual and takes on a catastrophic configuration, similar to those described by Jones (1977, in Rabinovich, 1981). There will be a drop to 'E' and if grazing is increased the system will reach state 'F'. If grazing is reduced then the system will move back to 'E' but further reductions in grazing pressure will not bring the system back to 'C' but will tend towards states 'G' or 'H'.

If the system has a resilient behaviour (Holling, 1973) going from 'A' to 'D' corresponds to typical responses of forest patches and savannas. However, if an added disturbance

FIG. 4. Dense and continuous shrub layer in a savanna of the Argentine Chaco. (Photo: J. Adámoli.)

factor shifts the system to forest and shrub patches the physiognomic structure of the system will be altered, resulting in a new set of relationships. In states 'A' to 'D' thick grasses strongly compete with shrubs for light and water but not necessarily for nutrients (Chaco soils are in general highly fertile). This balance favours grasses and is strengthened by periodic fires which weaken shrubs. Overgrazing, by interrupting the continuity of the grass cover and reducing its height, favours the germination of shrubs. Fruits are a large part of the cattle diet, and seeds of legumes passing through the digestive tract are scarified (many seeds have very hard coats). When these seeds are expelled with faeces, conditions are ideal for germination. In the end, the previous fire cycle stops because of low grass coverage and the discontinuous distribution of the herbaceous fuels.

If the system becomes dominated by shrubs, light becomes unfavourable for the heliophilous grasses. However, the mechanical removal of shrubs or renewed fires may cause a partial reduction of shrubs promoting grass in the openings. A low grazing pressure, together with fire, would promote a move from Forest and Shrub Patches back to Forest Patches and savannas. This leads to the re-establishment of competition favourable to the production of grasses.

Comparison with other regions

The processes leading to the virtual elimination of grasses and their replacement by thick shrub in Dry Chaco differs from the processes studied in other regions such as the Cerrados and Pantanal in Brazil, the Llanos in Venezuela or even the savannas in the Wet Chaco region. While in Dry Chaco the yearly rainfall varies between 500 and 700 mm, in Humid Chaco and Pantanal it reaches over 1200 mm and in Cerrados and Llanos over 1500 mm. In addition to the marked seasonality of rainfall in Dry Chaco there are other critical factors associated with seasonal rains. The Dry Chaco has increased day length and total radiation during the rainy season. High temperatures reduce water available to plants, especially when hot dry winds blow from the North. Under these environmental conditions, high grazing pressure has a synergistic effect during droughts (Frost *et al.*, 1986). One consequence is the displacement of the herbaceous species by shrubs, as discussed in the previous section.

In the other regions mentioned above, greater rainfall leads to a higher production of pasture biomass, which usually is not depleted by cattle grazing. Although overgrazing may affect the relationship between palatable and non-palatable grasses (Walker & Noy Meir, 1982; Walker, 1987), it rarely causes complete elimination of the herbaceous cover. As a consequence, competition for light and water differs, recruitment of shrubs is lower, and fires are more frequent.

GALLERY FORESTS: VEGETATION DYNAMICS

Role of fluvial modelling

The Chaco region lies on a sunken block which was filled with material provided mainly by the Pilcomayo, Bermejo and Juramento rivers. Fluvio-morphological processes are most important due to the absolute flatness of the region (with a regional slope of 20–40 cm km^{-1}) and the markedly seasonal, torrential nature of the river floods.

Owing to this peculiar low topographical relief, fluvio-morphological processes cause marked changes to local relief through successive river-bed migrations, river-bed filling, levee-bank formation and levee-bank disintegration (Fig. 7). These processes lead to intense pressure on vegetation composition and herbaceous/woody species relationships both at local and regional levels.

FIG. 5. Open herbaceous cover in savanna with shrub elements. (Photo: J. Adámoli.)

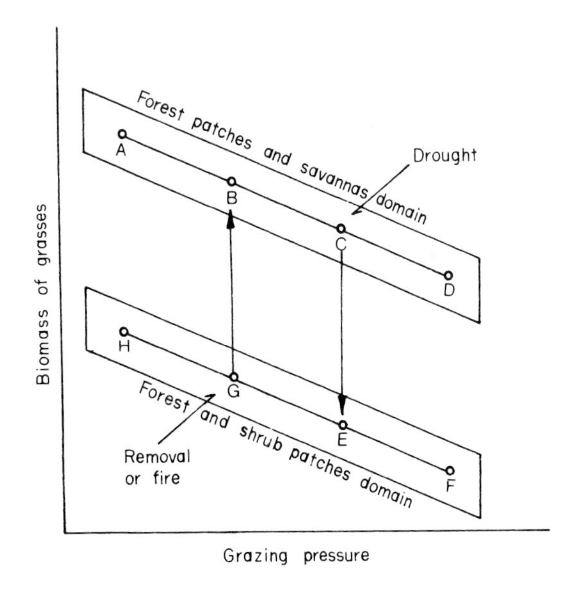

FIG. 6. Dynamics of the grasses in relation to grazing pressure.

When a river flows along a given path for a long period, several geomorphological units are formed in association with it, particularly side dykes or levee-banks resulting from sediment deposition caused by water overflow. These levee-banks constitute a peculiar environment with favourable water conditions since they receive an additional water supply and, at the same time, are well drained as they are raised with respect to the surrounding plain and adjacent river. During flood periods, rivers carry different propagules (seeds, fruits, branches and even complete trees), many of which originate in the montane forests of the upper basins. Rivers, then, generate the environmental conditions suitable for the development of gallery forest and also act as the main source of propagules for its typical species.

Analysis of the gallery forests

Because of the regional scope and the ecological relevance of the above-mentioned phenomena, research was carried out (Sennhauser, unpublished) into relationships between flood occurrence and forest structure and composition. The main hypothesis assumes a relationship between the largest and most frequent floods and the development of structurally complex forests, having great species diversity and floristically of Amazonian lineage (that is, composed of species alien to the Chaco region). On the other hand, ancient, abandoned river-beds whether filled in or not, are subject to deep physical, chemical and biological changes over time which finally lead to the development of a Chaco lineage forest adapted to the rainfall and seasonality of the region. These two extreme situations are represented respectively by 'H' and 'D' in Fig. 8(a). The diagonal joining the extremes represents the condition of maximum stability in the system, that is a long-term balance between environmental conditions and the biological responses of the forests, both in the herbaceous and the woody species. Obviously, intermediate flood situations which are stable will be represented by intermediate points along this diagonal.

The plain over which these rivers and their neighbouring streams wander is a large arc formed by the migration of the main river course. River-bed instability, then, is a determinant factor. Each of these river branches may alternate between active and inactive phases over time, with infilling during inactive phases and digression during active ones. In our study we found evidence of the superposition of up to four fluvial plains (Sennhauser, unpublished). As suggested by the data, forest structure and composition show a response to these ancient water pulses. That is, there may be disequilibrium between these gallery forest communities and the environmental conditions operating at present. Surface hydrology may change rapidly; for instance, a river may abandon its present course and start to flow along an ancient bed, all within a period measured in days (Fig. 9). In this situation, forest 'H' will keep all its structural and functional humid forest characteristics but will no longer have an additional supply of water, whereas forest 'D' will face the opposite situation, a dry forest (in structure and species composition) with water flowing along the adjacent river-bed. These extreme situations are represented in Fig. 8(b).

The diagonal joining both examples would represent

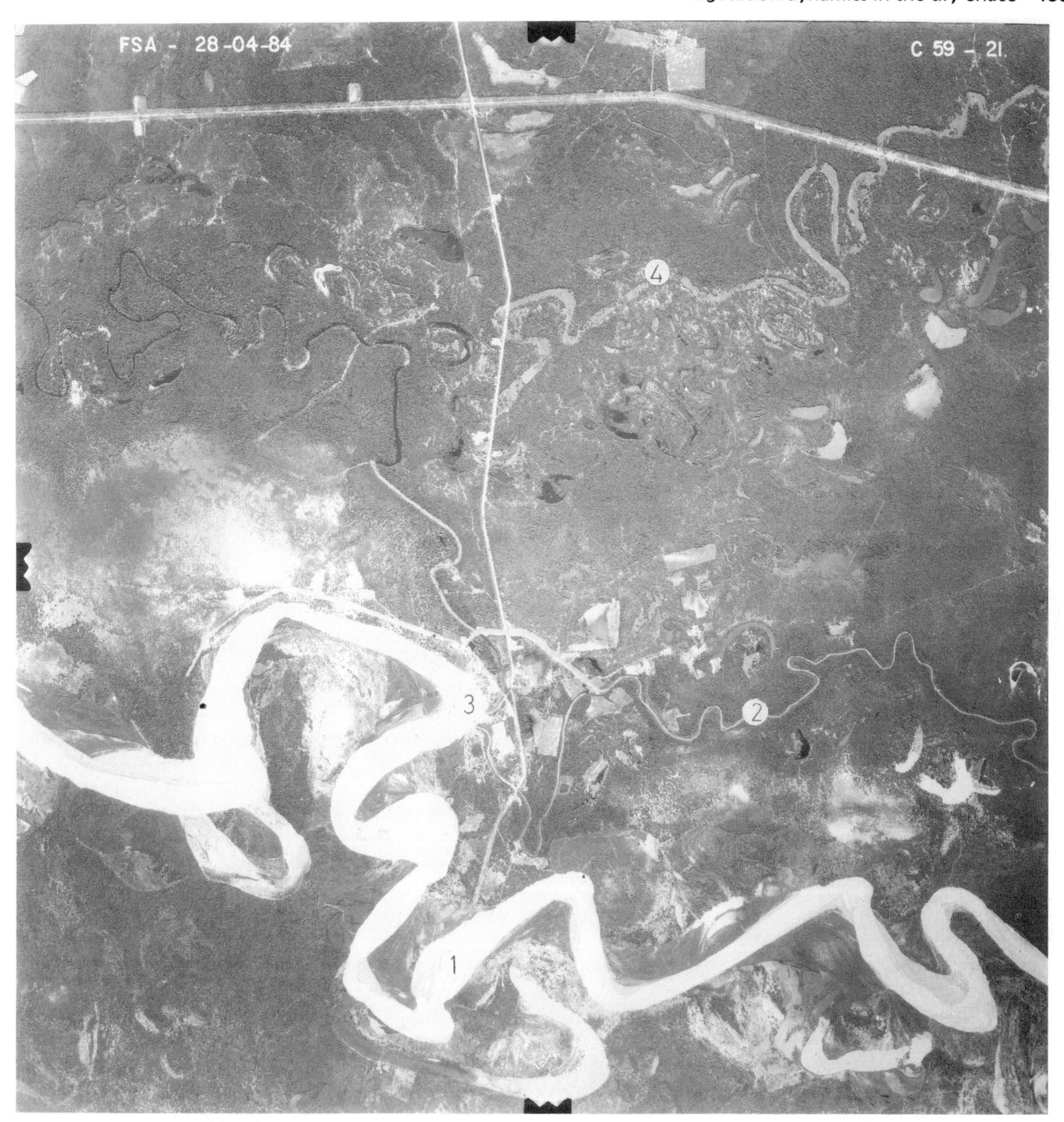

FIG. 7. Intense river-bed migration in Chaco region: 1, Bermejo river; 2, active river-bed captured by Bermejo river (point 3); 4, filled-in river-bed (see also Fig. 10). Evidence of the superpositions of fluvial plains can be seen in (2) and (4). (Scale 1:50,000. Photo: Paraná Air Brigade, 1984.)

maximum instability since it denotes maximum unbalance between the present water supply and the forests' biological response. Although changes in surface drainage patterns take only a few days, vegetation responses to these changes are slow: it takes years or decades for herbaceous components and centuries for woody species to adjust, owing to the population and community processes involved (Fig. 10).

At the moment maximum unbalance sets in, successional processes start which, if the new conditions persist, lead to the transformation of forest structure and composition until a new balance is reached. Sennhauser has studied several mechanisms underlying these changes, notably the following:

(a) In the initial stage of transformation of forest 'H' into forest 'D' (Fig. 8c) (i.e. the gallery forest growing on the levee-banks of an abandoned river-bed) biological processes – specially species replacement – prevail, starting with the species more dependent on the additional water supply provided by the river. During this first stage, bank structure is preserved. When forest structure disintegrates, total plant cover, annual primary production rates and litter supply decrease. Since the river bank is a positive relief structure with respect to the surrounding plain, physical degradation of the levee-bank commences. When water flow ceases, sediment deposition which may compensate for this erosion ceases.

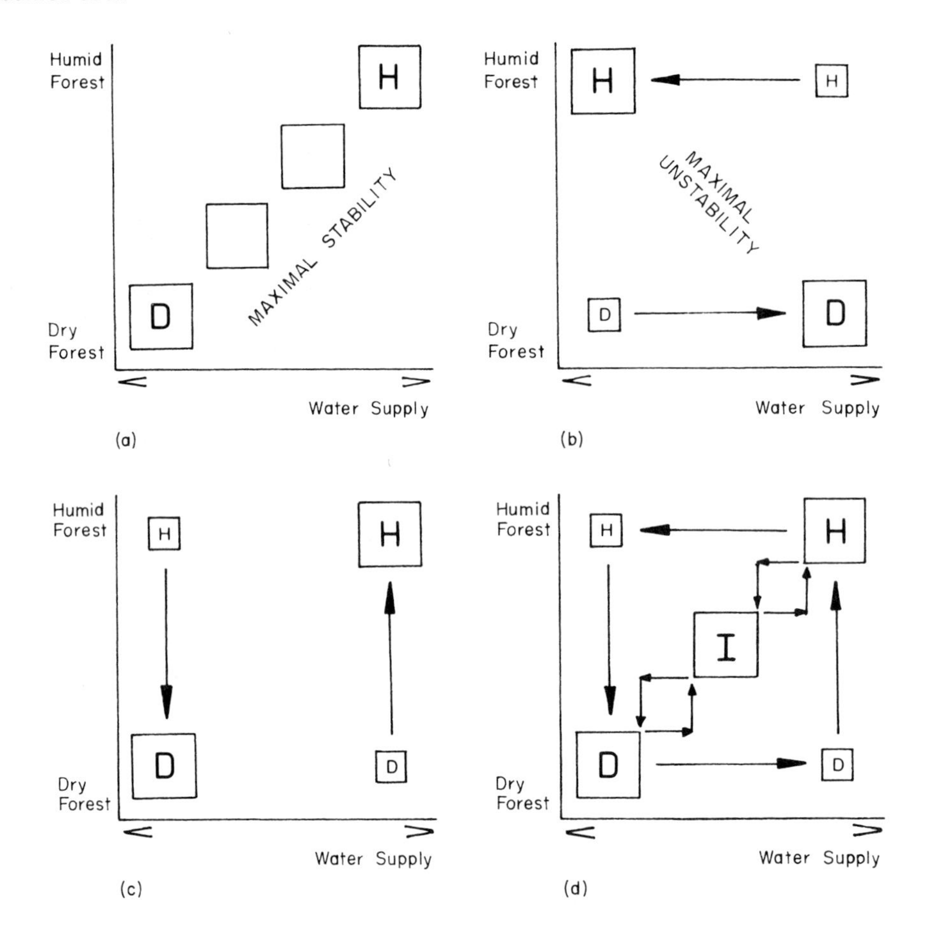

FIG. 8. Dynamics of forest galleries. (a) Condition of maximal stability. (b) Condition of maximal instability. Scale of time for changes: days. (c) Successional pattern. Scale of time for changes: centuries. (d) Direction and sense of global processes in the general model.

FIG. 9. A forest with structural and functional humid characteristics, unbalanced by a low actual water supply. (Photo: E. Sennhauser, 1988.)

FIG. 10. Ancient river-bed, Chaco Region. (Photo: J. Agraz, 1988.)

(b) The opposite process, that is, the transformation of dry forest 'D' into humid forest 'H' as a consequence of the re-activation of an ancient river-bed might be faster, since it would use bank structures already present. The recruitment of the biotic components would be favoured by fluvial transportation which is efficient at disseminating propagules.

Model's analogy with a circuit

The proposed model assigns directionality to these processes. This allows the establishment of an anology with a circuit where there are 'circulation rules'.

Rule 1. No circulation is possible along the diagonals. To gain access to any intermediate point the obligatory direction is counterclockwise.

In the diagram shown in Fig. 8(d), the previously described processes can be ecologically feasible only when they work in counterclockwise direction. An intermediate point such as 'I' (a forest subject to intermediate floods and consequently showing floristic and ecological conditions lying in between the two extremes) may result from a state constant equilibrium, but it may also be the last phase of a dynamic process, e.g., derived from dry forest 'D' through an increase in the additional water supply or from humid forest 'H', through a reduction in water availability. In both cases circulation will follow a counter-clockwise direction. Point 'I', in turn, may undergo an interruption of its flood cycle and evolve into a dry forest, or receive additional water and evolve into a humid forest. Again, in both cases the change will follow a counter-clockwise direction.

Rule 2. Reaching an extreme is not compulsory. There is 'permission to stop' at intermediate points.

This means that water supply instability does not work on an all-or-nothing basis, but includes all the intermediate cases. It also means that water availability may increase or decrease at any moment, which begins the dynamic process again.

Rule 3. There is a differential velocity along axes x and y.

As explained before, the rate at which water availability changes may be expressed in terms of days, whereas at constant water availability, biotic changes take years or decades in the case of the herbaceous cover and centuries in the case of woody species.

Community response will vary according to whether human exploitation affects forests situated along the maximum stability or the maximum instability diagonal. A forest exploited under conditions of equilibrium between water availability and the biota will return, over a given period of time, to the situation occurring before the exploitation. On the other hand, a forest exploited under conditions of maximum instability will not return to its initial state but rather move towards another state consistent with the new level of water availability.

Comparison with other regions

Gallery forests in the Chaco region grow mainly on river levee-banks, that is, on top of a positive relief structure with respect to the surrounding flat land (Fig. 11a). Similar processes operate in the two regions where hyperseasonal savannas reach greater development in Latin America: Pantanal in Mato Grosso (Brazil) and those areas of the Venezuelan Plain (Llanos) which are subject to flooding, particularly the Apure Plains. In northern Australia, some gallery forests also grow on elevated river levee-banks, although drainage patterns appear less mobile than described above.

In the Cerrados region, in contrast, gallery forests grow at bottom-of-the valley position, benefiting from the additional water supplied by the higher water table. Here rivers are incised so that their topographic position is at the bottom of a valley, below the level of the surrounding savanna vegetation (Fig. 11b). Similarly, gallery forests in northern Australia, for example Kakadu National Park, also differ from the Chaco gallery forests in that their topographic position generally lies below that of the savanna open forests. Consequently, Cerrados and all gallery forests growing at the bottom of valleys are more stable in time than those growing on elevated river levee-banks. Further, because these rivers have little opportunity to change course (as is the case in Chaco, Pantanal and Apure), the extreme range of successional process typical of Chaco is not present.

Finally, considering gallery forests as a source of woody species for colonizing the neighbouring savannas, another essential difference between Chaco and Cerrados (and also the other comparable regions mentioned) is evident. Cerrados (and Llanos, and northern Australian savanna open forests) cannot be colonized by woody species coming from their gallery forests whereas Chaco (and Pantanal and Apure) can.

If we consider the gallery forest, in any of the regional situations described, the species growing there will be adjusted to a relatively permanent water supply throughout the year. In Cerrados, owing to the permanent nature of the river courses and the impossibility of river-bed migration (since water flows along incised channels), the ecological isolation of the gallery forest species with respect to the savanna is very marked. In the Chaco, in contrast, intensive river-bed migration is predominant, with river banks and courses remaining inactive during centuries. In its final disaggregation stage, a river levee-bank will be subject to ecological conditions very similar to those of the surrounding savanna, which is why the woody species growing on it will prosper in the savanna if any disturbance allows their colonization.

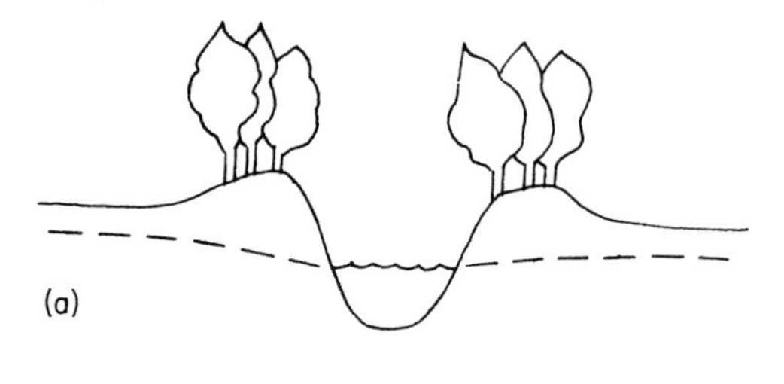

FIG. 11. Relative positions of forest galleries. (a) Forest gallery in Chaco, Pantanal and Apure (hyperseasonal 'Llanos'); (b) forest gallery in Cerrados, seasonal 'Llanos' and northern Australia.

Acknowledgment

This work is supported by CONICET.

REFERENCES

Adámoli, J., Neumann, R., Ratier de Colina, A. & Morello, J. (1972) El Chaco aluvional salteño. Revista de Investigación Agropecuaria. *INTA*, Ser 3, IX, (5), 165–237.

Frost, P., Medina, E., Menaut, J., Solbrig, O., Swift, M. Walker, B.H. (1986) Responses of savannas to stress and disturbance. *International Union of Biological Sciences, Special Issue*, **10**, 1–82.

Galmarini, A. & Raffo del Campo, J. (1964) *Rasgos fundamentales que caracterizan el clima de la región Chaqueña*. CONADE, Buenos Aires.

Goldstein, G. & Sarmiento, G. (1987) Water relations of trees and grasses and their consequences for the structure of savanna vegegation. In: *Determinants of tropical savannas* (ed. by B. H. Walker). IUBS Mon. Series 3, IRL Press.

Holling, C.S. (1973) Resilience and stability of ecological systems. *Ann. Rev. Ecol. Syst.* **4**, 1–23.

Isiche, A.O. (1988) The reversibility of changes in species composition, production and soil properties following defoliation, trampling and application of excreta. *Research procedure and experimental design for savanna ecology and management* (ed. by B. H. Walker and J. C. Menaut), pp. 64–67. Responses of Savannas to Stress and Disturbance, Publication No 1, CSIRO Printing Centre, Melbourne.

Jones, D.D. (1977) Catastrophe theory applied to ecological systems. *Simulation*, **29**, (1), 1–15.

Morello, J. (1968) Las grandes unidades de vegetación y ambiente del Chaco Argentino. I. Revista de Investigación Agropecuaria. *INTA, Ser. Fitogeográf.* (10), 1–126.

Morello, J. & Adámoli, J. (1974) Las grandes unidades de vegetación y ambiente del Chaco Argentino. II. Revista de Investigación Agropecuaria. *INTA, Ser. Fitogeográf.* (13), 1–130.

Morello, J. & Saravia Toledo, C. (1959) El bosque Chaqueño II: la ganadería y el bosque en el oriente de Salta. *Revista Agronómica Noroeste Arg.* **3**, (1–3), 209–258.

Rabinovich, J. (1981) Modelos y catástrofes: Enlace entre la teoría ecológica y el manejo de los recursos naturales renovables. *Interciencia*, **6**, (1), 12–21.

Saravia Toledo, C. (1984) Manejo silvopastoril en el Chaco Noroccidental de Argentina. Reunión de Intercambio Técnico de Zonas Aridas y Semiáridas. Catamarca, Argentina.

Sarmiento, G. (1984) *The ecology of neotropical savannas*. Harvard University Press, Cambridge, Mass.

Sarmiento, G. Goldstein, G. & Meinzer, F. (1985) Adaptative strategies of woody species in neotropical savannas. *Biol. Rev.* **60**, 315–355.

Sennhauser, E. (Unpublished) Estudio Ecológico sobre ambientes inundables (Area de influencia del río Bermejo en la prov. de Formosa). Informe final, CONICET, Buenos Aires, Argentina.

Shugart, H. (1984) *A theory of forest dynamics*. Springer, New York.

Walker, B.H. (1987) A general model of savanna structure and function. *Determinants of tropical savannas* (ed. by B. H. Walker), pp. 1–12. IUBS Mon Series 3, IRL Press.

Walker, B.H. & Noy-Meir I. (1982) Aspects of the stability and resilience of savanna ecosystems. *Ecology of tropical savannas* (ed. by B. S. E. Huntley and B. H. Walker), pp. 556–590. Springer, New York.

Section IV
Savanna Management for Pastoral Industries

The main economic enterprise in savannas is pastoralism, and rangelands cover significant areas of savannas on all continents. The two other major uses of savannas are for agricultural crops (with attendant human settlements) and for wildlife/wilderness reserves (often with significant tourism enterprises). The amount of land associated with each of these three uses varies greatly among (and within) continents. Further, socio-economic and cultural differences occur within each land use type. For example, in some savanna rangelands, especially those in Australia and South America which were settled by Europeans only in the past few centuries, human populations tend to be very low and individual properties very large, relative to savanna lands which have experienced pastoralism for millenia. Pastoral industries there tend to be large-scale ranching operations, and research to increase production concentrates on improved strains of livestock, manipulations of pasture plant species composition through introductions (e.g. legumes) or removals (e.g. trees), additions of fertilizers or mineral supplements, etc. The wise management of these enterprises requires knowledge, understanding, and decisions involving not only shorter-term processes affecting livestock and pastures, but also longer-term processes and the economic costs/benefits of the conservation of plants and soil.

The papers in this section have direct bearing on the management of savannas for ranging operations. Burrows *et al.*, McCown & Williams, and Moog discuss factors influencing productivity of livestock, mainly factors that impinge on fodder quality, quantity, and timing in relation to variability in rainfall. Burrows *et al.* discuss the relationship between woody plants and herbaceous plants in three savanna types in N.E. Australia, emphasizing an 'imbalance' of the two plant types in regard to livestock production, with the presence of woody plants reducing yield.

McCown & Williams and Moog show how legumes increase livestock production over native pastures in Australia and the Philippines, respectively. The tight linkages between water and nutrients ('nutrient drought') are dealt with by McCown & Williams, who point out that advances which reduce the effect of drought (e.g. genetically better adapted cattle, and legume introductions) increase the risk of overgrazing in severely dry years.

Winter makes a strong and compelling argument for a change in management philosophy in Australia's cattle industry which integrates fire management as a means of controlling animal movements and pasture quality, leading to a sustainable system of land use, including both increased yields and conservation of pasture.

Decision-support systems (DSS) are becoming increasingly important tools for land management. (See also McKeon *et al.*, Section I.) Stuth *et al.* have developed an integrated hierarchical planning system which is very inclusive and is targeted for service agency personnel, whereas the modular microcomputer-based DSSs of Stafford Smith & Foran and of Ludwig are aimed at the land managers of extensive grazing properties. The philosophy and approach to development of DSS for management presented by Stafford Smith & Foran is an excellent guide to ensure maximum effectiveness in the transfer of scientific knowledge and technology to land management in the types of savannas characterized by those in Australia.

P.A.W.

17/Management of savannas for livestock production in north-east Australia: contrasts across the tree–grass continuum

W. H. BURROWS, J. O. CARTER,* J. C. SCANLAN† and E. R. ANDERSON *Queensland Department of Primary Industries, P.O. Box 6014, Rockhampton Mail Centre, Queensland 4702, *Queensland Department of Primary Industries, P.O. Box 519, Longreach, Queensland 4730, †Queensland Department of Primary Industries, P.O. Box 183, Charters Towers, Queensland 4820, Australia*

Abstract. Three distinctive savannas of north-east Australia, originally dominated by unpalatable *Eucalyptus* L'Hérit spp., the valuable fodder *Acacia aneura* F. Muell. and *Astrebla* F. Muell. ex Benth. spp. grasslands are described. In each case an imbalance of woody and herbaceous plants now poses a threat to successful pastoralism. Tree/shrub–grass relationships and the impact of fire are examined for each community. Utilization of these areas for beef and wool production is discussed and some approaches to predictive modelling of the systems are outlined.

Key words. Savanna, tree–grass balance, *Eucalyptus*, *Acacia*, *Astrebla*, Australia.

INTRODUCTION

Of all the continental savannas those in Australia have suffered the shortest period of substantial modification by man, and the recent history of vegetation change is well documented. The generic term woodland has been used both for systems of structural classification and for vegetation mapping savanna-like formations in Australia (e.g. Carnahan, 1976). This emphasizes the woody upper stratum at the expense of the graminoid layer which many authors (Bourliere & Hadley, 1970; Huntley & Walker, 1982) consider as the principal functional element. The ground flora of savannas in tropical and subtropical north-eastern Australia is dominated by grasses of the tribe Andropogoneae. The soils are mainly weathered relics of earlier pedological processes (Hubble, Isbell & Northcote, 1983) resulting in their low nutrient status. Australian savannas have been described under various names by Moore (1970), Weston *et al.* (1981), Gillison (1983), Mott & Tothill (1984), Mott *et al.* (1985) and Weston (1988).

In this paper we describe three distinctive savanna types of north-eastern Australia: eucalypt (*Eucalyptus* L'Hérit spp.) woodlands, mulga shrublands (a 'disclimax' *Acacia aneura* F. Muell. community now dominated by *Eremophila* R.Br. shrubs) and mitchell (*Astrebla* F. Muell. ex Benth. spp.) grasslands (with an aggressive *Acacia* Miller invader) (Fig. 1). In each case an imbalance of woody and herbaceous plants poses a threat to successful pastoralism. Tree/shrub–grass relationships and the impact of fire are examined for each community. Utilization of these areas for beef and wool production is discussed and some approaches to predictive modelling are outlined.

PLANT COMMUNITIES

Eucalypt woodlands

These woodlands occur in the Black Spear Grass and Aristida–Bothriochloa Pastures of Weston (1988). They are characterized by a continuous grass dominated herbaceous ground cover with a variable upper stratum of *Eucalyptus* spp. (Fig. 2a).

The black spear grass pastures are the more extensive and productive native pasture community in terms of beef production in the humid and sub humid zones. They receive between 700 mm and 1200 mm mean annual rainfall. The main tree species are *Eucalyptus crebra* F. Muell., and *E. melanophloia* F. Muell. which range in height from 15 to 25 m. The original understorey was predominantly *Themeda triandra* Forsk. but this was modified by fire and grazing to *Heteropogon contortus* (L.) Roem. & Schutt (Tothill, 1969). Other important components include *Bothriochloa bladhii* (Retz.) S. T. Blake and *Bothriochloa ewartiana* (F. Muell.) C. E. Hubbard. Most of the area is made up of infertile texture contrast soils (natrixeralfs: Soil Survey Staff, 1975) and earths (alfisols and oxisols). These soils are characterized by low N and P levels. The land is used for cattle breeding and fattening.

The Aristida–Bothriochloa Pastures occur in semi-arid woodlands to the west of the Black Spear Grass Pastures.

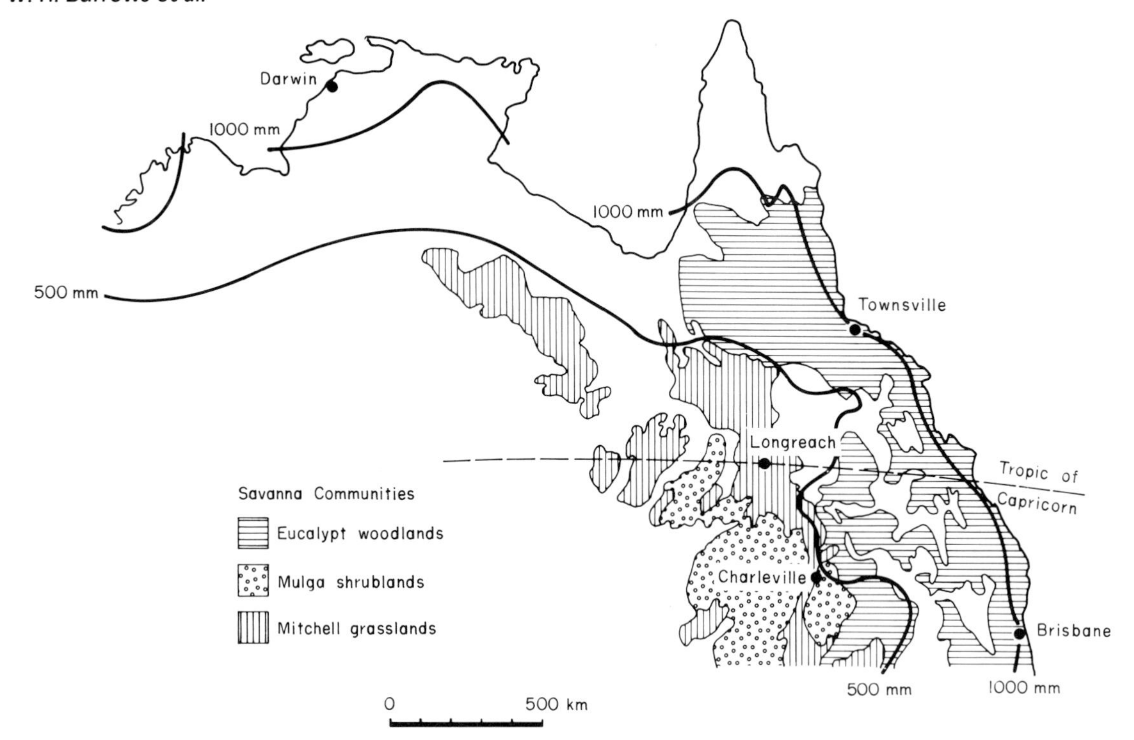

FIG. 1. Distribution of three distinctive savanna communities in north-east Australia. Rainfall isohyets for 500 and 1000 mm are indicated.

The main eucalypts here include *Eucalyptus populnea* F. Muell. and *E. melanophloia*. In the southern part of this region the grass understorey is mainly *Aristida* L., *Bothriochloa* Kuntze and *Chloris* Sw. spp. In the north *Chrysopogon* Trin. becomes a significant component. Infertile earths (alfisols and oxisols), texture contrast (natrixeralfs) and sandy soils (torripsamments) comprise most of the area. Sheep and cattle graze these poor quality pastures at low stocking rates: 19–40 ha per head for cattle and 2–3 ha per head for sheep.

Eucalypts are evergreen trees with sclerophyllous leaves and a high capacity to withdraw nutrients prior to leaf shedding. This produces leaf litter of low nutrient content and high C:N ratio, making decomposition slow. Most leaf material is shed in summer in synchrony with new leaf growth so that trees are rarely leafless (except for some northern species, e.g. *Eucalyptus alba* Blume-Specht and Brouwer 1975; Walker, 1979). Some tree legumes (mainly *Acacia* spp.) are present in the original community and may become prominent following disturbance.

Mulga shrublands

Structurally, mulga (*Acacia aneura*) associations range from open forests to sparse tall open shrublands (Boyland, 1973). Characteristic grass genera are *Aristida*, *Thyridolepis* S. T. Blake, *Eriachne* R.Br., *Digitaria* Haller and *Eragrostis* Wolf. Annual rainfall ranges from 250 to 500 mm.

Most of the soils are infertile earths (alfisols and oxisols), texture contrast soils (natrixeralfs) or sands (torripsamments). They have a particularly low P status. Most of the nutrients are in the surface few centimetres of soil and are lost if erosion occurs. These are traditional wool growing areas though cattle are often run conjointly with sheep.

Overclearing of mulga shrublands may occur as a result of felling the mulga for drought fodder or clearing to promote grass growth. This creates a niche for the rapid build-up of unpalatable shrubs (mainly species of *Eremophila*, *Dodonaea* Miller and *Cassia* L.). Shrub invasion in the mulga lands (Fig. 2b) has been documented by Burrows & Beale (1969), Burrows (1972, 1973) and Burrows *et al.* (1985). Shrubs reduce pasture productivity through competition for water and nutrients (Beale, 1973; Walker, Moore & Robertson, 1972), lower animal production (Burrows, 1986) and make property management more difficult.

Mitchell grasslands

The mitchell grasslands are the most resilient of the rangeland pastures in Australia (Orr & Holmes, 1984). Rainfall ranges from 250 to 500 mm annually. These grasslands occur on gently undulating downs where slopes are less than 2%. The soils are red, grey and brown alkaline cracking clays (vertisols) with a uniform profile with self-mulching surfaces. Soil depths up to a metre are common. Soil fertility is generally high, with adequate P and moderate levels of N.

Four species of *Astrebla* (*A. lappacea* (Lindl.) Domin, *A. elymoides* F. Muell. ex F. M. Bail., *A. pectinata* (Lindl.) F. Muell. ex Benth. and *A. squarrosa* C. E. Hubbard) are widespread and occur as tussock grasslands of low basal cover (2–4%). The interspaces are occupied by a range of annual grasses and forbs.

Most of the area is well supplied with artesian and sub-artesian water suitable for livestock. Where water is available from free flowing bores it is distributed by surface drains.

FIG. 2. Savanna communities of north-east Australia: (a) poplar box (*Eucalyptus populnea* F. Muell.) woodlands; (b) mulga (*Acacia aneura* F. Muell.) trees (background) felled for stockfeed and replaced by unpalatable *Eremophila gilesii* F. Muell. shrubs (foreground); (c) *Acacia nilotica* subsp. *indica* (Benth.) Brenan invading previously treeless mitchell (*Astrebla* F. Muell. ex Benth. spp.) grasslands.

These grasslands are most suited to sheep grazing though cattle numbers increase during favourable seasons and in response to market pressures. High temperatures and lack of shade are husbandry problems. The spread of the thorny tree, *Acacia nilotica* subsp. *indica* (Benth.) Brenan, *inter alia*, during the 1970s is leading to serious management problems in some central and northern areas. Two groups of woody plants are important in the grassland transformation:

(1) Native species which exist on the margins or at very low densities throughout and which are expanding in area. The most important native species are *Acacia cambagei* R. T. Bak. (gidgee) and *A. farnesiana* (L.) Willd. (mimosa). The main changes influencing the invasion of undesirable natives have been a change in fire regime and an increase in grazing pressure.

(2) Exotic species imported from Africa and America. *A. nilotica* is by far the most important of the exotic invading species (Burrows *et al.*, 1986; Burrows, McIvor & Andrew, 1986; Fig. 2c). The invasion has followed an exponential pattern with stepwise increases in density and spread driven by above average summer rainfall and changes in sheep/cattle ratios (see later).

TREE–GRASS RELATIONSHIPS

Eucalypt woodlands

Biomass effects. At low densities of large trees, low soil fertility status and low rainfall, the exotic grass *Cenchrus ciliaris* L. produces higher biomass under *Eucalyptus populnea* than native species in associated interspaces (Christie, 1975). This reflects the nutrient build-up beneath the tree canopies (Ebersohn & Lucas, 1965; Silcock, 1980). At higher densities of *E. populnea* with a woody understorey, low fertility, low rainfall and native pasture species, there is a strong negative curvilinear pattern of herbaceous dry matter production with increasing woody biomass (Walker *et al.*, 1972).

Conflicting patterns of herbaceous productivity also exist with *E. creba* communities. Scanlan (1986) and Scanlan & Burrows (1990) found a range of curvilinear relationships similar to Walker *et al.* (1972) (see Fig. 3a). By contrast, Walker *et al.* (1986a) reported a linear decrease in herbaceous production as basal area of *E. crebra* increased in a more favourable environment.

Thus eucalypts generally decrease the production of grass wherever such vegetation coexists. The magnitude of the decline depends on site fertility and total rainfall (i.e. potential site productivity). As site productivity increases, the effect of a given level of trees in depressing grass yield decreases (in relative terms).

Species composition. Trees modify the potential composition resulting from the particular soil, rainfall, temperature and grazing management histories of the site. Scanlan & Burrows (1990) noted that grasses within the Andropogoneae show decreasing dry matter production as tree density increases while grasses within the Paniceae display the opposite trend. However, forbs exhibited much less variation in actual production than did the grasses over a range of tree densities (cf. Ball, Hunter & Swindel, 1981; Clary & Jameson, 1981; Fig. 3c).

Community dynamics. In many parts of north-eastern Australia there has been a continuing cycle of timber treatment (firstly ringbarking or girdling, more recently stem injection with aboricide), followed by woody plant regrowth, followed by retreatment (Burrows, Scanlan & Anderson, 1988). A large proportion (50–90%) of all woody plants are too small for treatment by ringbarking or stem injection techniques (Scanlan, 1988). These multi-stemmed suppressed 'seedlings' grow rapidly when released from competition, showing higher growth rates than plants growing within the extant woodland (see Luken, 1988).

Leaving scattered mature trees (producing a derived savanna) ensures a seed source for continued regeneration of *Eucalyptus* spp. (Burrows *et al.*, 1988). A few trees can also have a relatively large impact on herbaceous productivity (Walker *et al.*, 1972; Scanlan, 1986). For example, if a site can produce 3000 kg/ha in treeless areas, then 70 trees/ha (20% of a dense stand) will reduce herbaceous production to 2000 kg/ha.

Overgrazing following tree clearing can result from attempts to recoup development expenditure rapidly. This results in lower soil organic matter, reduced nitrogen for grass growth and consequently reduced pasture productivity. Unwanted woody plant growth and soil erosion are the major causes of reduced pasture production seen in some eucalypt dominated regions of eastern Australia (Wilson, Tongway & Tupper, 1988).

Modification of a woodland to produce a savanna is undesirable for wildlife as it represents fragmentation on a microscale. The problem of habitat fragmentation is a major concern to wildlife biologists (Burgess & Sharpe, 1981; Harris, 1984).

The undesirable impact of converting woodlands into park-like savannas led Burrows *et al.* (1988) to propose that any clearing should be on the basis of complete tree removal on some areas while leaving trees undisturbed in remaining areas. This maximizes herbaceous production (given that a fixed proportion of trees are to be retained), minimizes the problem of seedling regeneration from seed trees and maintains some intact habitat for wildlife.

Mulga shrublands

Biomass effects. For this community, typical tree–grass yield relationships were shown for the original *A. aneura* tree cover by Beale (1973). He recorded sharp declines in grass yield with low tree densities and this pattern is repeated when the 'disclimax' *Eremophila gilesii* F. Muell, replaces *A. aneura* (Fig. 3b). The strong depressant effect the woody plants exert on grass production reflects the infertile and semi-arid environment of these communities.

Species composition. The mulga shrublands occur in areas with a reasonable expectation of some cool season rainfall. Nevertheless the predominant grass flora are summer growing (C_4) species. A close association between

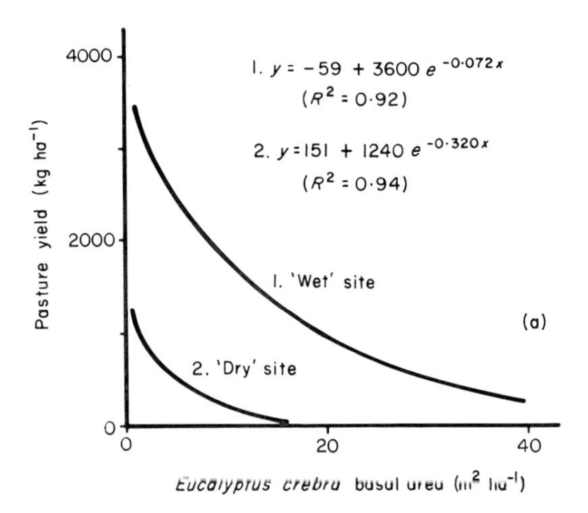

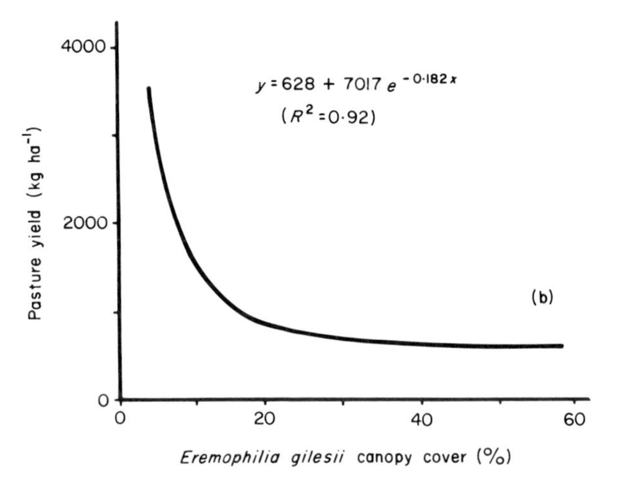

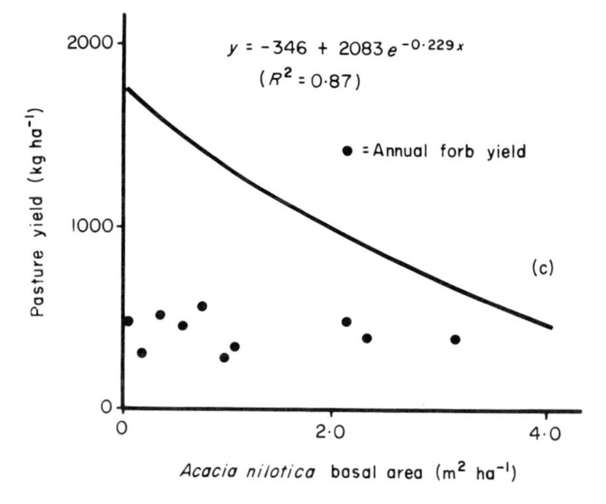

FIG. 3. Tree/shrub–grass relationships for north-east Australian savannas: (a) *Eucalyptus crebra* F. Muell. on texture contrast soils (alfisols: Soil Survey Staff, 1975); (b) *Eremophila gilesii* F. Muell. on red earth soils (alfisols); (c) *Acacia nilotica* (Benth.) Brenan on grey cracking clay soils (vertisols).

E. gilesii and *Aristida* spp. has been noted, suggesting an improved microhabitat beneath the shrub canopy and protection from grazing provided by the shrub branches (Sullivan & Pressland, 1987).

Community dynamics. Because the trees/shrubs exert such a major depressing effect on grass yield it is common for overgrazing to take place, especially when pastoralists underestimate the competition and consequently overestimate fodder availability, in poor rainfall years. Also, the competitive advantage of unpalatability is accentuated by the better adaptation of woody plants to the prevailing infertile soil conditions (Burrows, 1986).

The net result is for the woody shrubs to be increasing in density (Burrows *et al.*, 1985) and this has major management ramifications for the continued use of these communities by domestic stock. Unpalatable grasses can also fill the niche vacated by useful pasture species and some, such as *Aristida* spp., do so, but woody plants remain the dominant increaser plants in the mulga shrublands.

Mitchell grasslands

Biomass effects. Reduction in herbage production with increase in tree density follows a similar pattern to the other savanna types, with decreased productivity of perennial grasses. Growth of annual species in mitchell grasslands does not appear to be affected by tree density (Fig. 3c). Loss of available feed for animal production will increase as the invasion by native and exotic species progresses and will probably outweigh gains from the provision of shade and a high quality protein source during the dry season.

Dynamics. A. nilotica was introduced in the 1890s from India or the Middle East and was deliberately planted as a shade tree along drains distributing water from artesian bores. Little natural increase in populations occurred until high rainfall years in the 1950s. Areas of greatest increase in density were the town commons carrying large numbers of cattle and goats. Legislation enacted in 1957, prohibited growing of the plant. Limited descriptive data suggest the total area of *A. nilotica* was less than 0.5 ml/ha at this time. A run of dry years through the 1960s saw no increase in the species and little or no control by landowners. The 1970s saw large increases in population on many properties.

The reasons for the increases were (1) an accumulation of seed reserves, (2) above average wet years, and (3) a change from sheep to cattle grazing.

(1) *A. nilotica* is a prolific seeder, producing up to 175,000 seeds per tree if moisture is not limiting. Hard seed coats impose dormancy and seed is long lived (Blunt, 1948; Anon., 1955). One kilometre of bore drain with trees growing on one side of the drain only can produce an estimated 6–8 million seeds per year (Bolton, Carter & Dorney, 1987). Under conditions of low rainfall large seed banks accumulate.

(2) There were several periods in the 1970s when rainfall was much above average. Surface soils were often wet for a month or more allowing germination of hard and soft seed. Good rain in subsequent years ensured seedling survival.

(3) Changes in the type of grazing animal aided the spread of the species. Domestic animals eat most of the annual seed crop. Traditionally the area was grazed only by sheep. Sheep pass 1–2% of seed with about 30% being spat out in the eating process and a further 15–17% regurgitated during the day (Carter, unpublished data). Economic

circumstances led to increasing numbers of cattle being grazed in the region over the last 20 years. Cattle pass about 80% of seed eaten, more than half of which is viable (Harvey, 1981) and in an ideal medium for germination. Spread of this species via domestic stock parallels the spread of *Prosopis* species in the southern U.S.A. (Archer, this issue) and the spread of woody species in Africa (Walker *et al.*, 1981).

Distribution. Presently *A. nilotica* exists in over 25% of the 24 million hectares of mitchell grass, mostly scattered and at low density. Isolated plants occur in other grassland types. A bioclimatic analysis using all known locations of the plant in Australia and the BIOCLIM program of Busby (1984) showed that most of north-east Australia's natural grasslands are climatically suitable for *A. nilotica*. Pronounced increases in density can be expected in the future, as soil seed banks are large.

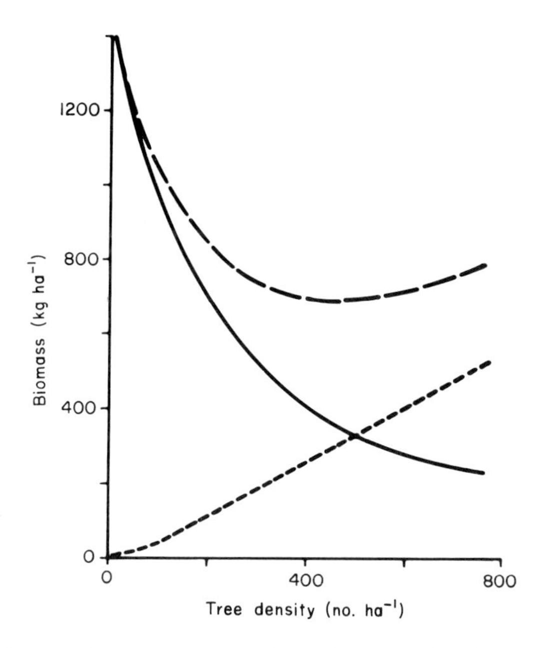

FIG. 4. Contributions to ground layer biomass (fuel loads: — — —) by herbage (grass and forbs:——) and the tree litter (-----) at the end of the dry season in *Eucalyptus populnea* F. Muell. woodlands. (Based on data from Scanlan & Burrows (1990) and D. M. Burrows, unpublished.)

FIRE EFFECTS

Eucalypt woodlands

Fire has played a key role in the evolution of Australia's dry tropical savanna communities. Recent studies in the Northern Territory (Bowman, Wilson & Hooper, 1988) have shown how protection against fire can result in development of a dense woody understorey in woodlands. The impact of fire on regrowth species in eucalypt communities has been demonstrated by Burrows *et al.* (1988) and Paton & Wilson (unpublished).

In north-east Australia total fuel loads of 800–1000 kg/ha are necessary for ground fire to carry under 'normal' dry season conditions. In grazed eucalypt communities such fuel loads are commonly achieved despite the inverse relation between tree cover and herbage yield. Fuel loads are boosted by large contributions from tree leaf litter (Fig. 4). Low tree populations also produce a significant seed 'rain' each year, with fire providing the establishment opportunity (sterile seedbeds) and sapling control (fire kill) (Burrows *et al.*, 1988) that maintains the woodland structure. Because the woody vegetation of eucalypt woodlands is largely unpalatable to domestic stock, a synergistic fire-grazing interaction is only likely to occur at high stocking rates, but by its very nature this combination is incompatible with continuous use.

Mulga shrublands

Fire is seen as one of the few management tools available to economically control woody plants in semi-arid environments (Hodgkinson *et al.*, 1984). Non-resprouting species *E. gilesii*, *Dodonaea attenuata* A. Cunn. and *D. tenuifolia* Lindl. are easily killed (Sullivan & Pressland, 1987). Seed banks for the species are often high (Orr & Sullivan, 1986) and re-establishment occurs after late summer rains. Those shrubs which regrow from basal or aerial shoots, e.g. *E. bowmanii* F. Muell. and *E. sturtii* R.Br. are suppressed by fire but not killed. Mortality is high for all species if burnt when less than 2 years old (Harrington & Hodgkinson, 1986). Herbage composition is little affected by single fires.

Carter & Johnston (1986) used herbage production models and long-term rainfall records to simulate fire frequency in mulga shrublands. While fire frequency was reduced with increasing grazing pressure it was most notably affected by increased canopy cover. Change in shrub cover from 0 to 10% reduced the potential fire frequency from 64% to 13% of years respectively. Thus fire should be used opportunistically to control shrub build-up before it begins to depress grass (and hence fuel) yields.

Mitchell grasslands

Little is known about the impact of fire on ephemeral forbs in mitchell grasslands as their presence is determined mainly by the timing of rainfall (Lorimer, 1978; Orr, 1981). Fire increased tillering in *Iseilema* spp. (a major annual grass), but reduced plant number. *Astrebla* spp. tussock numbers and seedheads per plant increased while basal areas remained unchanged following dry season wildfires (Scanlan, 1980). Dry matter on offer was similar on burnt and unburnt sites, 2 years after burning, even though tiller age structures were markedly different (Scanlan, 1983). Thus the amount of litter remaining and the timing and amount of rainfall determine the response of mitchell grasslands to burning.

As noted earlier, the main woody species present in northern mitchell grasslands are *Acacia farnesiana*, a pantropical species and the exotic *A. nilotica*. Burning does not kill *A. farnesiana* in south Texas, but reduces canopy cover and increases stem number (Scifres, Mutz & Drawe, 1982). Few data exist for fire effects on *A. nilotica*, although it appears to behave in a similar manner to other *Acacia* spp. and *Prosopis glandulosa* Torrey. Defoliation below cotyledons is required to kill *P. glandulosa* (Scifres *et al.*, 1985) and *A. nilotica* reacts similarly (Brown & Booysen, 1967). Severe drought kills a large percentage of *A. nilotica* and may kill a small percentage of *P. glandulosa* (Carter, 1964).

It appears that the timing of fire is critical. Absence of fire following years of above average rainfall allows woody species to establish and grow to a stage where susceptibility to fire is reduced. Grass production is also suppressed to a level where the chance of producing sufficient fuel to carry a fire is low.

There are suggestions that withdrawal of fire may encourage more vigorous grass growth and result in death of *A. nilotica* (van der Schiff, 1957; Pienaar, 1959) although the latter believes that during the early stages of invasion, fire can help prevent woody increase. This hypothesis is not supported by Qadri (1955), Pratt & Knight (1971) or Trollope (1980) who suggest that *A. nilotica* can persist in frequently burnt savannas.

Although frosts (Parker, 1918), grass competition (Brown & Booysen, 1967) and grazing by goats (Pratt & Knight, 1971) all reduce regeneration, the rapid maturation and seed longevity of *A. nilotica* suggest that, once established, it will remain a persistent feature of the landscape.

For most regrowth problems, the use of fire and browsers offers best hope for control. Vegetation management involves maintaining vigorous grasses to compete with seedlings and facilitate periodic fires, while reducing seed-set. Animal management should consider avoiding seed dispersal which results from consumption of *A. nilotica* seed pods. For example, ripe seed pods form at the end of the dry season and their pendulous arrangement on the plant, high nutritive value and palatability contribute to their selection by stock at this time. At other times of the year the pods are on the ground and less attractive to animals, while associated pasture species provide adequate animal diets.

UTILIZATION

Stocking rate has been shown world-wide to exert a very powerful influence on pasture production and stability. The savannas of north-eastern Australia are continuously grazed by domestic stock but there have been few studies on the effects of such grazing on the vegetation and animal performance.

Most savanna pastures are of low quality except for the first few months of the wet season. Stocking rates have traditionally been determined by the ability to carry stock through the ensuing dry season. The large imbalance between 'wet' and 'dry' season carrying capacity usually results in overstocking in the dry season and loss in animal condition. For example, from 87 data years McLennan *et al.* (1988) observed an average annual liveweight gain of 102 kg per head for *Bos indicus–B. taurus* cross steers continuously grazed under normal management conditions. Yet on conservatively stocked pastures in the eucalypt woodlands of north-eastern Australia annual liveweight gains in excess 200 kg/head have been recorded and have averaged 155 kg/head from 1977 to 1988 (C. H. Middleton, pers. comm.). Middleton's data suggest that these liveweight gains are being obtained at pasture utilization levels of 30–40%, whereas commercial stocking rates often exceed 50% utilization.

TABLE 1. Derived stocking rates (ha/adult beast) for *Eucalyptus populnea* F. Muell. communities before and after tree removal. Data are presented for good (pasture yield 3000 kg/ha in the open) and poor (pasture yield 2000 kg/ha in the open) seasonal conditions and based on consumption of 30% of the annual pasture growth.

	Basal area of trees (m^2/ha)	Pasture yield (kg/ha)	Derived stocking rate (ha/beast)
(i) Good season			
'Treed'	15	1040	13
'Cleared'*	3	2050	5†
(ii) Poor season			
'Treed'	15	260	50
'Cleared'	3	1300	10

* 'Cleared' areas have 20% of the original tree population retained.
† This compares with a two-fold increase in stocking rate as a result of tree clearing in *E. melanophloia* F. Muell. woodlands (Tothill, 1983).

A series of trials (Orr & Holmes, 1984; Wilson, Harrington & Beale, 1984) in the drier mulga shrublands and mitchell grasslands have also shown that low pasture utilization rates (20% and 30% respectively) maximize individual liveweight gain and wool production in merino sheep.

Failure to adjust stocking rates to available feed is a common management problem in continuously grazed savanna pastures. Managers do not appreciate that the competitive effect of the woody plants is magnified as growing conditions become harsher (Table 1). The most notable feature of these calculations is that the carrying capacity ratio between 'cleared' and 'treed' sites changes from about 2.5:1 in good seasons to almost 5:1 in poor seasons. Generally, landholders are not so responsive in changing their stocking rates, as a result of changing seasonal conditions. Similarly, they often fail to take into account the strong depressant effect of regrowing shrubs and trees on pasture production following initial clearing (Harrington & Johns, 1990).

Most damage to grass pasture occurs when stocking pressure is too high when the season breaks, rather than in the dry season itself (Mott & Tothill, 1984). These conditions occur in savannas when the effect of competing unpalatable trees and shrubs (such as *Eucalyptus* spp. and *Eremophila* spp.) is underestimated, or the availability of browse or nitrogen supplements enable grazing pressures to be maintained, when they might otherwise have been inevitably reduced.

DISCUSSION

The three savanna types detailed in this paper have different floristics climate and soils. A major similarity is that they all experience a highly variable rainfall regime. In each system the aim of the manager should be to maximize

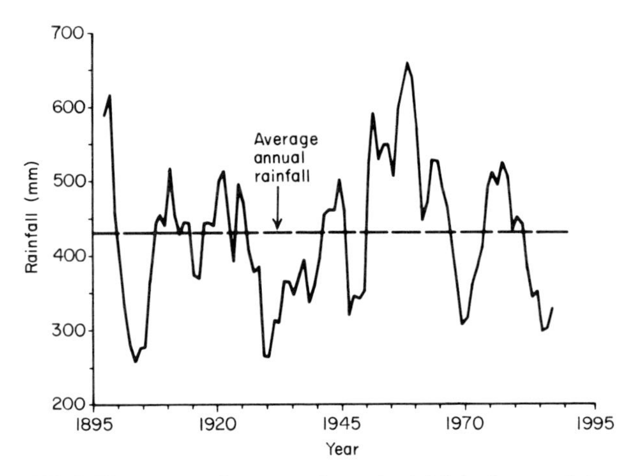

FIG. 5. Five-year running mean of annual rainfall for Longreach (see Fig. 1 for location) showing extended periods of above and below average rainfall on mitchell grasslands. Woody weed invasions occurred after above average rainfall in the 1950s and 1970s.

animal production commensurate with the long-term maintenance of the resource. This can be achieved by reducing or controlling woody plant density so production of herbage is maximized, while conserving adequate shade, shelter and wildlife habitat.

In all situations, trees have some stimulatory and some competitive influences on the understorey (Wu *et al.*, 1985; Walker *et al.*, 1986b). Thus individual trees may have no effect (Teague, 1984, for *Acacia karroo* in Africa), a net positive effect (Christie, 1975, for *E. populnea* in Australia) or, more commonly, a net negative effect (Burrows *et al.*, 1988, for *E. crebra* in Australia).

Nevertheless, on an individual tree basis some generalities exist at the community level. For example, eucalypts decrease herbaceous production with increasing density. The rate of this decrease declines as the potential site productivity increases, i.e. given poor soils and/or low rainfall, trees have a major effect while on fertile sites with high rainfall trees have less impact.

Northern areas of Australia (<20°S) may differ from more southerly areas as Mott & Tothill (1984) suggest there is little competition for water during the wet season, but rather the trees compete for nutrients which have been shown to be the more important limitation for plant growth on infertile sites (see Mott *et al.*, 1985).

Modification of the savanna systems in north-east Australia, whether by clearing of woodlands or invasion of undesirable woody species, is occurring rapidly. The role of the applied ecologist is to predict resource degradation, changes in economic returns and the risk element for a large number of management options. This is difficult because of the variable rainfall regime in which there are extended periods of above and below average rainfall (Fig. 5). Short-term experimental data are of limited value unless they can be extrapolated over a wide range of climatic conditions. One way this can be achieved is to combine pasture production models with historic climatic data for different zones to predict likely feed availability under different management scenarios.

A simulation model has been developed (G. M. McKeon and J. C. Scanlan, pers. comm.) to incorporate the effects of trees on forage production into a woodland development strategy, evaluation package. A water balance and pasture production submodel (McKeon *et al.*, 1982) estimates pasture production in the absence of trees from soil data and daily climate, using empirical relationships established from field trials. Equations from Scanlan & Burrows (1990) are then used to simulate the effect of trees on pasture production. Thus, given soil, rainfall, initial tree basal area and grazing utilization level, it is possible to compare alternative development strategies for north-eastern Australian eucalypt woodlands.

Development of reliable models has depended on establishing empirical relationships from field data. For example, in the savanna systems examined in this paper any prediction of animal productivity must take into account tree/shrub populations and their effects on associated pasture yield. The need for model validation and tuning has led to research aimed at collecting a minimum standardized data set for a wide range of savanna types (McKeon *et al.*, this issue).

The pattern of tree or shrub invasion of north-east Australia grasslands and the formation of induced savannas is similar to that in Africa and America. A woody plant–herbage imbalance exists or poses a threat to current management for each of the savanna types considered in this paper. In eucalypt woodlands application of 'new' technologies has the potential to switch this balance in favour of grass over 30 million hectares. In the mulga shrub lands utilization of the palatable *A. aneura* has removed tree competition, only to see the trees replaced by unpalatable shrubs. On the other hand, the introduction of shade/fodder trees into treeless mitchell grasslands has led to rapid invasion by *A. nilotica* at the expense of grass forage. In all areas optimizing the tree/shrub-grass balance for livestock production remains a major challenge for land managers.

ACKNOWLEDGMENTS

Savanna studies reported in this paper have received financial support from the Australian Meat and Livestock Research and Development Corporation and the Australian Wool Corporation on the recommendation of the Wool Research Trust Fund.

REFERENCES

Anon. (1955) *Forest research in India 1950–51*, Part 1. The Forest Research Institute, Indian Forest Service, Delhi.

Ball, M.J., Hunter, D.H. & Swindel B.F. (1981) Understorey biomass response to microsite and age of bedded slash pine plantations. *J. Range Managemnt*, **34**, 38–42.

Beale, I.F. (1973) Tree density effects on yields of herbage and tree components in south-west Queensland mulga (*Acacia aneura* F. Muell.) scrub. *Trop. Grassl.* **7**, 135–142.

Blunt, H.S. (1948) Further notes on forestry in the Sudan. *J. Oxford Univ. For. Soc.* (3rd Series), No. 3, pp. 28–31.

Bolton, M.P., Carter, J.O. & Dorney, W.J. (1987) Seed production in *Acacia nilotica* spp. *indica* (Benth.) Brenan. *Proc. Weed Seed*

Biol. Conf., Orange, N.S.W., pp. 29–34.

Bourliere, F. & Hadley, M. (1970) The ecology of tropical savannas. *Ann. Rev. Ecol. Syst.* **1**, 125–152.

Bowman, D.M.J.S., Wilson, B.A. & Hooper, R.J. (1988) Response of *Eucalyptus* forest and woodland to four fire regimes, Munmarlary, Northern Territory, Australia. *J. Ecol.* **76**, 215–232.

Boyland, D.E. (1973) Vegetation of the mulga lands with special reference to south-western Queensland. *Trop. Grassl.* **7**, 35–42.

Brown, N.C. & Booysen, P. de V. (1967) Seed germination and seedling growth of two *Acacia* species under field conditions in grassveld. *S. Afr. J. agric. Sci.* **10**, 659–666.

Burgess, K.L. & Sharpe, D.M. (1981) *Forest island dynamics in man-dominated landscapes*. Springer, New York.

Burrows, W.H. (1972) Productivity of an arid zone shrub (*Eremophila gilesii*) community in south-western Queensland. *Aust. J. Bot.* **20**, 317–329.

Burrows, W.H. (1973) Studies in the dynamics and control of woody weeds in semi arid Queensland. 1. *Eremophila gilesii*. *Qld J. Agric. Anim. Sci.* **30**, 57–64.

Burrows, W.H. (1986) Potential ecosystem productivity. *The Mulga Lands* (ed. by P. S. Sattler), pp. 7–10. Royal Society of Queensland, Brisbane.

Burrows, W.H. & Beale, I.F. (1969) Structure and association in the mulga (*Acacia aneura*) lands of south western Queensland. *Aust. J. Bot.* **17**, 539–552.

Burrows, W.H., Beale, I.F., Silcock, R.G. & Pressland, A.J. (1985) Prediction of tree and shrub population changes in a semi-arid woodland. *Ecology and management of the world's savannas* (ed. by J. C. Tothill and J. J. Mott), pp. 207–211. Australian Academy of Science, Canberra.

Burrows, W.H., Carter, J.O., Anderson, E.R. & Bolton, M.P. (1986) Prickly acacia (*Acacia nilotica*) invasion of Mitchell grass (*Astrebla* spp.) plains in central and northern Queensland. *Proc. 4th Bienn. Conf. Aust. Rangel. Soc., Armidale*, pp. 104–106.

Burrows, W.H., McIvor, J.G. & Andrew, M.H. (1986) Management of Australian savannas. *Proceedings of the third Australian conference on tropical pastures* (ed. by G. J. Murtagh and R. M. Jones), pp. 1–10. Tropical Grasslands Society of Australia, Occasional Publication No. 3.

Burrows, W.H., Scanlan, J.C. & Anderson, E.R. (1988) Plant ecological relations in open forest, woodlands and shrublands. *Native pastures in Queensland: the resources and their management* (ed. by W. H. Burrows, J. C. Scanlan and M. T. Rutherford), pp. 72–90. Queensland Department of Primary Industries Information Series Q187023, Queensland Government Printer, Brisbane.

Busby, J.R. (1984) Bioclimate Prediction System. *User's Manual Version 1.0* (mimeo.). Bureau of Flora and Fauna, Canberra.

Carnahan, J.A. (1976) Natural vegetation. *Atlas of Australian Resources, 2nd Series.* Department of National Resources, Canberra.

Carter, M.G. (1964) Effects of drought on mesquite. *J. Range Managemnt*, **17**, 275–276.

Carter, J.O. & Johnston, P.W. (1986) Modelling expected frequencies of fuel loads for fire at Charleville in western Queensland. *Third Queensland Fire Research Workshop, Gatton*, pp. 55–67.

Christie, E.K. (1975) A note on the significance of *Eucalyptus populnea* for buffel grass production in infertile semi-arid rangelands. *Trop. Grassl.* **9**, 243–246.

Clary, W.P. & Jameson, D.A. (1981) Herbage production and shrub removal in the pinyon-juniper type in Arizona. *J. Range Managemnt*, **34**, 109–113.

Ebersohn, J.P. & Lucas, P. (1965) Trees and soil nutrients in south-western Queensland. *Qld J. Agric. Anim. Sci.* **22**, 431–435.

Gillison, A.N. (1983) Tropical savannas of Australia and the southwest Pacific. *Tropical savannas* (ed. by F. Bourliere), pp. 183–243. Elsevier, Amsterdam.

Harrington, G.N. & Hodgkinson, K.C. (1986) Shrub–grass dynamics in mulga communities of eastern Australia. *Rangelands: a resource under siege* (ed. by P. J. Joss, P. W. Lynch and O. B. Williams), pp. 26–28. Australian Academy of Science, Canberra.

Harrington, G.N. & Johns, G.G. (1990) Herbaceous biomass in a *Eucalyptus* wooded grassland after removing trees and/or shrubs. (In press).

Harris, L.D. (1984) *The fragmented forest.* University of Chicago Press

Harvey, G.L. (1981) Recovery and viability of prickly acacia (*Acacia nilotica* ssp. *indica*) seed ingested by cattle. *Proc. 6th Aust. Weeds Conf., Gold Coast, Queensland*, pp. 197–201.

Hodgkinson, K.C., Harrington, G.N., Griffith, G.F. Noble, J.C. & Young, M.D. (1984) Management of vegetation with fire. *Management of Asutralia's rangelands* (ed. by G. N. Harrington, A. D. Wilson and M. D. Young), pp. 141–156. CSIRO, Canberra.

Hubble, G.D., Isbell, R.F. & Northcote, K.H. (1983) Features of Australian Soils. *Soils: an Australian viewpoint*, pp. 17–46. Division of Soils, CSIRO. CSIRO Melbourne/Academic Press, London.

Huntley, B.J. & Walker, B.H. (eds) (1982) *Ecology of tropical savannas*. Springer, Berlin.

Lorimer, M.S. (1978) Forage selection studies. 1. The botanical compostion of forage selected by sheep grazing *Astrebla* spp. pasture in north-west Queensland. *Trop. Grassl.* **12**, 97–108.

Luken, J.O. (1988) Population structure and biomass allocation of the naturalized shrub *Lonicera maackii* (Rupr.) Maxim. in forest and open habitats. *Amer. Midl. Nat.* **119**, 258–267.

McKeon, G.M., Rickert, K.G., Ash, A.J., Cooksley, D. & Scattini, W.J. (1982) Pasture production model. *Proc. Aust. Soc. Anim. Prod.* **14**, 201–204.

McLennan, S.R., Hendricksen, R.E., Beale, I.F., Winks, L., Miller, C.P. & Quirk, M.F. (1988) Nutritive value of native pastures in Queensland. *Native pastures in Queensland: the resources and their management* (ed. by W. H. Burrows, J. C. Scanlan and M. T. Rutherford), pp. 125–159. Queensland Department of Primary Industries Information Series Q187023. Queensland Government Printer, Brisbane.

Moore, R.M. (1970) *Australian grasslands.* Australian National University Press, Canberra.

Mott, J.J. & Tothill J.C. (1984) Tropical and subtropical woodlands. *Management of Australia's rangelands* (ed. by G. N. Harrington, A. D. Wilson and M. D. Young), pp. 255–269. CSIRO, Melbourne.

Mott, J.J., Tothill, J.C. & Weston, E.J. (1981) Animal production from native woodlands and grasslands of northern Australia. *J. Aust. Inst. agric. Sci.* **47**, 132–141.

Mott, J.J., Williams, J., Andrew, M.H. & Gillison, A.N. (1985) Australian savanna ecosystems. *Ecology and management of the world's savannas* (ed. by J. C. Tothill and J. J. Mott), pp. 56–82. Australian Academy of Science, Canberra.

Orr, D.M. (1981) Changes in the quantitative floristics in some *Astrebla* spp. (mitchell grass) communities in south-western Queensland in relation to trends in seasonal rainfall. *Aust. J. Bot.* **29**, 533–545.

Orr, D.M. & Holmes, W.E. (1984) Mitchell grasslands. *Management of Australia's rangelands* (ed. by G. N. Harrington, A. D. Wilson and M. D. Young), pp. 241–254. CSIRO, Melbourne.

Orr, D.M. & Sullivan, M.T. (1986) The role of seed banks in limiting control of *Eremophila gilesii* (green turkey bush) by fire. *Proc. 4th Bienn. Conf. Aust. Rangel. Soc., Armidale*, pp. 230–233.

Parker, R.N. (1918) *Acacia arabicas* Wild. A forest flora for the punjab with Hazara and Delhi. Superintendent of Government Printing, Lahore, India.

Pienaar, A.J. (1959) Bush encroachment not controlled by veld burning alone. *Fmg Sth Afr.* **35**, 16–17.

Pratt, D.J. & Knight, J. (1971) Bush-control studies in the drier areas of Kenya. V. Effects of controlled burning and grazing management on *Tarchonanthus/Acacia* thicket. *J. appl. Ecol.* **8**, 217–237.

Qadri, S.M.I. (1955) Ecological study of the riverine forest of Sind, Pakistan. *J. For.* **5**, 241–249.

Scanlan, J.C. (1980) Effects of spring wildfires on *Astrebla* (mitchell grass) grasslands in north-west Queensland under varying levels of growing season rainfall. *Aust. Rangel. J.* **2**, 162–168.

Scanlan, J.C. (1983) Changes in tiller and tussock characteristics of *Astrebla lappacea* (curly mitchell grass) after burning. *Aust. Rangel. J.* **5**, 13–19.

Scanlan, J.C. (1986) Influence of the tree basal area on pasture yield and compositon. *Rangelands: a resource under siege* (ed. by P. J. Joss, P. W. Lynch and O. B. Williams), pp. 66–67. Australian Academy of Science, Canberra.

Scanlan, J.C. (1988) Managing tree and shrub populations. *Native pastures in Queensland: the resources and their management* (ed. by W. H. Burrows, J. C. Scanlan and M. T. Rutherford), pp. 91–111. Queensland Department of Primary Industries Information Series Q187023, Queensland Government Printer, Brisbane.

Scanlan, J.C. & Burrows, W.H. (1990) Herbage productivity and composition in *Eucalyptus* spp. communities in central Queensland. *Aust. J. Ecol.* (in press).

Scifres, C.J., Mutz, J.L. & Drawe, D.L. (1982) Ecology and management on the Texas coastal prairie. *Texas Agricultural Experiment Station Report* B-1408. College Station, Texas.

Scifres, C.J., Hamilton, W.H., Connor, J.R., Ingliss, J.M. Rasmussen, G.A., Smith, R.P., Stuth, J.W. & Welch, T.G. (1985) Integrated bush management systems for south Texas: Development and implementation. *Texas Agricultrual Experiment Station Report* B-1493. College Station, Texas.

Silcock, R.G. (1980) Seedling growth on mulga soils and the ameliorating effects of lime, phosphate fertilizer and surface soil from beneath poplar box trees. *Aust. Rangel. J.* **2**, 142–150.

Soil Survey Staff (1975) *Soil taxonomy – a basic system of soil classification for making and interpreting soil surveys.* United States Department of Agriculture, Washington, D.C.

Specht, R.L. & Brouwer, Y.M. (1975) Seasonal shoot growth of *Eucalyptus* spp. in the Brisbane area, Queensland (with notes on shoot growth and litterfall in other areas of Australia). *Aust. J. Bot.* **23**, 459–474.

Sullivan, M.T. & Pressland, A.J. (1987) The effects of fire on woody weed control and compostion of ground flora in south western Queensland. Queensland Department of Primary Industries, Report to the Australian Wool Corporation (mimeo.).

Teague, W.R. (1984) The management of thornveld. *Dohne Agric.* **6**, 21–23.

Tothill, J.C. (1969) Soil temperatures and seed burial in relation to the performance of *Heteropogon contortus* and *Themeda australis* in burnt native woodland pastures in east Queensland. *Aust. J. Bot.* **17**, 269–275.

Tothill, J.C. (1983) Comparison of native and improved pasture systems on speargrass. CSIRO Australia, Division of Tropical Crops and Pastures, Annual Report 1982–83, p. 105.

Trollope, W.S.W. (1980) Controlling bush encroachment with fire in the savanna areas of South Africa. *Proc. Grassld Soc. Sth Afr.* **15**, 173–177.

Van der Schijff, H.P. (1957) Bush encroachment in South Africa. *Handbook for farmers in South Africa*, Vol. III. Government Printer, Pretoria.

Walker, B.H., Ludwig, D., Holling, C.S. & Peterman, R.M. (1981) Stability of semi-arid savanna grazing systems. *J. Ecol.* **69**, 473–498.

Walker, J. (1979) *Aspects of fuel dynamics in Australia.* CSIRO Division of Land Use Tech. Memo. No. 79/7.

Walker, J., Moore, R.M. & Robertson, J.A. (1972) Herbage response to tree and shrub thinning in *Eucalyptus populnea* shrub woodlands. *Aust. J. agric. Res.* **23**, 405–410.

Walker, J., Robertson, J.A., Penridge, L.K. & Sharpe, P.J.H. (1986a) Herbage response to tree thinning in a *Eucalyptus crebra* woodland. *Aust. J. Ecol.* **11**, 135–141.

Walker, J., Sharpe, P.J.H., Penridge, L.K. & Wu, H. (1986b) *Competitive interactions between individuals of different size: the concept of ecological fields.* CSIRO Div. Water Land Resour. Tech. Memo. 86/11.

Weston, E.J. (1988) Native pasture communities. *Native pastures in Queensland: the resources and their management* (ed. by W. H. Burrows, J. C. Scanlan and M. T. Rutherford), pp. 21–33. Deprtment of Primary Industries Information Series Q187023, Queensland Government Printer, Brisbane.

Weston, E.J., Harbison, J., Leslie, J.K., Rosenthal, K.M. & Mayer, R.J. (1981) *Assessment of agricultural and pastoral potential of Queensland.* Agric. Br. Tech. Rep. No. 27, Queensland Department of Primary Industries, Brisbane.

Wilson, A.D., Harrington, G.N. & Beale, I.F. (1984) Grazing management. *Management of Australia's rangelands* (ed. by G. N. Harrington, A. D. Wilson and M. D. Young), pp. 129–139. CSIRO, Melbourne.

Wilson, A.D., Tongway, D.J. & Tupper, G.J. (1988) Factors contributing to differences in forage yield in the semi-arid woodlands. *Aust. Rangel. J.* **10**, 13–17.

Wu, H., Sharpe, P.J.H., Walker, J. & Penridge, L.K. (1985) Ecological field theory: a spatial analysis of resource interference among plants. *Ecol. Mod.* **29**, 215–243.

18/The water environment and implications for productivity

R. L. McCown and John Williams* *Division of Tropical Crops and Pastures, CSIRO Davies Laboratory, Private Mail Bag, Townsville, Queensland 4814, and *Division of Soils, CSIRO Davies Laboratory, Private Mail Bag, Townsville, Queensland 4814, Australia*

Abstract. Water deficits constrain productivity of herbivores on savanna grassland in two ways. The lack of soil water in the dry season causes a nutritional stress due to the low protein concentration and low dry matter digestibility of dead tropical perennial grass tissue. Less frequently, lack of rain in the growing season causes feed shortages in the subsequent dry season. The first part of this paper deals with simulation of these effects using weather data and describing the variation which occurs in northern Australia.

Options for reducing the impact of the annual dry season include sowing a pasture legume, since dry legume leaf is relatively nutritious. The second part of the paper examines variation imposed by the water climate on legume production potential and on the risk of spoilage due to out-of-season rainfall.

Thirdly, the paper examines the effects of nitrogen and phosphorus deficiencies and the consequences of mismanagement of the soil surface on the water constraint.

It is concluded that the substantial progress in reducing nutritional stress of the annual dry season has changed the nature of the problem of periodic droughts. Whereas formerly the main impact of nutritional stress was high cattle mortalities, now genetically better adapted cattle survive by consuming a greater proportion of the low-quality herbage with the aid of mineral supplements. The sowing of legumes increases average carrying capacity and increased pasture utilization.

Both innovations increase the risk of overgrazing in years of below-average pasture production. Avoidance of accelerated land degradation as a result of loss of buffering capacity inherent to the former extensive system will require skilful management.

Key words. Savanna, climate, animal nutrition, intensfication, stability, northern Australia.

INTRODUCTION

Water deficiencies are a major determinant of the productivity of savanna vegetation and of any replacement vegetation. However, nitrogen and phosphorus deficiencies are at least equally important. Operationally, an important distinction is that it is feasible to supply nitrogen and/or phosphorus; however, when this is done, the importance of water as a constraint increases. This paper attempts to examine the water constraint in relation to the productivity of the economically most important producer of the Australian savanna – the beef animal. It looks first at constraints in the system where cattle graze native grassland and then at the modification where nutritional deficiencies are reduced by introduction of a pasture legume and P by supplementation or fertilization. In both cases four questions are asked: How does the water environment constrain cattle productivity? How can the effects be quantified using weather data? How does this constraint vary in time and space? and What are the options for management?

THE NATIVE GRASSLAND

A characteristic of the savanna is the long dry season which constrains animal production in every year; during this period, nutritional value of grass falls below maintenance requirements. Natural grass herbage in Australian tropical savannas is, in the main, highly sclerophyllous. Only young leaf has sufficient protein concentration and sufficiently high digestibility to support weight gain of cattle. This means that active pasture growth is a prerequisite for animal growth and couples animal growth closely to soil water supply. Fig. 1 shows the decline in diet quality and weight gains with decline in growth index (G.I.).

G.I. is defined as the ratio of growth rate under current environmental conditions to optimum growth rate and was calculated as the product of a water balanced-derived moisture index and a thermal index as per Fitzpatrick & Nix (1970). Using data on live weight changes and weather, G.I. has been calibrated to reflect pasture quality (McCown, 1980). In Fig. 1 it is evident that cattle began to lose weight when the G.I. dropped to approximately 0.1

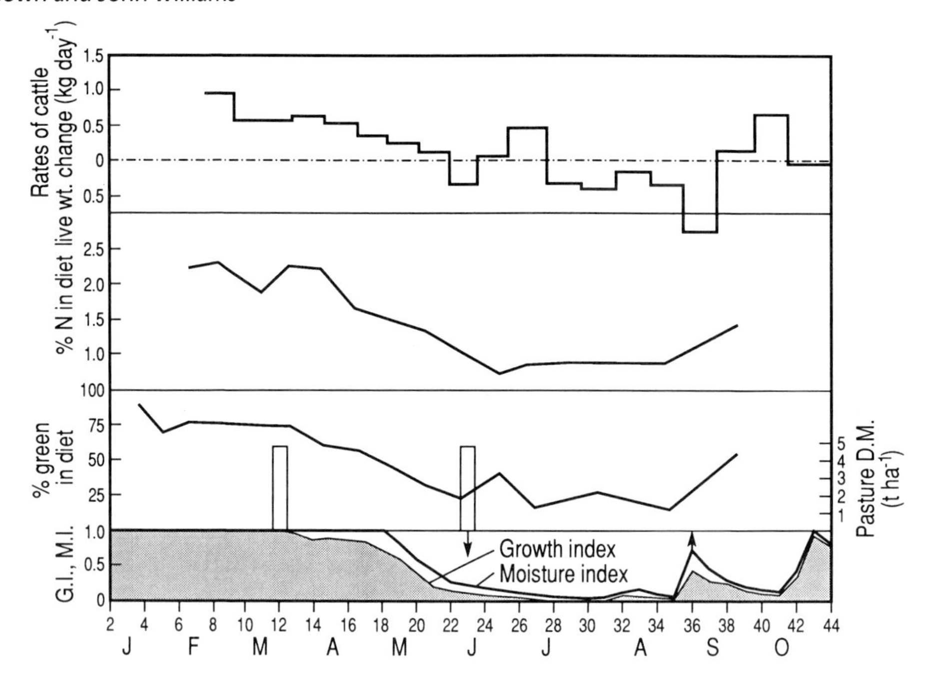

FIG. 1. Trends in liveweight, diet attributes, pasture yield, and pasture environment of cattle grazing native pastures near Woodstock, Queensland. (Liveweight and %N data from Hunter, Siebert & Breen, 1976; remaining data, McCown, unpublished.)

and, in the main, lost weight until rain in September caused a substantial new growth period. McCown (1980) used this point as the termination of a 'green season' (STOPWK) and the start of the 'dry season'. The initiation of green season, the main weight gain period, was found to coincide with a sequence of 8 weeks in which G.I. is greater than 0.1 for three of four and six of the eight weeks GOWK. (The event in Fig. 1 just barely qualifies as the start of the green season.) In this paper, 'green' and 'dry' seasons so defined will appear as Green Season and Dry Season and durations as Green Weeks and Dry Weeks.

Diet quality during a given week was classified as being either high enough to support liveweight gain (a Green Week) or not (a Dry Week). Variation in liveweight change among years or stations was explained by variation in number of Green Weeks or Dry Weeks. McCown *et al.* (1981a) found Dry Weeks in the Dry Season to account for 75% of variation in liveweight loss in the dry season and the relationship is approximated by the 'greater than 15' curve in Fig. 2.

Whereas water deficits limit production via diet quality in every year, drought years occur only periodically. In these years insufficient rainfall in the Green Season results in a shortage of feed in the dry season. Dry matter production can be predicted using the G.I.

For a specified vegetation–phenological stage–soil fertility situation, some maximum increment in dry matter production can be expected during a week with G.I. = 1.0.

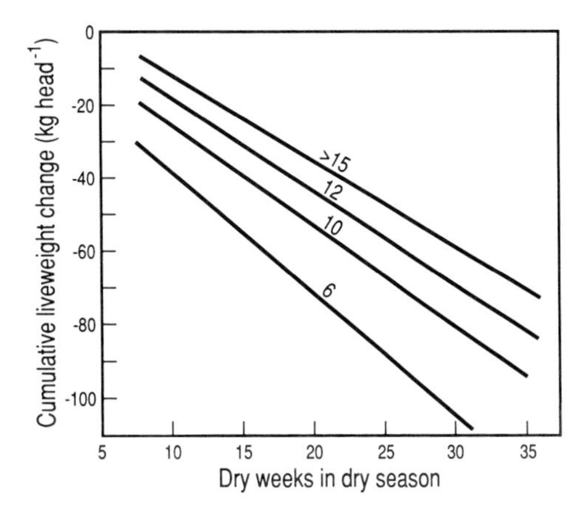

FIG. 2. The dependency of liveweight loss in the dry season on length of the dry season as influenced by Growth Weeks in the Green Season. (Generalized from McCown *et al.*, 1981a.)

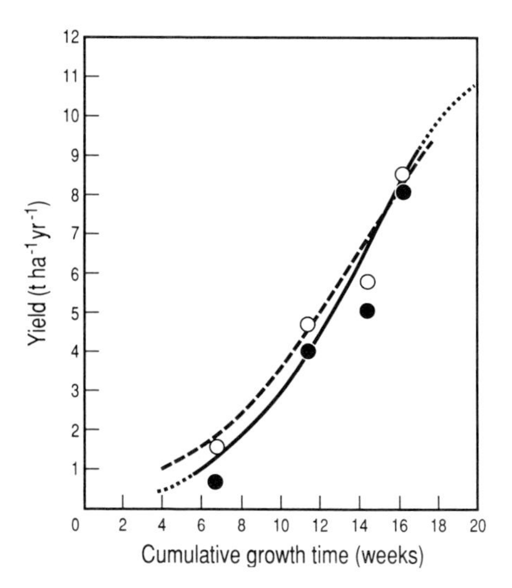

FIG. 3. Production of tropical legume swards as a function of Growth Weeks. (○, Total yield of *Stylosanthes humilis*-dominated swards (McCown, 1973); ●, yield of pure *Stylosanthes hamata* (Williams & Gardener, 1984).)

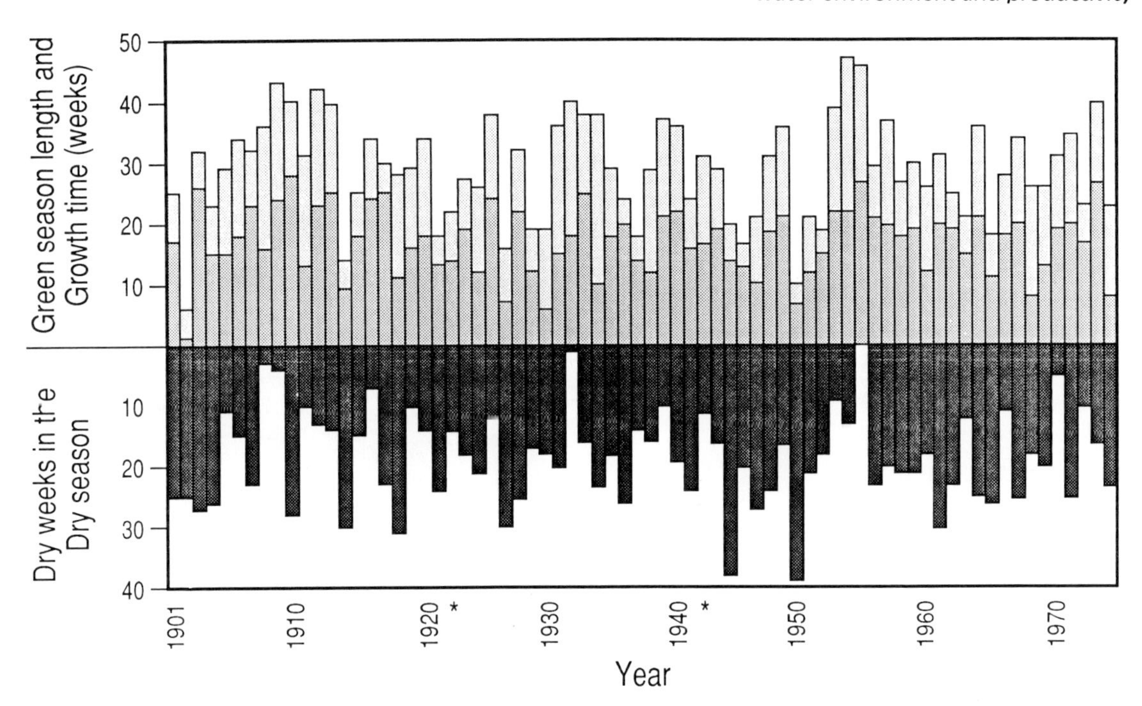

FIG. 4. Variation in simulated lengths of Dry Seasons, Green Seasons and Growth Weeks at Woodstock, Queensland.

Whereas 2 weeks at G.I. = 1.0 should produce twice as much, 2 weeks at G.I. = 0.5 might be expected to produce an amount equivalent to 1 week at G.I. = 1.0 (Williams & Gardener, 1984). This suggested a simple independent variable that combines the 'goodness' of conditions for growth and the duration at that level of 'goodness'. The 'growth time' (weeks or days) has proved valuable in being readily integrated over any real time period, and in the ease of interpretation of cumulative values by comparison to the real time duration, i.e. the maximum growth time, had conditions been optimum. (Williams & Probert (1983) set out a theoretical analysis which demonstrates a functional relationship between dry matter production and the growth time as determined by water and temperature regime.) A great deal of the variation in dry matter production among years and sites due to variation in rainfall and soil water storage capacity can be explained by relating production to a growth time variable (Fig. 3). This variable should relate closely to carrying capacity and should explain variation in cattle performance when herbage quantity is limiting.

In the data used by McCown *et al.* (1981a), there were years in which *quantity* effects on liveweight loss could be distinguished from low *quality* effects. Losses in the dry season in years in which Growth Weeks are greater than 15 are linearly related to Dry Weeks in the dry season, whereas larger losses occurred when Growth Weeks are less than 15 especially at high stocking rates (Fig. 2).

Mott *et al.* (1986) used this approach to compare the herbage productivity of six major savanna types in Australia. They found a close relationship between vegetation types, productivity, and the Growth Index.

With two variables that relate individually to forage quantity and quality (i.e. Growth Weeks and Dry Weeks), and having shown how they are derived from standard climatic data, it is possible to estimate how cattle performance varies, both from year to year and from place to place. Variation among years for growth weeks, the duration of the Green Season (potential Growth Weeks), and Dry Weeks in the Dry Season are shown for Woodstock, Queensland, for a 72-year period in Fig. 4. Year-to-year variation is high. Using generalized relationships from Fig. 2, liveweight loss for each year can be estimated. The resulting frequency distribution of losses is shown in Fig. 5. Variation among years for Katherine is low by comparison (Fig. 5). In seven years in 100, winter rainfall at Woodstock, representing the Townsville–Burdekin region, is sufficient to actually enable weight gain in the dry season, whereas cattle at Katherine would be expected to always lose at least 10 kg. At Woodstock, six years in 100 have low green season production (few Growth Weeks) preceding a long dry season, resulting in more than 80 kg loss. At Katherine this did not occur in the thirty-four years analysed.

Geographic and temporal variation in annual Growth Weeks, Green Weeks and Dry Weeks has been described using eighty stations with an average rainfall record of 40 years (McCown, 1982).

Increased production in the northern beef industry requires reduction of the close dependency of protein supply on green feed supply. One approach has been to feed non-protein nitrogen supplements in the dry season which result in increased intake of desiccated, low quality, perennial grass tissue. Because digestibility of this herbage remains low, production responses are modest; however, because the costs are relatively low and cattle mortality is greatly reduced, this practice has been widely adopted.

An approach with a greater potential to increase production is replacement (or supplementation) of native grasses with plants less sclerophyllic, and consequently, with more nutritious mature leaves. Introduced grasses,

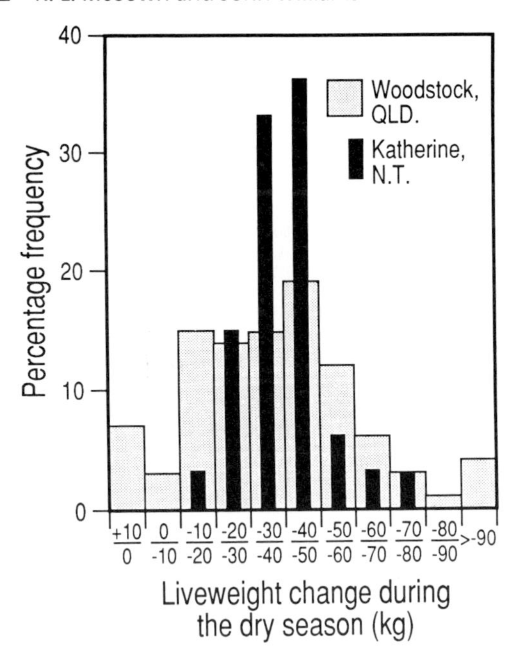

FIG. 5. A comparison of frequences of simulated Dry Season liveweight losses at two stations with similar mean annual rainfalls.

mainly from Africa, provide this to some degree, but these grasses have higher requirements of N and P than native grasses. Because legumes have much higher dry season nutritive value than any grasses but also fix their own nitrogen, most of the research on improved pastures in northern Australia has focused on finding and assessing legumes adapted to this climate. Several legumes, mainly in the genus Stylosanthes, have been found to grow and persist well (Edye, 1987). One of the most successful of these is Caribbean stylo (*Stylosanthes hamata* cv. Verano) which is shown in Fig. 6.

FIG. 6. An example of Caribbean stylo (*Stylosanthes hamata* cv. Verano) in the dry season showing edible leaf litter. A legume which has been found to grow and persist well in the wetter regions of the Australian savanna.

SOWN LEGUME-GRASS PASTURES

How does the water environment influence productivity on these pastures?

Climatically-induced variation in pasture legume production potential has been described by Williams & Gardener (1984) and by Williams & Probert (1983) using the methods outlined earlier for calculating growth time and relating this to dry matter production of Caribbean stylo. The cumulative probability-yield functions for Caribbean stylo with P non-limiting for a range of sites are shown in Fig. 7. The contrast between the Townsville–Burdekin sites of Pentland and Woodstock and that of Katherine is marked. The highly variable plant production at Pentland and Woodstock which is characteristic of the Townsville–Burdekin region is in strong contrast to the more reliable climate of Katherine which is representative of the more strongly monsoonal areas of northern Australia.

The means by which pasture legumes contribute to dry season nutrition of cattle differ. One group, currently typified by Caribbean Stylo, sheds leaf readily as the soil dries. Dry leaf, while less nutritious than green leaf, is much more nutritious than mature grass leaf, and cattle lick leaf litter from the ground. A second group, typified by Shrubby stylo (*S. scabra*) retains leaf in a greens state for much of the dry season even though growth is negligible.

The respective efficacies of the 'dry leaf strategy' versus the 'green leaf strategy' depend strongly on dry season rainfall and evaporation. Rain following leaf shed results in microbiological spoilage and decline in nutritive value and acceptability to cattle (Norman, 1967; McCown, Wall & Harrison, 1981b). Rain prior to leaf shed serves to prolong the period of green leaf retention. A recent study quantifies the geographic and temporal variation in dry season weather with regard to the dry leaf strategy (Wall & McCown, 1989). Fig. 8 shows the large variation among regions in the average contribution of legume in alleviating the 'nutritional drought'. In the far northern locations spoilage is of minor importance, and indeed much of the impetus for pasture legume research came from demonstration of the success of this strategy at Katherine (Station 7) (Norman, 1967).

Elsewhere, especially in the Townsville–Burdekin region, the average proportion of the dry season before spoilage occurs is modest. Fig. 9 shows the differences between the cumulative probabilities of having periods of sound legume greater than a given number of weeks for Katherine and Woodstock. Whereas at Katherine more than 18 weeks of sound legume can be expected in eight of ten years, at Woodstock, the same can be expected in only two in ten. Although in general, it can be presumed that the places least suited for a dry leaf strategy are better suited for a green leaf strategy, some areas appear to be poorly

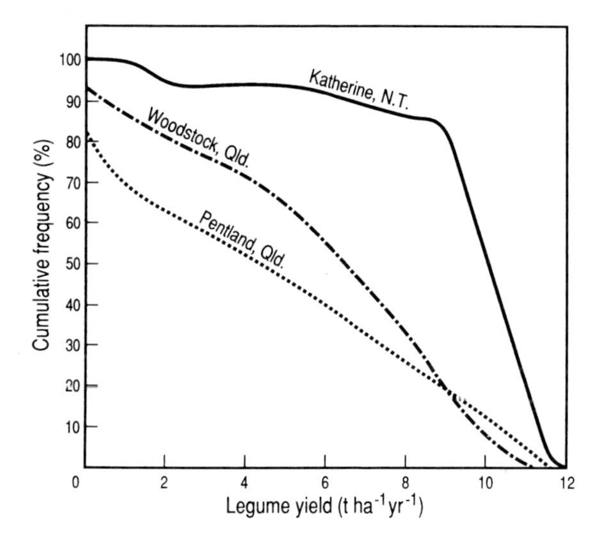

FIG. 7. Cumulative frequencies of *Stylosanthes hamata* cv. Verano yield at three stations. (From Williams & Probert, 1983.)

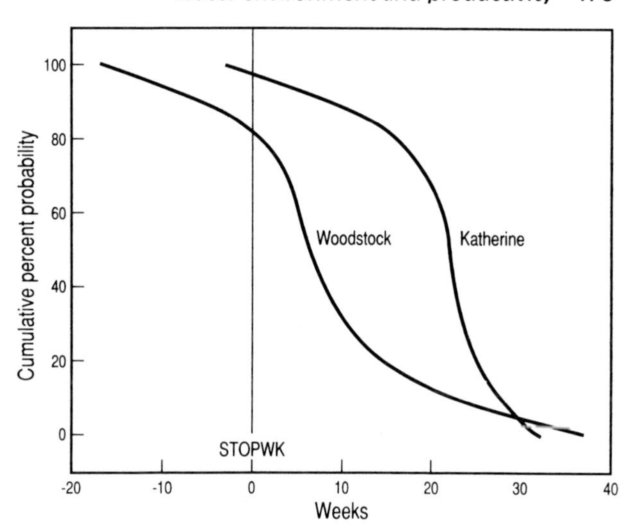

FIG. 9. Probability of exceeding a specified value of the time from the end of the Green Season of native grass pasture (STOPWK) until dry legume spoilage occurs.

endowed to support either reliably. In the Townsville–Burdekin and the West Kimberly regions, dry seasons are not dry enough for good dry legume preservation, but are often too long and dry for the strongly perennial legumes to make sufficient contribution of green leaf.

NON-CLIMATIC INFLUENCES ON THE WATER CONSTRAINT TO PRODUCTIVITY

Although savanna environments are usually characterized on the basis of the severity of the water constraint, any discussion of management of productivity of savannas without consideration of mineral nutrient constraints would be artificial. In the studies reviewed by Mott *et al.* (1986), the native pasture production potential predicted by their climatic analysis was often greater than measured herbage biomass production (Fig. 10). The potential production was estimated by computing the growth time which is not constrained by temperature, water or nutrient and multiplying this time by the potential growth rate of the grass species under conditions where nutrients are non-limiting.

The shortfall is most apparent on the massive sesquioxidic soils (Isbell, 1983) represented by the

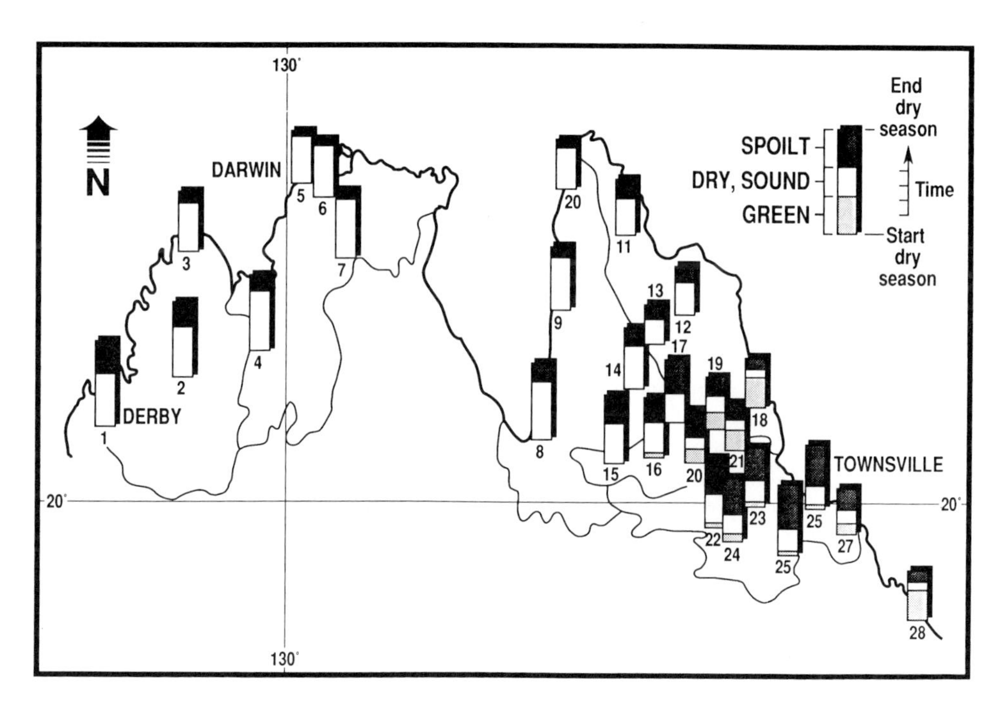

FIG. 8. Simulated states of leaf during the Dry Season at twenty-eight stations in northern Australia (mean length of Dry Season, mean times for which leaf is green, dry and sound, and spoilt). (From Wall & McCown, 1989.)

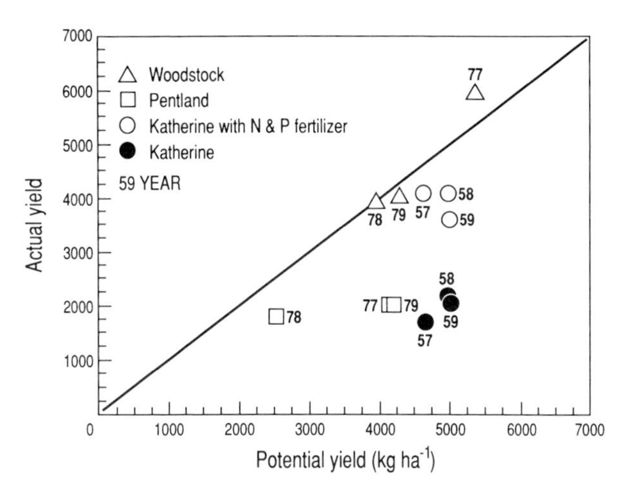

FIG. 10. Actual yield of native grass pastures versus simulated yield in years indicated, assuming nutrients non-limiting. (From Mott *et al.*, 1986.)

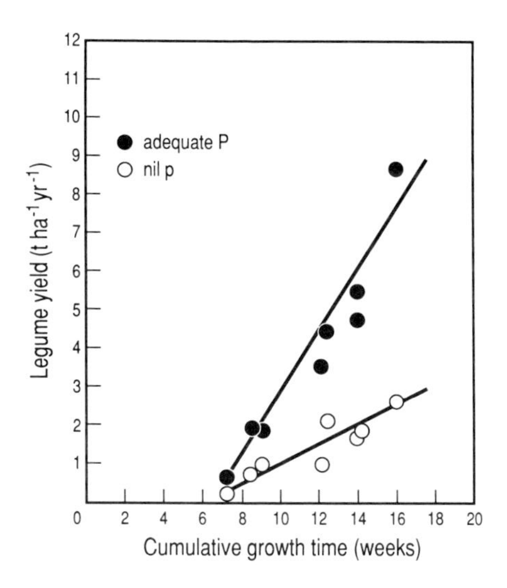

FIG. 11. The effect of P supply on the yield versus growth weeks relation for *Stylosanthes hamata* cv. Verano growth at 'Redlands' near Charters Towers on a red earth soil. (From Williams & Probert, 1983.)

Katherine and Pentland sites. At these sites the actual production of the native grasses falls well below potential production and shows small variation between the high and low rainfall years. When deficiencies of N and P were eliminated at Katherine, actual production increased to approximate the estimated potential production for the season. On the somewhat more fertile texture contrast soils at Woodstock, actual grass production approximated potential production implying that in these seasons grass yield was not nutrient constrained.

In a case where the vegetation was a sown legume (Williams & Probert, 1983) supply of adequate P to a severely P-deficient soil resulted in 3-4-fold increases in Caribbean stylo yield (Fig. 11). Yield is proportional to growth time (water supply) at both low and high P levels with difference in yield between the driest and wettest years exceeding 10-fold. However, the addition of P results in additional yield of only 1 t/ha in a season of 8 Growth Weeks, the same input results in an additional 6 t/ha in a season with 16 Growth Weeks.

In a grazing industry in which it is difficult to rapidly acquire or dispose of stock in response to feed supply, there is a problem of capitalizing on greatly improving production in good seasons except by increasing the normal stocking density. This, however, increases the risk of overstocking in dry years. In the nutrient-limited system, overgrazing in dry years is less of a problem because nutrient deficiencies keep normal stocking density low.

The effectiveness of rainfall is influenced by surface processes that affect the proportion of rain that enters the soil, i.e. rainfall interception, surface detention, overland flow, and infiltration. Much of the research specific to northern Australia has been conducted recently at study sites in little-disturbed savanna near Torrens Creek in central north Queensland. Water balance work by Williams & Coventry (1979, 1981) established that the water losses were by evapotranspiration and deep drainage alone with the implication that runoff was very small. Rainfall was shown to infiltrate and penetrate to depths in excess of 6 m in the deep massive sesquioxidic profiles. The eucalypt woodland extracted the water from similar depths during the dry season to yield a profile water store of approximately 480 mm.

In an undegraded state, the field saturated hydraulic conductivity is moderately high (50–80 mm h^{-1}) (Bonell & Williams, 1986; Williams & Bonell, 1988). These authors showed that under the high intensity rainfall of 60–120 mm/h^{-1} for storm durations of 20–30 min that, although Hortonion overland flow was rapidly generated, less than 2% was redistributed and this subsequently infiltrated. During high intensity rainfall the ponding depth increases, exposing higher points of the surface around tussocks to ponding. Bonell & Williams (1986) found that the conductivity of the higher surfaces near vegetation was often much greater than on the lower bare surfaces. The consequence of this was that the deeper surface ponding in the woodland the higher the effective hydraulic conductivity of the land surface. The tendency of runon to equal runoff is shown in Fig. 12. Here we see the net change in overland flow at locations down the hillslope being quite small and essentially cancelling each other out.

Water entry is sensitive to disturbance at the soil surface in these savanna woodlands. Overgrazing alters the hydrology by first reducing infiltration by reducing the surface roughness and the hydraulic resistance of tussock vegetation. Death of vegetation can lead to the formation of 'scalds' whose permeability is very low (Mott, Bridge & Arndt, 1979). With similar soil, slopes and rainfall intensities as studied by Williams & Bonell (1988) but for heavily utilized sown pasture on land previously cropped, Ive *et al.* (1976) found upwards of 37% of rainfall became overland flow which was lost as runoff. Preliminary results from the Croplands Erosion Research Project at Douglas Daly Experimental Farm also show the close relation of runoff and soil erosion to degree of disturbance in this savanna zone (Mohammed Dilshad, pers. comm.).

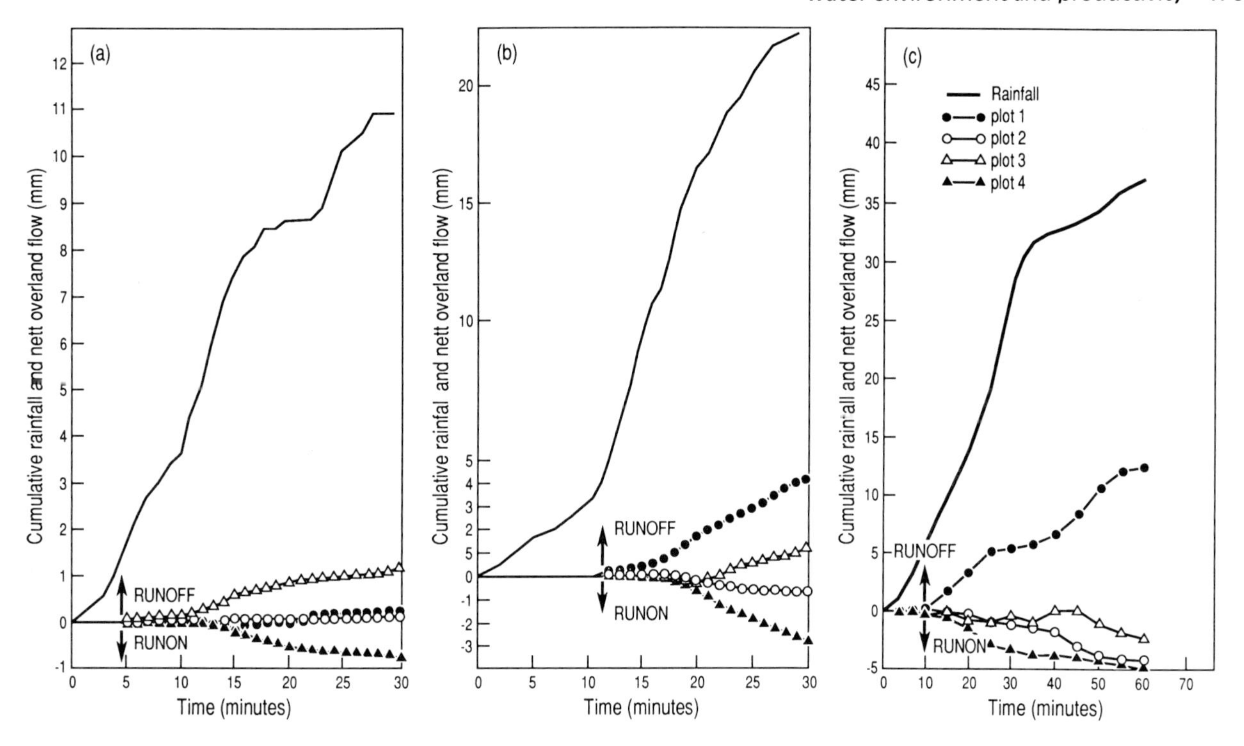

FIG. 12. Cumulative rainfall and net overland flow as a function of time for three storms (a) 10.8 mm, (b) 22.3 mm, and (c) 38.0 mm in Eucalypt Woodland on red earth soil near Torrens Creek. (From Williams & Bonell, 1988.)

CONCLUSION

Water deficits that severely limit plant growth are normal in the savanna zone, and an extensive beef industry which is adapted to this environment has existed in northern Australia for over 100 years. In recent years a revolution has been taking place that greatly reduces the inherent dependency of animal production on soil water supply via green grass. The first step has been the infusion of *Bos indicus* germplasm, resulting in cattle with greater tolerance of environmental stresses including the poor nutritional regime. The second step has been the widespread adoption of feeding non-protein nitrogen supplements in the dry season, which results in a further increase in herd productivity and a more complete utilization of dry grass herbage. Although not widely adopted, the sowing of legumes can greatly increase carrying capacity and individual production performance.

The recent succession of years with below-average rainfall has shown up the increased vulnerability of the savanna lands to overgrazing and soil erosion as a result of these technologies which enable more complete utilization of herbage. Although with a continuing cost-price squeeze, such intensification is necessary for continued economic viability, urgent changes in management that prevent ecologically damaging increases in herd size, e.g. greater turnoff of cows, are required. Much remains to be learned about management of the more intensively utilized pastoral ecosystem, but there seems to be little prospect for a return to the traditional low input/low productivity enterprise in spite of the appeal of its comparative ecological stability.

REFERENCES

Bonell, M. & Williams, J. (1986) The two parameters of the Philip infiltration equation: their properties and spatial and temporal heterogeneity in a red earth of tropical semi-arid Queensland. *J. Hydrol.* **87**, 9–31.

Edye, L.A. (1987) Potential of *Stylosanthes* for improving tropical grasslands. *Outlook on Agric.* **16**, 124–130.

Fitzpatrick, E.A. & Nix, H.A. (1970) The climatic factor in Australian grassland ecology. *Australian grasslands* (ed. by R. M. Moore), pp. 3–26. Australian National University Press, Canberra.

Hunter, R.A., Siebert, B.D. & Breen, M.J. (1976) Botanical and chemical compositon of the diet selected by steers grazing Townsville Stylo-grass during a liveweight gain period. *Proc. Aust. Soc. Anim. Prod.* **11**, 457–460.

Isbell, R.F. (1983) Kimberley–Arnhem–Cape York (III). *Soils: an Australian viewpoint.* pp. 189–199. CSIRO/Academic, Melbourne/London.

Ive, J.R., Rose, C.W., Wall, B.H. & Torssell, B.W.R. (1976) Estimation and simulation of sheet run-off. *Aust. J. Soil Res.* **14**, 129–138.

McCown, R.L. (1973) An evaluation of the influence of available soil water storage capacity on growing season length and yield of tropical pastures using simple water balance models. *Agric. Meteor.* **11**, 53–63.

McCown, R.L. (1980) The climatic potential for beef cattle production in tropical Australia. I. Simulating the annual cycle of liveweight change. *Agric. Systems*, **6**, 303–317.

McCown, R.L. (1982) The climatic potential for beef cattle production in tropical Australia. IV. Variation in seasonal and annual productivity. *Agric. Systems*, **8**, 3–15.

McCown, R.L., Gillard, P., Winks, L. & Williams, W.T. (1981a) The climatic potential for beef cattle production in tropical Australia. II. Liveweight change in relation to agro-climatic variables. *Agric. Systems*, **7**, 1–10.

McCown, R.L., Wall, B.H. & Harrison, P.G. (1981b) The influence of weather on the quality of tropical legume pasture during the dry season in northern Australia. I. Trends in sward structure and moulding of standing hay at three locations. *Aust. J. agric. Res.* **32**, 575–587.

Mott, J.J., Bridge, B.J., & Arndt, W. (1979) Soils seals in tropical tall grass pastures of northern Australia. *Aust. J. Soil Res.* **30**, 483–494.

Mott, J.J., Williams, J., Andrew, M.H. & Gillison, A.N. (1986) Australian savanna ecosystems. *Ecology and management of the world's savannas* (ed. by J. C. Tothill and J. J. Mott), pp. 65–82. Australian Academy of Science.

Norman, M.J.T. (1967) The critical period for beef cattle grazing standing forage at Katherine, N.T. *J. Aust. Inst. agric. Sci.* **33**, 130–132.

Wall, B.H. & McCown, R.L. (1989) The influence of weather on the quality of tropical legume pasture during the dry season in Northern Australia. IV. Geographic variation in risk of spoilage of standing hay. *Aust. J. agric. Res.* **40**, 579–90.

Williams, J. & Bonell, M. (1988) The influence of scale of measurement on the spatial and temporal variability of the Philip infiltration parameter – an experimental study on an Australian savannah woodland. *J. Hydrol.* **104**, 3–22.

Williams, J. & Coventry, R.J. (1979) The contrasting soil hydrology of red and yellow earths in a landscape of low relief. *Hydrology of areas of low precipitation symposium, Canberra, 1979*, pp. 385–395. IAHS-AISH publication No. 128.

Williams, J. & Coventry, R.J. (1981) The potential for groundwater recharge through red, yellow and grey earth profiles in central north Queensland. *Proceedings of the Groundwater Recharge Conference, James Cook University of North Queensland, 1980*, pp. 169–181. Australian Water Resources Council Conference Series No. 3, Canberra.

Williams, J. & Gardener, C.J. (1984) Environmental constraints to growth and survival of *Stylosanthes. The biology and agronomy of Stylosanthes* (ed. by H. M. Stace and L. A. Edye), pp. 181–201. Academic Press, Sydney.

Williams, J. & Probert, M.E. (1983) Characterization of the soil–climate constraints for predicting pasture production in the semi-arid tropics. *Research to resolve selected problems of soils in the tropics*, pp. 61–75. Proceedings of the International Workshop on Soils, Townsville, 1983, ACIAR Proceedings Series No. 2.

19/Philippine grasslands: liveweight gains in cattle and buffaloes, with and without introduced legumes

F. A. MOOG *Department of Agriculture, Visayas Avenue, Diliman, Quezon City, Repubic of the Philippines*

Abstract. Grasslands in the Philippines have been reduced from 11% of the total land area to 6% over the past few decades, mainly due to the conversion of these ecosystems into cropland, including pasturage. Because the soils are poor in nutrients, the introduction of legumes has been used for pasture improvement. The liveweight gains of cattle and buffaloes are different in various regions of the Philippines, and depend on both the rainfall of the region and the introduced legume. In some areas gains of livestock on improved pastures were up to 3 times that on native pastures.

Key words. Grasslands, introduced legumes, cattle production, buffalo, Philippines.

INTRODUCTION

Grasslands in the Philippines have been traditionally referred to as 'cogonal' lands mainly because *Imperata cylindrica* (L.) Raeusch, locally called 'cogon', is the predominant species. In fact Philippine grasslands comprise four community types (Sajise *et al.*, 1976). These are the *Imperata cylindrica* grasslands and those dominated by *Themeda triandra* Forsskal, *Chrysopogon aciculatus* (Retz.) Trin. or *Capillipedium parviflorum* (P.Br.) Stapf. Each community type is characterized by a particular dominant species, but *Paspalum dilatatum* Poir. and *Fimbristylis* spp. occur generally in most of the communities.

For several decades, these grasslands, taken together, have been considered to comprise over 3 million hectares or about 11% of the total land area of the Philippines. However, a recent (1987) report shows that grasslands comprise only 6.1% of the country's land area (Table 1). As a consequence of population pressure, former grassland and bushland areas have been converted into cropland (including pasturage), with the result that now cultivated regions mixed with bushland and grassland occupy 10.1 million hectares or 34.2% of the land area.

TABLE 1. Current land use in the Philippines.

	Area (km^2)	Percentage
Grassland	18,131	6.13
Cultivated area mixed with bushland and grassland	101,143	34.24
Pine forest	812	0.27
Mossy forest	2,455	0.83
Dipterocarp forest	24,344	8.24
Mangrove vegetation	1,493	0.50
Cultivated and other areas in forest	307	0.10
Coconut plantation	11,327	3.83
Other plantations	908	0.31
Cropland mainly cereals and sugar	43,922	14.87
Cropland mixed with coconut plantation	37,480	12.69
Cropland mixed with other plantations	3,648	1.23
Fishponds derived from mangrove	1,971	0.67
Other fishponds	102	0.03
Eroded areas	6	<0.01
Quarries	88	0.03
Riverbeds	816	0.28
Other barren land	104	0.03
Marshy areas	1,035	0.35
Lakes	2,051	0.69
Built-up areas	1,317	0.44
Other	42,056	14.24
Total	295,546	100.0

Modified from: NAMRIA – Department of Environmental and Natural Resources (DENR), 1987.

INTRODUCTION OF LEGUMES IN GRASSLANDS

Philippine grasslands are mostly rolling to hilly, and generally have nutrient-poor soils. The introduction of legumes

TABLE 2. Liveweight gains on *Imperata* and *Imperata*/legume pastures.

Location (animal)	Pasture	Stocking rate	ADG (kg)	Live weight gain per head	Live weight gain per hectare
Masbate[1] (cattle)	*Imperata*	0.5	0.12	43.2	21.6
	Imperata	1.0	0.07	26.6	26.6
	Imperata/stylo	1.0	0.32	116.6	116.6
	Imperata/centro	1.0	0.25	91.8	91.8
Bukidnon[2] (cattle)	*Imperata*	1.0	0.21	77.4	77.4
	Imperata/centro	1.0	0.26	94.1	94.1
Bohol[3] (carabao)	*Imperata*/*Themeda*	0.5	0.24	85.4	42.7
	Imperata/*Themeda*	1.0	0.22	78.9	78.9
	Imperata/*Themeda*/stylo	0.5	0.35	127.0	63.0
	Imperata/*Themeda*/stylo	1.0	0.25	92.2	92.2
Bohol[4] (carabao)	*Imperata*	0.75	0.22	68.1	51.1
	Imperata/*Leucaena*	1.5	0.35	111.9	167.9
	Imperata/*Leucaena*	2.0	1.28	87.4	174.8

[1]Siota *et al.* (1977); [2]Magaden (1974); [3]Castillo *et al.* (1987); [4]Moog *et al.* (1981).

has been shown to be the quickest and most practical approach to large-scale improvement of these areas. Overseeding is done after burning or heavy grazing of non-accessible areas or after disking the more accessible areas. Shrub or tree legumes can be introduced by transplanting seedlings during the rainy season.

Most grassland soils are acidic and deficient in phosphorus, and therefore present a problem for introducing legumes. Establishment of legumes can be improved by liming and phosphorus fertilization. Alternatively, the Stylos (*Stylosanthes* spp.) have been found to be tolerant of acidic and low phosphorus soils.

ANIMAL LIVEWEIGHT GAIN

Introduced legumes in pasture have been found to increase productivity by providing a greater quantity of higher quality herbage for grazing animals even during the dry season. The liveweight gains of cattle and buffaloes are

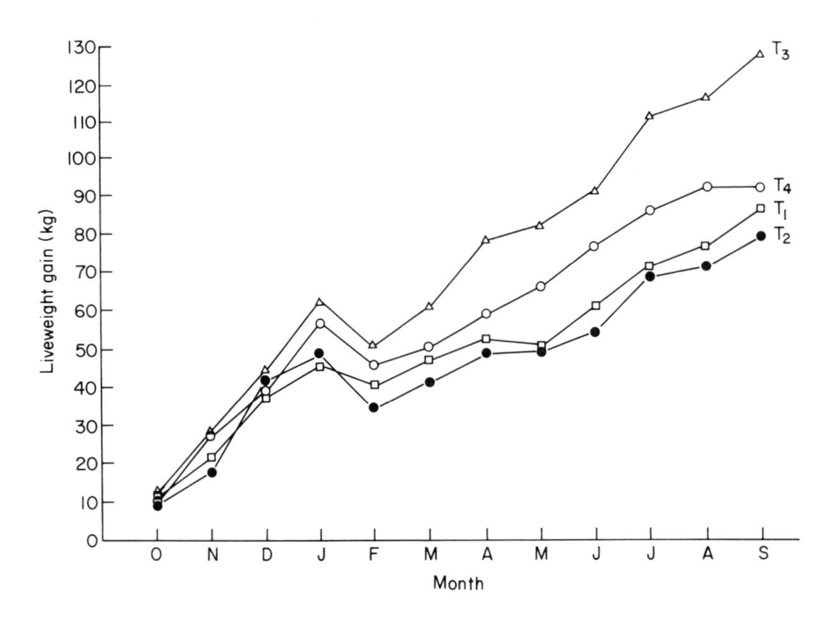

FIG. 1. Cumulative liveweight gains per head of buffaloes grazing native and native/stylo pastures at different stocking rates (from Castillo *et al.*, 1987). T_1 = native pasture at 0.5 a.u./ha; T_2 = native pasture at 1.0 a.u./ha; T_3 = native/stylo at 0./5 a.u./ha; T_4 = native/stylo at 1.0 a.u./ha.

different in different locations in the Philippines (Table 2). In Masbata province (Siota *et al.*, 1979), *Imperata* pastures produced 22–27 kg ha^{-1} yr^{-1} at stocking rates of 0.5 and 1.0 animals per hectare respectively, while *Imperata*/Centro and *Imperata*/Stylo pastures at 1.0 animals per hectare produced 92 to 117 kg ha^{-1} yr^{-1}, respectively. Periodic herbage samples taken during the grazing period indicated that the *Imperata*/legume pastures could still support an additional 0.5 animals per hectare. Animals on *Imperata* pastures lose weight during the dry season, but those on *Imperata*/ legume pastures do not.

In Bohol province, Castillo *et al.* (1987) reported that buffaloes on both *Imperata–Themeda* and *Imperata–Themeda*/Stylo pastures at stocking rates of 0.5 and 1.0 animals per hectare lose weight during the dry season (Fig. 1). Liveweight gains obtained from native pasture with stylo were higher than those on pasture without stylo. In the same location, Moog *et al.* (1981) working with *Imperata*/ *Leucaena* pasture, observed that buffaloes did not lose weight during the dry season, even at slightly higher stocking rates than those used in the other experiment. Liveweight gains of buffaloes grazed on *Imperata*/*Leucaena* pastures were 3 times those of animals on *Imperata* pastures because of the capacity of the former to support more animals due to higher yield and quality of herbage, with *Leucaena* comprising over 50% of the herbage on offer throughout the grazing period.

In Bukidnon province (higher elevation with higher rainfall more evenly distributed than in Masbate and Bohol), liveweight gains of cattle grazed on *Imperata* pastures were higher than those observed elsewhere (Table 2).

FUTURE OF PHILIPPINE GRASSLANDS

Grassland plays, and will continue to play, a major role in the development of the Philippine livestock industry. The country's growing population will increase pressure to transform livestock production in these areas from extensive to semi-intensive or intensive systems. The introduction of legumes will be an intermediate phase in attaining the above objective.

REFERENCES

Castillo, A.C., Macalandag, E.C., Moog, F.A. & Salces, C.B. (1987) Pasture and carabao (*B. bubalis*) productions from native and native/stylo pasture at two stocking rates in Bohol province. Abstracts: Paper presented at the 25th PSAS Annual Convention, PSCARRD, Los Baños, Laguna.

Magadan, P.B. (1974) Beef production on native *Imperata cylindrica* and Para grass (*Brachiaria mutica*) pastures. M.S. thesis, University of the Philippines, Los Baños.

Moog, F.A., Castillo, A.C., Tibayan, R.G. & Tombocon, H.P. (1981) Carabeef production on native and native/ipil-ipil pastures at different stocking rates. *Philippines J. Animal Industry*, **36**, (1–4), 24–33.

Sajise, P.E., Orlido, N.M., Lales, J.S. Castillo, L.C. & Atabay, R. (1976) The ecology of Philippine grasslands: 1. Floristic compostion and community dynamics. *Philippines J. Animal Industry*, **59**, 317–334.

Siota, C.M. Castillo, A.P. Moog, F.A. & Javier, E.Q. (1977) Beef production on native, native stylo and native centro pastures. Cited in Moog *et al.* (1981).

Valenzuela, F.G., Moog, F.A., Tibayan, R.G. & Tombocon, H.P. (1982) Productivity of guinea grass/cock style mixed pasture at different stocking areas. *Philippiness J. Animal Industry*, **37**, (1–4), 1.

20/Australia's northern savannas: a time for change in management philosophy

W. H. WINTER *Cunningham Laboratory, Division of Tropical Crops and Pastures, CSIRO, 306 Carmody Rd, St Lucia, Queensland 4067, Australia*

Abstract. Much of Australia's northern savannas have undergone substantial change and degradation in the century of grazing by sheep and cattle. The extent of change is variable – from relatively minor changes in perennial grass species to more serious and irreversible woody weed invasion and sheet and gully erosion, depending upon characteristics such as soil type, original species and year–year rainfall variability. The root causes of this degradation have been the increase in stock numbers in response to the cost/price squeeze in recent years and the uncontrolled grazing of stock.

Concern for the long-term effects of these changes have, until recently, only been of interest to the urban population when the effects of change have affected their lifestyles by, for instance, polluting water supply dams. Nowadays graziers and scientists alike are faced with the demands of the urban populations to develop sustainable systems of land management. Such systems will need to offer higher profit levels than current management methods otherwise they will not be adopted and the conflict between rural landholders and urban philosophy will worsen.

A system of management of tropical tallgrass savanna is described in which the grazing area is rotationally burnt on an annual basis and stock are supplemented to overcome nutrient limitations. The rotational burn has the dual advantages of attracting animals away from previously heavily grazed areas and so halting pasture decline and of improving the quality and accessibility of the grasses in the late dry and early wet seasons. Annual animal liveweight gains are more than doubled over those obtained under normal commercial management. This is the type of system which may have some chance of adoption and which will prove to be sustainable.

Structural adjustment in the industry is required. Fortunately some grazier groups recognize the problems and are responding favourably to the new demands for sustainability. Unfortunately, there is a paucity of long-term research from which to draw information.

Key words. Degradation, grazing, savanna, fire, supplements, Australia.

INTRODUCTION

Biologists must recognize that land use, and ultimately land condition, is affected as much by social and economic conditions as by the subjective attitudes of land users toward the land.

Research and development undertaken by individuals and organizations should consider these factors which must ultimately be taken into account when providing advice to land users.

There is little doubt that the grazing lands of northern Australia have been changed dramatically in the last 100 years. There is some strong opinion that this process has been accelerated in recent years, particularly in the more closely settled areas on North Queensland, but there is little supportive evidence. It may well be that we are now more aware of the problem because of our ability to document changes more easily on a grand scale from one location.

Until the past 20 years or so the pastoral industries of northern Australia have been of little significance to the national economy. Even today, when about 20% of the national beef herd is north of the Tropic of Capricorn, the region still contributes less than 5% of the national agricultural wealth (ABS, 1988). Concern for the state of the land has been in proportion to the economic value of the region because the effects of land management have not affected people other than those living in those regions. However, in recent years the Australian community had become increasingly aware of its environment and the management of resources has become a social, political and economic issue. In some instances this awareness is out of balance with the resource value – whether from lifestyle or exploitive use. At present there is little attention given to the north-west, but in the more populous north-east the condition of the land, particularly its degradation, is becoming more of a social issue because of its effects on the non-rural population, e.g. sedimentation of the water supply dams. This sequence of events is similar to that pertaining to the recognition of the importance of soil erosion in developing countries (Blaikie, 1985).

Although there will inevitably be disagreements about the relative roles of human intervention and natural processes as causes of change, this ultimately will be of little consequence as the 'national wish' will be imposed on the system to bring about further change, presumably for the better. Long-term documentation of change which can be interpreted to understand the biology of the systems, and then to aid the development of plans for future management, are in short supply. Nevertheless, the 'state' will become involved at a number of levels – political, scientific, and at the district or regional level to bring about the desired changes. The livelihood of the land-user may be affected through such things as changes in conditions of land use, taxation laws and access to credit, and the value of production from the industry may be reduced, at least in the short-term. The possibility of conflict of interest between the 'state' and the current land users is highly likely. However, landholders will have to accept some degree of external control if urban taxpayers' money is used to correct and halt land degradation.

I believe that if recommendations for change to current land management are to have any chance of success they must be couched within an economic framework which will minimize this potential conflict. Joint involvement of the 'state' and land users in setting the guidelines and timetable for such a programme is essential.

The aim should be to improve the 'quality' of livestock production embracing the philosophy of sustainable land use and a production system which is consistent with it.

AUSTRALIA'S SAVANNAS

The savannas of northern Australia have been adequately described by Mott *et al.* (1985). Beef cattle production is the predominant industry and utilization of the native forages is influenced by a number of non-biotic factors such as accessibility, water supply, and proximity to markets, as well as the biotic factors such as pasture type and seasonal conditions. The most intensively developed area is the tropical tallgrass region of the NE Queensland, much of which forms the catchment of the Burdekin River. Cattle populations are also high on the midgrass areas of the base of Cape York Peninsula in Queensland, in the Northern Territory and in the Kimberleys in Western Australia. Generally, the soils in these regions are of relatively low nutrient status and, after water supply, the potential grassland productivity is considered to be limited by deficiencies of nitrogen and/or phosphorus supply (Mott *et al.*, 1985).

SAVANNA CONDITION

Signs of disturbance and degradation are most evident in the catchments of the major rivers. In the Kimberleys, the Fitzroy River catchment has experienced uncontrolled grazing of first sheep then cattle for over a century. Stock have concentrated on the river frontages causing loss of vegetation then loss of topsoil. Over 30% of the 10,000 km^2 catchment was degraded to poor or very poor condition in 1976 (Payne *et al.*, 1978) and the situation has deteriorated in the ensuing decade. Revegetation or reclamation of much of this land is improbable given the sodic nature of the exposed subsoils.

The condition of the Ord River catchment first came to prominence when a dam was proposed in the 1960s. The annual sediment load of 24 million tonnes threatened the life of the dam (Anon., 1976), so reclamation and controlled grazing of some of the catchment was implemented. Despite the success of some of the revegetation work the sediment loads in the river are little changed as the site of erosion has shifted from widespread sheet erosion to gully and stream-bank erosion, due to the continued high run-off from the landscape (Winter, 1987a).

A large proportion of the most valuable land within the 120,000 km^2 Victoria River catchment has been degraded to some degree. While only 3–5% of the land is eroded (Condon, 1986), there is widespread replacement of the preferred species with less palatable species such as *Aristida* and the thicket forming shrubs *Calotropis procera* (Wild.) R.Br. ex Aiton f. and *Acacia farnesiana* (L.) Wild. There is a paucity of factual information on the condition of this valuable resource and only meagre data on the biology of some of the species (Mott & Andrew, 1985) and the potential recovery of a degraded system (Foran, Bastin & Hill, 1985).

This northwestern region of Australia has the benefit of reasonably reliable wet season rainfall from the viewpoint of native grass production (McCown, 1981). This reliability is not shared by the tropical tallgrass region of the Burdekin catchment in NE Queensland where, although the median length of the wet season is longer, and that of the dry season shorter, the variation about that median is greater than in the NW (McCown, 1982). In this region increased grazing pressures and fire frequency caused a shift in dominance from *Themeda triandra* Forsskal to *Heteropogon contortus* (L.) Beauv. ex Roemer and Schultes within the last 100 years (Shaw & Bissett, 1955), while in the last 15 years *Bothriochloa petusa* (L.) A. Camus has replaced *Heteropogon* in many situations. The recent run of 5 years of dry conditions has seen large areas completely denuded, and what rain has fallen has caused substantial sheet and gully erosion.

SAVANNA MANAGEMENT

In every situation the root causes of degradation can be traced to uncontrolled overgrazing. In these circumstances animals are able to graze favoured areas until the original vegetation is removed. The processes of species replacement and accelerated soil losses are then set in train. The concept of a safe grazing pressure for an uncontrolled situation has little meaning, as even if stock numbers are reduced stock will still concentrate in, and degrade, favourable areas.

Graziers have used uncontrolled grazing because of its economic advantages. The costs of land degradation have not been included in the long-term budget and in certain circumstances there have been financial inducements to overgraze, e.g. drought subsidies. Generally short- to medium-term economic considerations dominate the style of property management and the main ways of combatting the economic cost/price squeeze have been to increase herd size and/or increase the efficiency of the herd by using supplements. Both of these strategies increase the utilization of the forage resource. The adoption of Brahman cattle has been blamed for the apparent acceleration in degradation, as death rates are lower in poor years when compared with Shorthorns they replaced, and they have a wider grazing range. While this is true, Brahmans should only be seen as a more efficient tool in the hands of the land managers rather than the root cause of the problem.

Seasonal variability has also contributed to the process of degradation. In the NW the poor quality of the pastures during the dry season has protected the pastures to some degree so that degradation has primarily been associated with the better country first and then a later spread to less favoured country. The situation is different in the NE where 'It would seem that although production in this region is extraordinarily unpredictable and catastrophies relatively frequent, the probability of having a good year is sufficiently high to more than compensate' (McCown, 1982). In this sense McCown was referring to monetary compensation (which is aided by drought relief fund availability) but not biotic compensation. The subject of stability and resilience of the tropical savanna species is discussed thoroughly by Hodgkinson & Mott (1987) and it is clear that these plants cannot withstand continuous heavy defoliation. It is also worth noting that there is a considerable economic lag during a drought, i.e. in a drought lasting 2 years income may not be substantially reduced in those years as the marketable stock are generally the least affected; the effects of lower fertility and poorer calf growth are not recognized financially until the following years. This effect probably has a dampening effect on the grazier's management response to drought. The consequences for the pastures are all too often neglected.

In the more intensively developed regions of NE Australia land prices have risen to the extent that the removal or killing of trees and/or the replacement or augmentation of native vegetation with grasses and legumes is considered as an option to the purchase of more land when increases in income are required. Pasture sowing is occurring at a rate of about 3000 km^2 annually in the higher rainfall regions where the risks associated with pasture establishment are less and surety of annual production is greater. There is little widescale pasture sowing in the NW because of the different economic and management status of the industry. Animal production from improved pastures accounts for 10–15% of the total, with the remainder from native pastures. The remainder of this discussion will concentrate upon improving animal production from the latter resource *in situ*.

Property management is generally oriented toward the cattle, with little concern for the pastures. If animals are moved from one area to another the prime reason is likely to be to reduce mustering costs or to have the animals more accessible for potential markets. Similarly, if pastures are burnt the purpose will be to make mustering easier or to provide a firebreaker rather than for pasture management. Naturally, economic concerns predominate in this process and in many instances practices which would be beneficial to the landscape are contrary to good economic management, e.g. fencing off the highly favoured but sensitive riparian areas is not only costly but generates the need for artificial waters. Recommendations aimed toward developing sustainable systems will ineviably contain some capital and variable costs and, unless they also result in higher profitability, they will have little chance of adoption.

A POTENTIAL MANAGEMENT SYSTEM

The perennial tallgrass pastures at Katherine (330 km southeast of Darwin) in north-west Australia are known to degrade through the process of plant death and scalding (Mott, 1987). Stocking rates are low but animals concentrate their grazing and create patches where defoliation is severe. If the animals are not removed from the pasture, grazing will continue in these patches until the perennial plants die and, since recruitment rates are so low, these areas become bare and scalded. Cattle production from these pastures is low regardless of whether they are in poor or good condition. The primary limitation is the quality of the forage for most of the year.

Thus, when considering options for the development of a sustainable production system at Katherine the two major considerations were firstly to develop a system of pasture management which would break the sequence leading to plant death, and secondly to improve cattle growth by reducing the nuritional limitations through supplementation

TABLE 1. The mean ($n = 4$ years) liveweight gain of Brahman-cross steers for each seasonal period when grazing native perennial grasses. From Winter (1987b).

Treatment	Per season (g head day^{-1})				Per year
	Early wet	Late wet	Early dry	Late dry	(kg head yr^{-1})
Control	318^{a*}	287^{a}	−144^{ab}	−292^{a}	21
Salt	346^{a}	385^{b}	−209^{b}	−211^{ab}	47
Fire	401^{a}	160^{c}	−106^{a}	−103^{c}	46
Salt+Fire	785^{b}	345^{ab}	−219^{b}	−135^{bc}	66
Salt+Fire+ Supplement	863^{b}	575^{d}	70^{c}	−94	134
LSD P<0.05	97	71	77	94	—
No. days	85	105	88	65	—

* Within columns different letters denote significant differences (P<0.05).

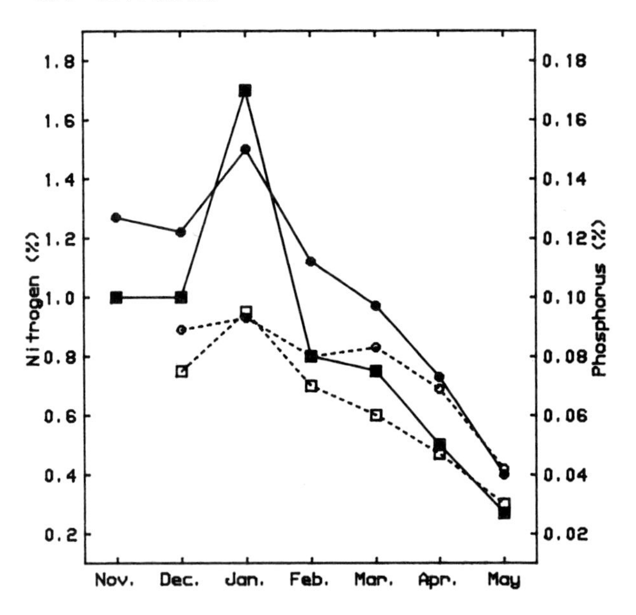

FIG. 1. The effect of burning the pasture during the dry season on the mean concentrations of N (○) and P (■) in the native perennial grasses during the wet season. Burnt = closed symbols; unburnt = open symbols. Data from Winter (1987).

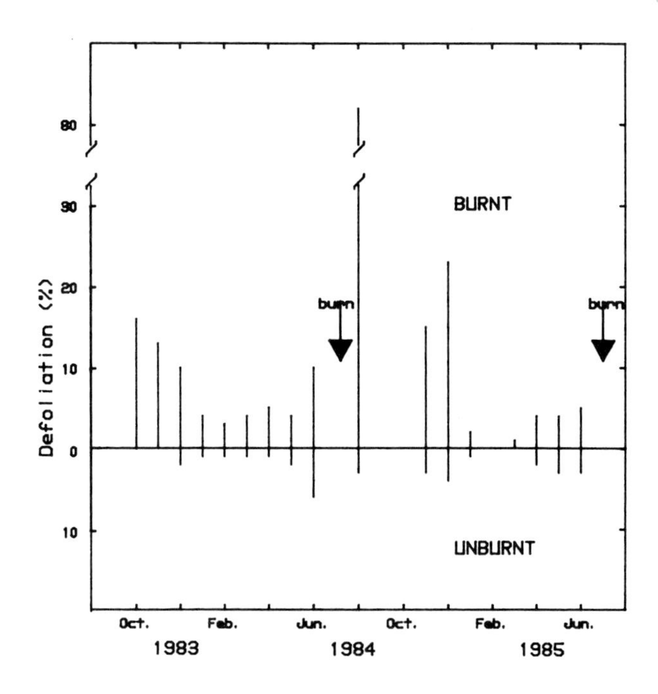

FIG. 2. Cattle grazing behaviour expressed as mean defoliation (%) of native grasses in recently burnt and unburnt halves of the paddocks. Data from Andrew (1986).

and possibly through pasture management. Two additional elements also needed consideration, i.e. that pastures were regularly burnt by graziers and that sodium was likely to be a limiting nutrient for cattle in addition to P and N.

An experiment with five treatments was conducted over 4 years; the treatments were (i) control, (ii) Na supplement, (iii) half of the pasture burnt annually in June–July, (iv) Na supplement and pasture burning, (v) Na supplement plus pasture burning plus P and N supplement year-round. Details of this work are presented by Winter (1987b).

The mean liveweight gains of cattle from these treatments during seasonal periods of the year are presented in Table 1. Access to previously burnt pasture was advantageous during the late dry season (LDS), which includes the period of early storms, and early wet season (EWS). The EWS response was, however, dependent upon the supply of Na which was the prime limitation during that period. The response to burning was undoubtedly due to the better quality of the forage in the previously burnt area (Fig. 1) and the greater accessibility of the forage to the grazing animal. In the previously burnt pasture all the tillers were from the crown and were easily prehended, while in the unburnt plants early growth was from secondary tillering or was less accessible among the old tillers. Diet selection was not measured in this experiment but elsewhere it has been shown that cattle continued to prehend stem material for some time after the early rains during the transitional period between the LDS and EWS on unburnt pasture (McCown & McLean, 1983).

Cattle were attracted to the area of burnt pasture both immediately after the burn and throughout the following wet season (Fig. 2). Regrowth of pasture during the dry season from burning in June/July was meagre, usually less than 150 kg/ha, but of higher quality than dry feed. Animal production was similar on both pastures during the early dry season (EDS), so it seems likely that there was a trade-off between quality and quantity, so that nutrient intakes were similar on both pastures. The attraction of animals to the burnt areas had substantial animal production benefits during the LDS and EWS, as mentoned above, but of equal importance was the modification of grazing pressure on the pasture. Heavily grazed patches which had been formed in the previous wet season were abandoned in favour of the new post-burn regrowth and this spelling enabled complete recovery of the individual plants. In the next cycle of burning these areas burnt as well as the surrounding pasture and reformation of a heavily grazed patch at a particular site in the next year was a chance event (Andrew, 1986).

The nutritional limitations of the native grasses were most apparent during the late wet season (LWS) and EDS. Liveweight grains were doubled with a complete supplement during the LWS and moderate losses turned to moderate gains during the EDS.

The net effect of rotational burning, provision of salt blocks and feeding a complete mineral supplement on annual growth of steers was substantial. While it is difficult to establish an industry 'control' production level, given that pastures are burnt on an erratic basis and animals might obtain salt from natural sources such as seepages, the complete supplement doubled the next best production level which would be the upper level of industry productivity. The effects on the pasture were two-fold. Firstly, the rotational burning broke the sequence of pasture degradation which results from patch grazing and so acted as a stabilizing influence. Secondly, supplementation undoubtedly increased the utilization of the native grasses by removing the previous nutritional limitations which acted to both lower the efficiency of digestion and to reduce intake.

Overall, this package has the potential to fulfill the requirements of new technology being introduced into an existing industry as, for instance, it required little change in existing management and capital outlay is minimal. It has the potential to improve pasture stability while at the same time increasing income to the land user. However, there are likely to be problems of implementation which will only come from testing the proposal at a realistic scale. The integration of fire and supplementation, or any other technological innovation, into a whole property management scheme, taking account of the needs of the various animal classes, the much greater geographic variability that is normally dealt with in formal experiments, and the economic realities which a land user faces, is a task that has not been widely addressed in northern Australia. This is the next logical progression of the work and will be conducted in conjuction with co-operating graziers in the next few years.

CONCLUSIONS

The grazing industry in north-west Australia in particular is at a cross-roads. It can no longer operate as a hunter-gatherer of near-feral stock as the state has already imposed the restriction of disease-free herds. This has resulted in a capitalization previously unknown and unneeded. The old method of increasing herd size and turn-off to meet new financial demands will no longer suffice – particularly as the area of land available is in most cases less than before the disease-free status was required. In the short term many of the marginal land users will try the old strategy, but inevitably they will fail if other changes are not made. Structural adjustment of the industry is inevitable, and the sooner the land users and the 'state' reach this conclusion and begin the process the better for the land.

REFERENCES

ABS (1988) *Livestock and livestock products*. Australian Bureau of Statistics, Australian Government Printer, Canberra.

Andrew, M.H. (1986) The use of fire for spelling tropical tallgrass pasture grazed by cattle. *Trop. Grassl.* **20**, 69–78.

Anon. (1976) *Ord Irrigation Project*. Australian Government Printer, Canberra.

Blaikie, P. (1985) *The political economy of soil erosion in developing countries*, pp. 89–95. Longman Scientific and Technical, Essex.

Condon, R.W. (1986) A reconnaissance erosion survey of part of the Victoria River District, N.T., Hassall and Associates, Canberra.

Foran, B.D., Bastin, G. & Hill, B. (1985) The pasture dynamics and mangement of two rangeland communities in the Victoria River district of Northern Territory. *Aust. Rangel. J.* **7**, 107–113.

Hodgkinson, K.C. & Mott, J.J. (1987) On coping with grazing. *Grazing-lands research at the plant–animal interface* (ed. by Floyd Horn, John Hodgson, John J. Mott and Ray W. Brougham), pp. 171–192. Winrock International, Arkansas.

McCown, R.L. (1981) The climatic potential for beef cattle production in tropical Australia. III. Variation in the commencement, cessation and duration of the green season. *Agric. Systems*, **7**, 163–179.

McCown, R.L. (1982) The climatic potential for beef cattle production in tropical Australia. IV. Variation in seasonal and annual production. *Agric. Systems*, **8**, 3–15.

McCown, R.L. & McLean, R.W. (1983) An analysis of cattle liveweight changes on tropical grass pastures during the dry and early wet seasons in northern Australia. 2. Relations to trends in the pastures. *J. agric. Sci., Camb.* **101**, 25–31.

Mott, J.J. (1987) Patch grazing and degradation in native pastures of the tropical savannas of northern Australia. *Grazing-lands research at the plant–animal interface* (ed. by Floyd Horn, John Hodgson, John J. Mott and Ray W. Brougham), pp. 153–162. Winrock International, Arkansas.

Mott, J.J. & Andrew, M.H. (1985) The effect of fire on the population dynamics of native grasses in tropical savannas of north-west Australia. *Proc. Ecol. Soc. Aust.* **13**, 231–239.

Mott, J.J., Williams, J., Andrew, M.H. & Gillison, A.N. (1985) Australian savanna ecoystems. *Ecology and management of the world savannas* (ed. by J. C. Tothill and J. J. Mott), pp. 56–82. Australian Academic Press, Canberra.

Payne, A.L., Kubicki, D.G., Wilcox, D.G. & Short, L.C. (1978) A report on erosion and range condition in the West Kimberley area of West Australia. *W.A. Dept of Agric Tech. Bull.* No. 42.

Shaw, N.H. & Bissett, W.J. (1955) Characteristics of bunch spear grass (*Heteropogon contortus* (L.) Beauv.) pasture grazed by cattle in subtropical Queensland. *Aust. J. agric. Res.* **6**, 539–552.

Winter, W.H. (1987a) Forage resource management in north-west Australia. *Herbivore nutrition research* (ed. by Mary Rose), pp. 219–220. Australian Society of Animal Production.

Winter, W.H. (1987b) Using fire and supplements to improve cattle production from monsoon tallgrass pastures. *Trop. Grassl.* **21**, 71–81.

21/RSPM: a resource systems planning model for integrated resource management

J. W. STUTH, J. R. CONNER, W. T. HAMILTON, D. A. RIEGEL, B. G. LYONS, B. R. MYRICK and M. J. COUCH *Department of Range Science, Texas A&M University System, College Station, Texas 77843-2126, U.S.A.*

Abstract. Computer-aided planning of grazed ecosystems is gaining more attention among scientists, action agency personnel, and consultants concerned with management of natural resources. An integrated system, RSPM (Resource Systems Planning Model) has been developed by the Ranching Systems Group at Texas A&M University for nationwide distribution in USDA–Soil Conservation Service field offices. The hierarchical planning system is targeted for service agency personnel working directly with the individual property owner or landholder. The strategic level subsystem allows the user to make an inventory of forage and animal resources, conduct enterprise bioeconomic optimization analyses and determine economic viability of long-term investments in land management technologies. Embedded expert systems allow for improved technology selection and assessment of managerial risks. The tactical subsystem allows analysis of forage and nutrient balance associated with scheduled use of paddocks by specified herds. Least cost effective nutrient mediation tactics may be determined to correct imbalances. On-site monitoring systems of forage supply and animal nutritional status are coupled with a hydrologic-based forage production model to form the foundation of the operational subsystem. Detailed information on RSPM is provided for each module of the various subsystems.

Key words. Decision support system, resource inventory, grazing planning system, nutritional mediation, investment analysis.

INTRODUCTION

With the advent of computer technology and the exponential expansion of information about natural resource management, interest in the use of computerized decision support systems (DSS) has increased. These systems assist managers in dealing with complex planning problems and in selecting appropriate technology. DSS in natural resources are similar to those in other businesses in that they are designed for a specific problem area, incorporate specified planning horizons and guide decision makers through a process of logical planning and technology selection (Taylor & Taylor, 1987).

Decision support systems are a blend of structured problem solving and analytical procedures solved by hard data and semi-structured procedures which may use heuristics as the primary information source (Negoita, 1985; Scifres, 1987). Heuristic information is obtained from expert opinion when precise data and/or algorithms are not available. Generally, DSS are models that help the user understand complex management environments and select feasible technologies to solve problems.

The purpose of this paper is to provide an overview of a hierarchical DSS which facilitates planning and analysis of grazed ecosystems. A brief overview of the entire system is provided first, then a more detailed description is provided for each system component.

SYSTEM OVERVIEW

The Ranching Systems Group (RSG) in the Range Science Department at Texas A&M University began development of a decision support system in 1986 to address managerial needs for the ranching industry in Texas. The system was recently adopted by USDA-Soil Conservation Service as the official planning tool for use with rangeland, grazable woodland, pastureland, grazable cropland and hayland (Fig. 1).

A hierarchical approach to resource management and decision support has been adopted. RSPM (Resource Systems Planning Model) has three levels of decision making (Fig. 2). The strategic level focuses on (1) the assessment of management goals, (2) forage and animal resource inventory, (3) enterprise analysis and optimization, and (4) long-term investment analysis of selected technology. The tactical level considers the allocation of resources for a specified period of time, usually 12 months. Since the primary users of grazed systems are domestic herbivores, the tactical level of RSPM supports stocking rate, scheduling and nutritional mediation decisions. Once the tactical plan has been implemented, RSPM supports operational

FIG. 1. USDA-Soil Conservation Service range conservationist Sid Brantly helps Florida ranchers test range and pasture planning using a decision support system developed by the Ranching Systems Group. Field transects, forage balance sheets and economic reports are computed, printed and delivered to the ranchers on site.

decisions, such as adjustments of forage and animal resources related to variable weather events and market conditions. Monitoring regimes in the system measure these variable conditions. While these regimes are considered most important at the operational level, they service the resource inventory subsystem as well.

Databases and applications of the system are divided into client-independent and client-dependent activities. Client-independent information is data that does not vary in a given management environment. When the uniqueness of a management environment must be considered in the development of data, it is considered a client-dependent activity in RSPM. Fig. 3 provides an organized view of the databases. The system is designed to provide the framework for planning in any localized situation. The system provides the structure, but the user must fill that structure with data.

It is important to note that RSPM is currently under extensive development and will continue to undergo revisions. The strategic level of RSPM will be distributed in 1989, followed by the tactical subsystem, and later by the operational subsystem. The system will use the C programming language for both MS-DOS and UNIX operating systems.

STRATEGIC LEVEL

Goal analysis

This portion of the system addresses the user's identified goals and the financial resources allocated to the user's needs. The primary focus is on hierarchical elements that include (1) retaining ownership, (2) setting standards of living, (3) minimizing risk of catastrophic losses, and (4) reinvesting profits. This subsystem helps identify the level of funds available for land management practices that may be analysed in other financial analysis subsystems.

Geographical Information System (GIS)

The forage inventory system will be linked to a GIS which allows the user to define layers of information about property boundaries, soils, elevation, ecological condition classes, woody plant cover, physical barriers, water locations and roads. The GIS will define the number of polygons, called response units, and calculate adjustment factors for defining stocking rates of given response units. The system will also determine optimum (biological and economic) placement of water locations and woody plant control in consideration of landscape features and various classes of domestic stock (cattle, sheep and goats). SCSGRASS is the primary GIS for the UNIX version and can be interfaced with the MS-DOS version of the system. However, RSG is currently developing a custom GIS to interface with the MS-DOS version of the system that is specifically designed for RSPM.

Plant attribute database

The user defines the common and scientific names and associated codes for species important in the area where

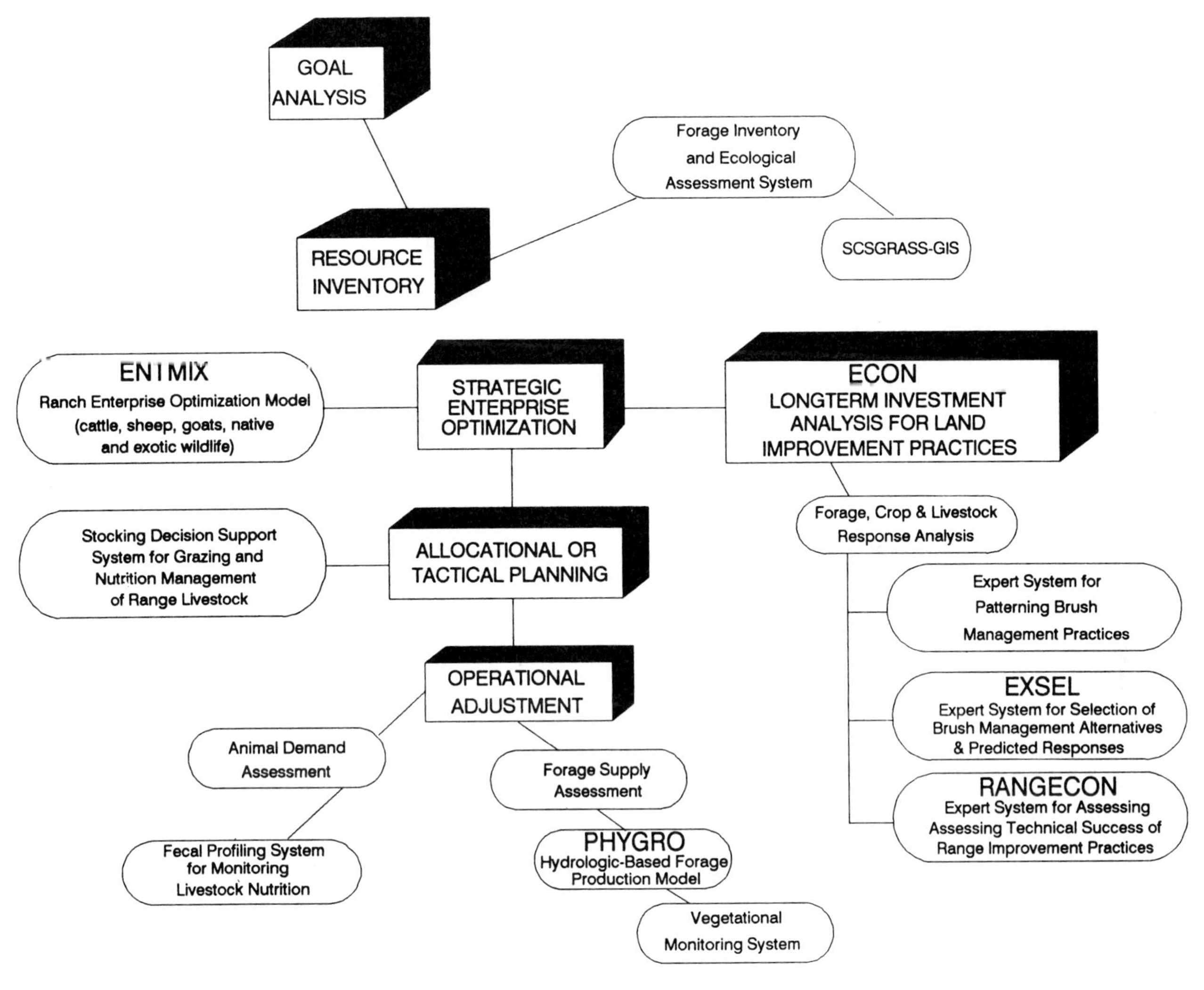

FIG. 2. Primary decision support components of the Resource Systems Planning Model (RSPM).

RSPM is being used. Along with the plant's name, attributes are stored which influence its preference value to herbivores. At this date the following attributes are being considered for parameterization: life cycle, primary growth season, habit (life form), size, canopy volume weight, coarseness, physical hindrance, secondary compounds and toxicity. Levels within an attribute are broad qualitative values, e.g. size = short, mid, tall. An embedded expert system is being developed which provides the user with a predicted preference category (preferred, desirable, undesirable) and associated herbivore specific harvest efficiency (per cent actual forage consumed at proper use). These data are used to assess habitat value of a site and estimate single species stocking rates and proper stocking of mixed herbivore grazing. The data are also used in the bioeconomic enterprise optimization model to set biological constraints on the analysis. Additional growth attributes will be added to the database to meet the data requirements of the hydrologic-based forage production model.

Animal attribute database

The database accepts inputs on physiological parameters which define nutrient requirements of a specified breed-type within an animal species. The primary parameters focus on herbivore class, a metabolic scaler, birth weight of the offspring, mature weight, maximum fleece weight, mill production coefficients and frame score. These data support nutrient requirements and animal performance calculations in the nutritional mediation subsystem and determination of habitat value and recommended stocking rate when data is not available.

Soils

The user can create the soils database from available soil survey information or from data imported or merged from a regional database. Land use classification is chosen from a selection presented to the user so that a single soil can be tied quickly to several land uses and assigned specific identifications. Pertinent information from the soils database and other databases used in strategic planning is linked to a specific client.

Plant growth curves

A plant growth curves database is provided that reflects

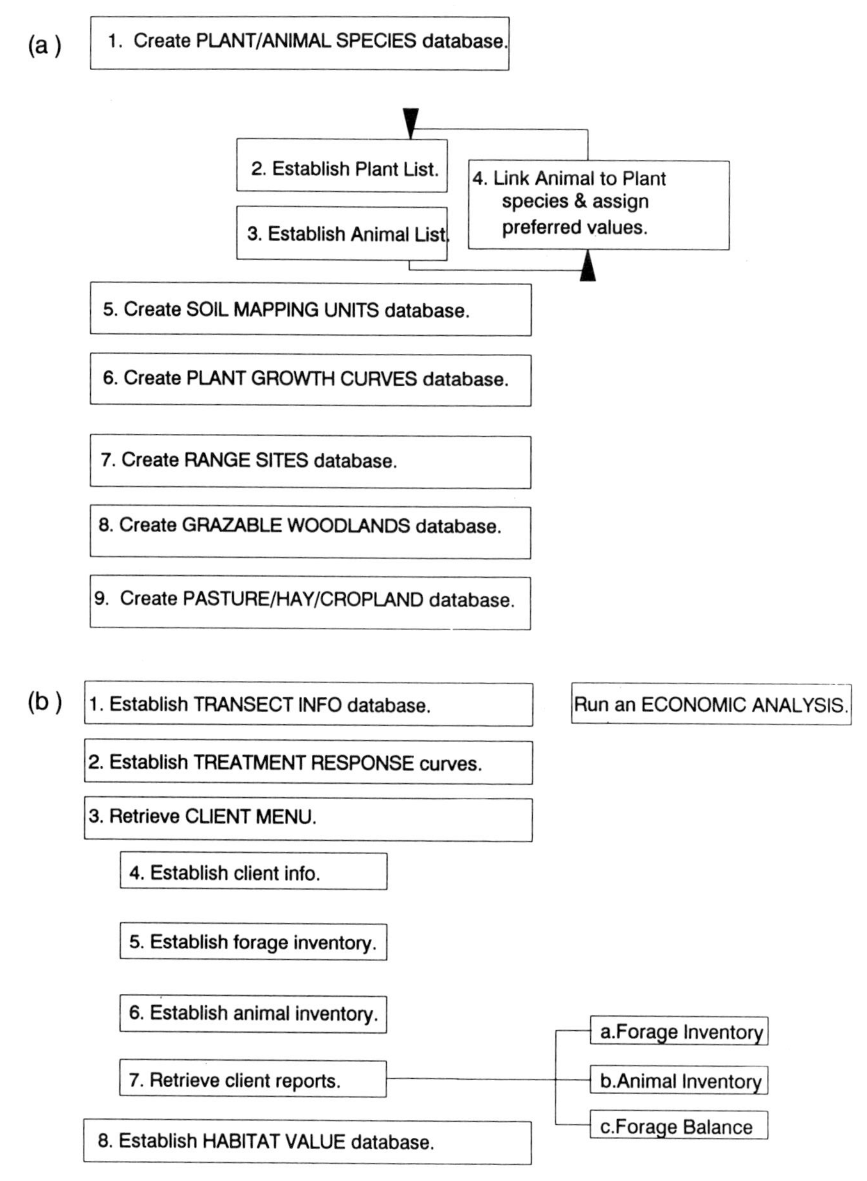

FIG. 3. Example of sequential setup of (a) client-independent and (b) client-dependent databases in the strategic subsystem of RSPM.

the percentage of the entire plant growth (production) expected for each month of the year. Growth curves may be defined for the same range site (see below) to reflect different species compositions and/or seasonal growth characteristics. The degree of differentiation of curves is a function of user input.

Anticipated monthly growth is entered as a percentage of the total for the year and tied to an amount of annual production to produce the client's monthly forage supply. This supply is then compared to an animal inventory to derive a balance between supply and demand.

Land use descriptions

RSPM provides for land uses of rangeland, grazable woodland, pastureland, grazable cropland, and hayland. These land uses are described and included in client independent databases.

Range sites. Range site information includes narratives on soils, the soil mapping units associated with the range site, landscape factors (topography and elevation), a table of climatic data pertinent to the geographic region in which the site is located, a description of the site, and ground cover and structure of the vegetation.

The range site description also includes such details as the vascular plant composition, production, and per cent allowable in determining range condition; plant growth curves associated with the site (this may include several curves for the same site); estimates of total annual production for favourable, normal and unfavourable years; a guide to recommended initial stocking rate based on ecological condition of the site; a list of the wildlife species common to the site; a narrative description of the process of site degradation following longtime overuse and a description of the typical site location.

Grazable woodland. The grazable woodland database is

currently composed of woodland suitability group information required to determine an initial stocking rate. Stocking rate is based on a combination of forage value ratings (very high, high, moderate, and low values) and tree canopy classifications (open, sparse, medium, and dense). Recommended initial stocking rates are expressed as both acres per animal unit (AC/AU) and animal unit months per acre (AUM/AC) for each combination of canopy class and forage value rating.

Pastureland, hayland and grazable cropland. The user selects a land use, defines it as irrigated or non-irrigated and identifies the dominant plant species providing the forage resource. Yield is estimated for grazing, hay or silage production for each land use entry. One entry in any of the harvest methods is expanded by a calculation to the other two methods. Grazing yields are expressed in AUM/AC, and silage in tons/acre. Values for harvest efficiency for each harvest method, grazing, hay or silage, and for moisture content are shown as default values that may be changed by the user.

Estimated yields are also entered by level of fertilizer input, using categories for no fertilizer and low, moderate and high levels. Adjustment factors are provided on the input screen for additional AUM of grazing that can be obtained by added nitrogen applications. Adjustments in stocking rate caused by imbalance of the phosphorus and potassium ratios to nitrogen are also calculated from user entries based on information about the specific soils and fertilizer recommendations for the area.

Interproperty defined databases

Range improvement scenarios, expressed as the response expected over a planning horizon, are entered and maintained in a treatment response database. These treatment scenarios may be created for a specific client but used from the database for other applications.

Transect information database. Currently, RSPM only allows input of summarized transect information either on a weight basis (lb/ac) or an estimated stand composition (%) and total forage production (lb/ac) basis. RSG is developing field data logger programs to allow the plant monitoring subsystem to interface with this database. The amount of keying time needs to be minimized when dealing with large amounts of field data.

Treatment response curves. Response curves reflect change from current production (expressed on rangeland as AC/AU) anticipated from introduction of an improvement practice or treatment set. To create a treatment response curve in the database, the user enters an annual projected per cent increase in range carrying capacity above current (year 0) carrying capacity. Planning horizons of 20 years or less are allowed.

The user can also project an increase or decrease in carrying capacity at the end of the planning horizon if no treatment was applied. The year in the planning horizon that the change (decrease or increase) will begin and end are identified. This feature allows change caused by planned treatment scenarios to be measured each year against both the assumption that carrying capacity will remain the same over the planning horizon and that an increasing or decreasing level of productivity will occur if no treatment is applied.

Client information

Each client in RSPM is identified by the enterprise (farm or ranch) name, contact person and address and telephone number. A mailing list of all clients can be entered in the database and printed for general mailing.

Forage inventory. Once a client is identified, previously stored client-specific information can be recalled. If the client is new, initial information can be entered on the forage inventory screens for all the land uses (rangeland, woodland, cropland, hayland and pastureland).

The acreage associated with each response unit becomes the basis for calculating forage supply and animal units of carrying capacity after stocking rate is determined in AC/AU or AUM/AC from additional entries. The land use screens for rangeland and grazable woodland differ from the cropland, hayland and pastureland screens. Woodland input requires entries for canopy class and forage value to calculate carrying capacity instead of an estimated ecological condition class. Entry methods for calculating stocking rate common to both rangeland and woodland include transect (use of actual field surveys to calculate per cent climax vegetation remaining), and total pounds of production. Rangeland transect entries use a calculated per cent climax vegetation remaining to interpolate the initial stocking rate from stocking rate ranges provided in range site descriptions. Total pounds of production is used with the harvest efficiency factor to calculate stocking rate based on the annual dry matter required of an animal unit.

The cropland, hayland and pastureland screens allow the user to identify irrigated or non-irrigated use, the soil capability class and the specific crop or pasture species to be included in the forage inventory. Level of fertilizer use is entered as well as harvest efficiency factors and moisture content associated with each harvest method; grazing, hay or silage. Entry methods for calculating stocking rate include actual pounds of nitrogen to be applied, total pounds of production, and fertilizer category (none, low, medium or high). Calculations for stocking rate vary with the entry method. In all methods, an adjustment factor to account for reduced efficiency of forage harvest caused by such problems as poor grazing distribution mediates the stocking rate. The user may also change default values for harvest efficiency and per cent moisture content of the forage which in turn will increase or reduce stocking rate.

Animal inventory. The animal inventory is a client-dependent database that identifies the kind, number and forage demand (AUE) of the animals within an enterprise. Demand is entered for each month of a complete production cycle in order to reflect physiological stage of breeding animals and age and weight changes for non-breeding animals. The animal inventory is used to create the monthly demand. This data is compared to supply from the available forage inventory to produce the forage balance report.

Habitat value. This module allows the user to attach a transect to a specified breed-type within a type of animal and determine the proper stocking rate for that animal. A user-defined index based on the composition of preferred, desirable and undesirable vegetation allows the system to assign not only an ecological condition score but also a habitat value score for a specified herbivore.

Forage/animal inventory and balance reports. Currently, RSPM provides the user with strategic information on forage inventory, animal inventory and forage balance for a given property or firm. The forage inventory report provides the user information about acreage, stocking rate, potential and available animal units and tons of hay and silage by management unit in the forage inventory report. The animal inventory report provides information on the number of animals by kind and class and an analysis of monthly demand which reflects the changes in dry matter intake as physiological stage of the animals change. The forage balance report contrasts monthly forage availability with monthly forage demand to determine those periods in the year where shortfalls in forage supply may occur. The balance report also provides the user with information on each month's accumulated supply of hay and silage. However, RSPM does not allocate grazing or stored feed sources at this stage of development. The strategic forage/animal balance report provides the initial evaluation of how forage is distributed relative to the demand of the animals on the property. The obvious next step is to allocate and schedule grazing once strategic analysis has occurred.

A grazing scheduler is currently under development that allows the user to create herds and allocate them to a given set of pastures or paddocks. Scheduled animal demand derived from either the animal unit values or predicted dry matter intake is contrasted to available forage supply on a daily basis to allow analysis of forage and nutrient balance.

Consolidated landscape treatment response curves. Treatment response curves that were built and stored in the client-independent database for different treatment scenarios can be combined in the client-specific part of the program. This capability allows development of combined carrying capacity from response units where differences exist due to sites, treatments or other variables. Response units are combined into a single carrying capacity curve to perform the economic analysis in the planning process.

Menu selections allow the user to combine management units, all response units within a management unit, or combine selected response units within multiple management units. A weighted average is also calculated for the stocking rate by years during the planning period and for the increase or decrease in carrying capacity if no management is applied. The combined curve may then be imported into the economic analysis portion of the program.

Enterprise optimization and analysis

The ENTMIX subsystem selects appropriate mixes and levels of livestock and/or wildlife enterprises subject to both seasonal availability and animal preferences and demand for grass, forbs and browse. Constraints on other resources, i.e. labour and capital, are also incorporated in the optimal solutions. Projected diet composition of major preference groups (preferred, desirable, undesirable) is the determined composition of available forage and associated harvest efficiency of associated animal species. Linkages are provided to the forage and animal databases to facilitate determination of forage availability and animal preferences/demand.

Long-term investment analysis

ECON is a principal component of RSPM. It was designed to assist resource managers in estimating returns on investments for range improvement and/or grazing management practices. It is structured to facilitate delineation of annual costs and revenues for a specific practice or set of practices over a 1–20-year planning period. The annual costs and revenues delineated are only the excess over those which would be expected for each year in the same period in the absence of the practice or set of practices. Thus, ECON is developed on the precept that the range improvement or grazing management practice will be applied to an existing ranch operation which will continue to operate throughout the planning period whether or not the practice or set of practices is put into effect.

An additional underlying precept is that one or more livestock or wildlife production enterprises are the basic source of revenue for the ranching operation and that enhancement of these same enterprises, i.e. more production per acre, will provide the added revenues which must offset the costs of implementing the practice. ECON accounts only for costs and returns directly accruing to the land owner/operator. It does not include off-site benefits accruing to other members of society.

Although ECON analyses a practice or set of practices applied to either a management unit or an entire ranch, grouping managerially unrelated practices, i.e. practices applied in different management units, permits an investment analysis for the grouped practices. However, providing information for an individual practice causes ECON to produce an investment analysis for the specific practice applied to each unit.

Brush control practices, fences, water facilities, etc., are all potential investments which require a large initial capital outlay and produce variable annual returns over several years. Other investment opportunities always exist. Money invested in a range improvement practice or program cannot be used for one of the alternatives. Thus a resource manager would want the range improvement practice to earn a return on the money invested in it at least as large as the return that could be earned in the alternative. This concept is termed opportunity cost.

ECON simply accounts for the added costs and revenues associated with the range improvement practice occurring in each year in the planning period, discounts them with the appropriate discount factor, sums the discounted costs and revenues, and subtracts the summed discounted costs from the summed discounted revenues. The difference is the net present value (NPV) of the investment in the range improvement practice. If the NPV is zero or positive, then the investment in the range improvement practice is estimated to be economically feasible.

The ECON procedure also produces an additional investment indicator, the internal rate of return (IRR). The IRR can be defined as the discount rate that results in a NPV of zero. Thus, it is a direct estimate of the average annual rate of return that the investment in the range improvement practice will produce over the entire planning period. This indicator is particularly useful in comparing the relative merit of alternative investments which may require different time periods and/or different initial outlays of capital because it reports returns as an average annual percentage. NPV, on the other hand, only reveals the present value of net returns in dollars.

Expert systems

Currently, two expert systems are under development for use within the strategic component of RSPM. One of these systems allows assessment of the probability for success of planned technologies based on managerial capability and environment. The other system matches brush and weed problems with technically feasible brush management alternatives.

Technology selection and response. EXSEL allows users to identify and characterize a brush or weed management problem. Queries include a description of the geographic region where the problem exists, soils, primary target species, secondary species, woody plant growth form, density, stem diameter, maturity and other pertinent attributes that can be viewed on the screen. The system also provides treatment alternatives developed by experts for all combinations of attributes. Mechanical, chemical and prescribed brush burning weed control methods are included in the alternatives. An added feature of the system is a database of suggested response curves linked to treatment alternatives. These curves can be accepted or modified by the user to fit specific situations.

Technology success. RANGECON establishes a level of managerial skills and management environment based on answers developed and weighted by expert opinion. Such items as the decision-making environment, educational level, capital availability, complexity of the technology being considered, past history of success in technology application, and source of technical information are used to predict the probability that the manager can capture the full benefits of the treatment(s) being evaluated in the planning process (Ekblad *et al.*, 1989). Capturing full benefits would mean that the predicted internal rate of return (IRR) on investment in the treatment(s) could be met. Results from RANGECON are reported as a numerical value on a negative to positive scale.

TACTICAL LEVEL

Animal class definition database

Because of the wide variation in naming of animals and associated husbandry practices, this module allows the user to define a client-specific set of names for each class of mature, nursing or growing animals on a property. The user establishes a set of default names and parameters to avoid repetitive data entry. The user also specifies age limits (month) to determine timing in changing animals from one animal category to another. Each established animal class will have a breed-type attached in the database to assure access to physiological information in later nutritional calculations.

Forage quality database

The estimated forage quality is expressed as percentage crude protein (% CP), percentage digestible organic matter (% DOM), digestible dry matter (DDM), organic matter digestibility (OMD), or total digestible nutrients (TDN), or mcal net energy (NE), metabolizable energy (ME), or digestible energy (DE) on a monthly basis. The user can store many different forage quality scenarios to allow proper assessment of potential nutritional consequences of various grazing situations and available feedstuff.

Environmental profile database

Monthly values of average daily minimum temperature, maximum temperature, humidity and mud factors are stored in this database to adjust for costs of thermal regulation among the animals in question. The user can define multiple examples to determine the potential impact of variations in the weather on forage and nutritional balance. The stochastic weather generator in the hydrologic based forage production model will be integrated with the grazing scheduler for users wishing more detailed analysis.

Herd profile database

This module allows the user to create a client-dependent database of the composition of a given herd of animals managed as a group. It allows specification of animal class, numbers of animals, and death and birth rates by month.

Grazing scheduler

This client-dependent subsystem allows the user to schedule access of various herds to specified pastures. The information allows analysis in forage balance relative to demand of the animals and forage supply in the pasture. The scheduler provides a daily, weekly or monthly analysis of forage and nutrient balance. The user can attach forage quality and an environmental profile to the scheduler. The scheduler can be implemented on an animal unit basis or a digestibility driven demand basis. Currently, the system acquires stocking rate information from the forage inventory database, converts the information to annual production and distributes that production by month using the linked forage growth curve (% by month). Future versions will have the capability to use an embedded simulation model to predict forage growth based on hydrologic characteristics of the site and the nature of vegetation occupying the area. [See section on hydrologic based forage production model (PHYGRO).]

Feedstuff dictionary

The user establishes a database of various feedstuffs that are available to mediate nutritional deficiencies or induce a higher level of animal performance. Values of crude protein (%) and mcal/lb of energy maintenance and gain (% TDN, DDM, DOM, OMD, or mcal/lb of DE, ME, NE) are stored in the database.

Nutritional mediation

Once a set of management units is scheduled and a forage/nutrient balance analysis is conducted, the user specifies those periods when intervention with supplemental or substitutional nutrients is desired. Any deficit periods specified are subject to an optimization analysis which determines the most cost effective combination of feedstuffs to mediate a nutritional deficiency. The user specifies candidate feedstuffs and inputs purchase, delivery, storage and feeding costs associated with each feedstuff. The linear programming algorithm determines the mix of feedstuffs, provides a recommended feeding schedule and gives a quantity report.

Enterprise budgeting

The module establishes and maintains annual cost and returns budgets for each animal enterprise and a cash flow statement for the entire firm. The budgets and cash flow statements can be developed as planning documents at the beginning of each year and then updated as the production year progresses. Also, as supplementation plans are modified, due to unexpected forage shortages, for example, the budget and cash flow information on feed costs will be automatically updated through linkage to the nutritional mediation module. This module will provide the user with a means of assessing the impact on financial objectives of marketing and/or production alternatives at any point before or during the production year.

OPERATIONAL LEVEL

Once a plan of action at the strategic and tactical level is established, the user must react to the weather and market conditions during the upcoming year. Generally, this task requires some ability in assessing the current conditions of vegetation and animals and in assessing impact of probable outcomes. To address effectively these needs, emphasis was placed on the development of simulation models for plant growth and measurement techniques that could provide accurate estimates of current states.

Hydrologic-based forage production model

The PHYGRO model is a process-oriented compartment model that simulates a water balance and maintains a soil water inventory on a daily basis at the range site level. The model uses a description of the plant ecosystem in conjunction with daily weather inputs. The ecosystem is described as a container with up to four compartments or soil layers in and/or on which a plant community functions. The plant community is defined by its potential standing live phytomass curve, root distribution and ability to utilize water and potential or long-term forage yield. Standard soil parameters by soil layer are used to define each compartment of the container. The total container is the aggregation of all compartments.

The model estimates daily potential evapotranspiration for rangeland (ETpr) as the product of daily potential evapotranspiration (ETp) estimated by the Jensen and Haise equation, and a site-specific water use coefficient. The model uses the relative growth curve and a site-specific transpiration parameter to partition ETpr into potential transpiration (Tp) and potential soil evaporation (Ep). Actual daily soil evaporation (Ea) is estimated as the quotient of Ep and the square root of the number of days since the soil surface was last wet. Actual daily transpiration (Ta) is the sum of actual daily transpiration of the different soil layers which are estimated as the product of the ratio of available soil water content of that day to available soil water capacity, the root density factor and soil temperature factor.

In operation, the model first checks for precipitation (P), and, if it occurs, it uses the Soil Conservation Service (SCS) curve number technique to predict runoff (R). The difference between P and R is added to one soil layer at a time. Water is first added to the surface layer until it reaches field capacity. Only then is it added to the next layer and so on until all water is accounted for or all soil layers have been filled. When either one of these situations occurs, the excess water is counted as deep drainage (D). Soil water extracted for plant transpiration also proceeds one layer at a time. If the surface layer cannot supply enough water to meet daily Tp, then it is extracted from the second layer and so on until the demand has been satisfied or all the layers have been sampled.

A weather input file and input parameters require eighty-five parameters to run a simulation. From those parameters, three define the type of printout desired, six define the simulation period, thirty-three are needed to run the weather generator options, and forty-three describe the range ecosystem, including species composition. The model will account for heterogeneity in composition of species as it affects potential site yield. The weather input file contains the year, julian date and daily values of precipitation, solar radiation and maximum and minimum air temperatures.

The model allows the estimation of forage production per growing season. The procedure used is not process-oriented (Wight, Hanson & Whitmer, 1984). Growing season herbage yield is calculated at peak standing crop as the product of site production potential and a yield index for that growing season. Currently the model provides two yield indexes. One is the cumulative actual to potential transpiration ratio. The other is the ratio of the area under the plotted seasonal, cumulative actual to potential transpiration curves.

ERHYM-II, PHYGRO's precursor, has been demonstrated to be effective for predicting soil water, evapotranspiration and forage production of sites in Montana, North Dakota and Idaho (Wight & Hanks, 1981; Wight, 1983). The model has been shown to be more accurate in predicting forage production response than more complex models such as SPAW and CREAMS (Cooley & Robertson, 1984; Wight, Hanson & Cooley, 1986). The combination of simplicity and ability to predict forage yield makes this model ideal for management applications.

Plant monitoring system

Three field methods were combined into an integrated analysis software package called HABITAT. The comparative yield, dry rank weight and crown volume weight methods were used to characterize herbage and browse available to animals in the specific sites. The system provides estimates of standing crop and composition by species. Data summaries from HABITAT are transferred to the transect database for use in other modules. Programs for small hand-held microcomputers are under development that allow the user to acquire information and download it for analysis in HABITAT. The transect methodology also provides analysis of structural attributes of a site for use in wildlife habitat scoring systems.

Animal monitoring system

Near infrared (NIR) spectroscopy technology has been successfully used to develop calibration equations for scanning fecal material of cattle and predicting the crude protein and energy content of their diet. Approximately 85–90% of observed variation in data sets of known diet quality could be accounted for using this technique. RSG is currently developing a system of collection, delivery, analysis and feedback to the producer. Estimated turn around time appears to be 4 days where telephone communication is good and 9 days where information must be transmitted by mail. The goal of the system is to provide the user with a mechanism to rapidly assess the nutritional quality of diets of animals on pasture. Information provided would allow assessment of relative quality of pastures, enhanced knowledge on adjusting rotation schedules and accurate input for the nutritional mediation subsystem.

Weather monitoring system

Many land managers collect information on precipitation. However, in the future DSS will require additional information on temperature and sunlight. A mechanism is being developed to update a weather record file or download weather information from either distributed weather databases or on-site portable weather monitoring stations. The information will be particularly useful in the hydrologic forage production model. Access to historical weather information is essential to assess weather-induced risk associated with long-term investment strategies.

CONCLUSIONS

Government agencies responsible for research and technology transfer have been accumulating a large amount of information on various attributes of natural systems for many years. Many of these databases were created with little definition about the application of the information. While these databases have served as mechanisms for cataloguing large amounts of important information, their use has been ill-defined. Decision support systems, however, have forced scientists and action agency personnel to organize information into a uniform format that is acutely defined for the resolution of knowledge that is required for effective decision making.

Generally, DSS can be categorized into policy and firm, or individual, level systems. The increased use of DSS in natural resource management will become more evident as policy level funding agencies will require greater depth of analyses in less time while firms will require greater resolution of their analyses, particularly as they relate to long- and short-term risks associated with weather and market conditions. As communications technology makes data available on a global basis, as marketing networks expand, and concerns for our world environment increase, the compilation of well-defined databases for use in multiple applications will require greater coordination among DSS developers who share common data sources.

The use of decision support systems in natural resource management is directly tied to the improved performance/costs ratios of computers capable of processing large amounts of data in a timely manner. As computing power increases, overcoming information requirements for more detailed analyses becomes less of a problem for developers of DSS. The advent of distributed databases, which represent a subset of information from large centralized databases, has been helped by improved hard and optical disk technology, the latter of which is most appropriate for large, static databases that are not updated more than once each year. Examples of these databases include soils, climate, elevation, property boundaries, cover types, aerial photo images, plant attribute and animal attribute data.

Application of embedded expert systems appears to be a viable means of providing for decision requirements today while data hungry embedded simulation models come on line tomorrow. Many DSS require the user to project the response of a given treatment, select an array of appropriate technologies or assign index values to a complex relationship. The embedded expert system puts the expert at the side of the user to assist in parameterizing the application. This trend will expand until technology changes our current problem solving paradigm or until information can be obtained and used in a more advanced simulation with minimal user input. Like most applications, the quality of the information depends on the quality of the system expert. To ensure that expertise will be properly captured, there will be greater demands on higher education to produce knowledge engineers for the natural resources arena.

Computer-aided decision making through DSS is forcing us to set better objectives and to direct the conduct of science as well as to create new paradigms about how natural systems function when man is an integral part of the equation. Organization and synthesis of information is our immediate challenge. Breaking current modes of research to address needs generated for resource decision making will prove to be our greatest challenge in the future. However, it also provides a great opportunity to improve the efficiency of research in support of application requirements. Developers should keep a DSS simple and question the need for information inputs. They should fight tendencies to include information because it might be important in the future. Flexible programming methods should be used that allow for system modifications and expansion. The end user must be clearly identified and his inputs considered early in the design and development process as in the concept of prototyping described by Mathieson (1988). A system must be used to be successful. Therefore, the DSS must be perceived as more valuable to the decision maker than current methodology, and its adoption of the system as a facilitator rather than as a major disruption of on-going activities. Most decision makers are time-limited and will make a decision regardless of the level of sophistication of the system. DSS should grow with the user. Therefore, flexibility in depth of information inputs required to generate an analysis from DSS is critical to early adoption of the system.

DSS is not the end of the story but only the beginning of a new era in which information technology will have a profound impact on the way resource managers access and analyse information. This trend will require a restructuring of the way science is conducted in the future requiring biological, economic and social sciences to merge with information science to produce individuals with the necessary competencies to react to the information age.

Acknowledgement

This paper was approved by the Deputy Director of the Texas Agricultural Experiment Station as TA-25312.

REFERENCES

Cooley, K.R. & Robertson, D.C. (1984) Evaluating soil water models on western rangelands. *J. Range Managemnt*, **37**, 529–534.

Ekblad, S.L., Hamilton, W.T., Stuth, J.W. & Conner, J.R. (1989) A knowledge-based management evaluation system for assessing success of selected range improvement practices. Absract 132. Presentation at the Annual Meeting of the Society for Range Management, Billings, Montana, U.S.A.

Mathieson, K. (1988) Prototyping expert systems. *AI Applications*, **2**, (2–3), 3–11.

Negoita, C.V. (1985) *Expert systems and Fuzzy Systems*. Benjamin/Cummings Publishing Company, Inc., Menlo Park, California.

Scifres, C.J. (1987) Decision-analysis approach to brush management planning: Ramifications for integrated range resources management. *J. Range Managemnt*, **40**, 482–490.

Taylor, J. & Taylor W, (1987) Searching for solutions. *PC Magazine* **15**, 311–322.

Wight, J.R. (1983) Application of a water-balance, climate model for research and management in a desert-shrub community. In: Wildland Shrub Symposium 'The Biology of Atriplex and Related Chenopods,' Provo, Utah.

Wight, J.R. & Hanks, R.J. (1981) A water balance, climate model for range herbage production. *J. Range Managemnt*, **34**, 307–311.

Wight, J.R., Hanson, C.L. & Cooley, K.R. (1986) Modeling evapotranspiration from sagebrush rangeland. *J. Range Managemnt*, **39**, 81–85.

Wight, J.R., Hanson, C.L. & Whitmer, D. (1984) Using weather records with a forage production model to forecast range forage production. *J. Range Managemnt*, **37**, 3–6.

22/RANGEPACK: the philosophy underlying the development of a microcomputer-based decision support system for pastoral land management

D. M. Stafford Smith and B. D. Foran *CSIRO, Division of Wildlife and Ecology, P.O. Box 2111, Alice Springs, Northern Territory 0871, Australia*

Abstract. CSIRO RANGEPACK is an evolving decision support system for the management of extensive grazing properties in the context of a variable climate and market. The system's modular tools are mainly microcomputer-based. They include: programs such as *HerdEcon*, an integrating herd dynamics and property economics model; easily-used databases such as *Climate*, which provides ready access to meteorological data; and design tools such as *Paddock*, which assists managers to develop the spatial layout of their property.

RANGEPACK philosophy includes the following: (i) decision support tools must be directed at significant management problems to be useful; users must be involved in establishing implementation priorities; (ii) a modular design allows important decisions to be targeted individually, and allows modules to be region specific: large process models of complex systems are obscure and difficult to interpret; (iii) it is important to make it easy for users to proceed by 'successive approximation': optimization techniques tend to obscure limitations in models; (iv) it must be possible to apply tools to specific properties and paddocks: managers do not like making decisions based on 'district averages'; (v) programming design should ensure ease of use, flexible communications between modules, and easy updating of modules; (vi) alternative media, such as wall charts and manuals, may be better than computers for some tasks, especially where microcomputer technology is not available.

Key words. Land management, decision support system, microcomputer, cattle, sheep, Australia.

INTRODUCTION

CSIRO RANGEPACK is an Australian project to develop a microcomputer-based decision support system for pastoral land management. RANGEPACK resulted from frustration at the slow adoption rate of ecologically-based management practices in Australia's rangelands, but its philosophy and many of its modules are applicable to any extensive grazing system. The causes of slow adoption include the rapid turnover of skilled advisory personnel in remote areas, and the common failure to integrate research results into real management systems. These problems led to the need to develop integrating tools, which would simultaneously communicate research results, provide continuity of advisory expertise, and so help managers make better decisions.

The methodology for analysing critical decisions has been developing for many years (e.g. Norton & Mumford, 1984; Norton & Walker, 1985; Scifres *et al.*, 1985). The term 'decision support system' has been coined for a system which develops from such decision analysis (e.g. Bennett, 1983; Mittra, 1986). These often use techniques of 'expert systems' (Starfield & Bleloch, 1983; Harmon & King, 1985; Davis, Hoare & Nanninga, 1986), but rarely meet the strict 'inference engine-knowledge base' design criteria of the latter. This article describes the past and future development of one such decision support system, RANGEPACK, and the design philosophy that has evolved through the experience of workshops and feedback from users.

RANGEPACK BACKGROUND

Our initial development of RANGEPACK, and much of the philosophy discussed in this paper, was based on questionnaire responses from about 100 people, and on a workshop held in October 1986 with another eighteen people. Both groups represented a wide range of skills, including managers, economists, rangeland advisors, researchers and computer programmers, from all around Australia. They were asked to identify the aspects of management which were most in need of assistance, to enable priorities to be set for RANGEPACK.

It soon emerged that there was no point providing ecological information unless this was presented in the necessary economic context, that is, the financial circumstances

FIG. 1. Farmers in Australia are increasingly using portable computers and associated decision support software in the day-to-day planning and running of their enterprises.

of the manager's own property. Furthermore, it was important to address the problems of the managers themselves rather than those of the advisory agencies (Fig. 1). If this was done successfully, the same tools would be useful to the advisory personnel, and in educational institutions.

There are many management decisions which are trivial; there is nothing to be gained by addressing these. General tasks with which computers could help managers were identified as: (i) performing large numbers of calculations, (ii) handling large data bases, (iii) making inaccessible process models available, (iv) presenting visual or graphic display, and (v) providing some form of useful expert system, i.e. where only a few people had some expert knowledge which was not readily quantified, nor accessible to the community at large.

The workshop also analysed what types of decisions are made in pastoral management in Australia. Decisions were classified as strategic – about long-term management goals – or tactical – about implementing the strategies. Some of the strategic options that are open to cattle enterprises in various states of development are shown in Fig. 2 (Stafford Smith & Foran, 1988). Tactical decisions relating to paddock design and stocking rates were also analysed.

When a set of options such as those shown in Fig. 2 is examined, biological researchers tend to think most readily about those options which are biological in nature. However, these may have less impact on short-term profitability than organizational changes. If managers are to take any notice of suggested improvements in biological management, therefore, these options need to be placed in a context where they can be assessed against other management changes. They also need to be assessed in the context of riskiness resulting from the unpredictability of climate and markets. These sorts of messages have been identified before, of course (e.g. Crouch, 1972; Bardsley, 1981), but seem destined to need to be re-worked by every person concerned with extension, since research results and advice are still sometimes promoted without the necessary overview or integration.

The workshop therefore gave top priority to the development of a module now called *HerdEcon*, which can integrate all aspects of biological and economic management on an enterprise into the common currency of cash flows.

RANGEPACK MODULES

HerdEcon (Stafford Smith & Foran, 1988) is a dynamic herd and property economics model. It allows the user to

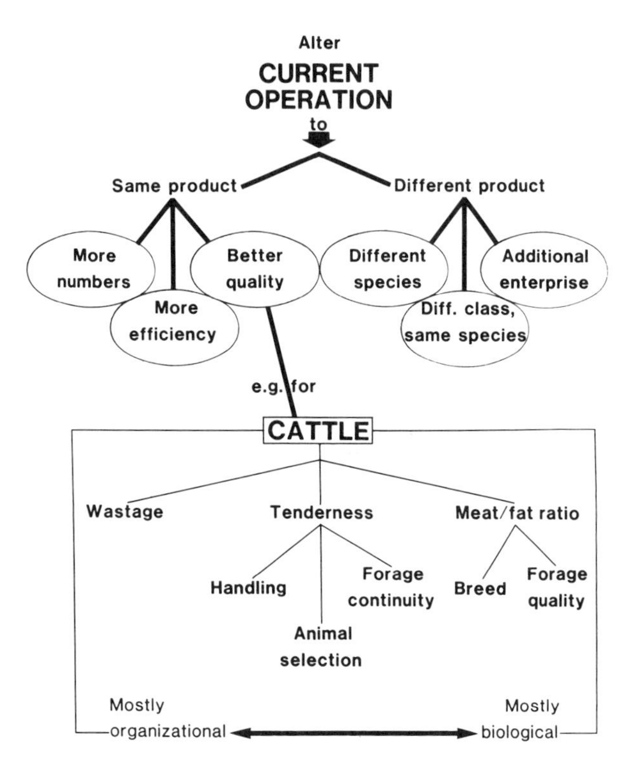

FIG. 2. A section of the tree of possible development options available to an Australian cattle grazier seeking to improve property viability. Factors which could be considered for cattle enterprises are shown for the 'better quality' branch (after Stafford Smith & Foran, 1988).

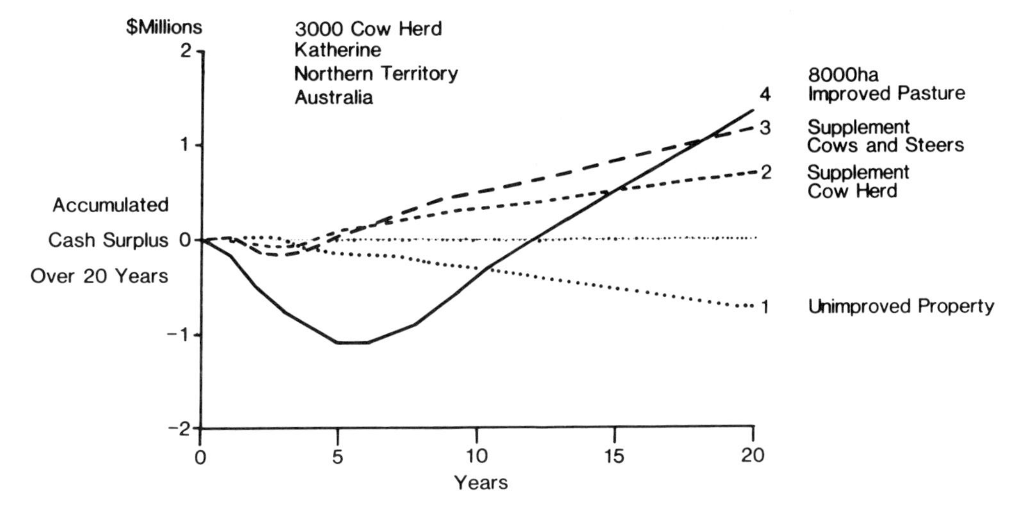

FIG. 3. An example of the use of *HerdEcon* to graphically compare four development options for a property in the Katherine region of the Northern Territory, showing how accumulated cash surplus changes during implementation (from Foran, Stafford Smith, Niethe, Michell & Stockwell, in prep.). The details of the options are unimportant, but involve comparing an unaltered management system (1) with improved breeder (2) and steer (3) management strategies, and a major planting of introduced pastures (4).

set up the herd structure of any property, with birth, death and growth rates specified for each animal class and age group. These rates can take account of differing seasonal conditions. Costs and receipts, including variable costs related to animal numbers, can be defined in considerable detail for the property. Buying and selling strategies can be specified. The modelled property can then be run for a series of years, for example to track the implementation of a new technology (e.g. Fig. 3), or to test alternative strategies in the face of climatic or market variability (e.g. Fig. 4 which shows the mean expected outcome, as well as the variance or riskiness, of different strategies of managing cattle in central Australia). An integrated graphics package permits easy comparison of alternative strategies, and a report-generator provides flexible output. Further details and examples are given in Stafford Smith & Foran (1988).

HerdEcon is easy to use, and flexible enough to be tuned to almost any grazing enterprise, regardless of the mix of animal species and classes. Typical properties have been described for many regions, so that new users can usually find an example in *HerdEcon* format which they can modify to suit their particular enterprise. Following the commercial release of *HerdEcon* (available from the authors), we are now developing modules to deal with the more biological aspects of management. In expected order of release, they include *Climate*, *Paddock*, *Forage*, *Animal* and *Fire*.

Climate is an interactive climate data base providing historical rainfall, temperatures and other climatic variables for stations in the pastoral lands, as well as the facility to add a property's own records. The module will provide probabilities of specific categorical climatic sequences, including conditional probabilities when these are justified, and provide realistic sets of yearly conditions against which management strategies in *HerdEcon* can be tested. The example in Fig. 4 is a case where the probabilities of different combinations of years can be estimated from the climatic record, to assess the overall expected outcome of a series of different runs with *HerdEcon*. The output of *Climate* may be as answers to questions, or in graphical form. A prototype of *Climate*, developed by CSIRO Division of Plant Industry as 'METACCESS', is already available (Donnelly, 1988).

Paddock is a graphics-based module to help with designing paddocks, in particular to promote even grazing patterns (Stafford Smith, 1988); an example of its use is shown in Fig. 5. It will also include an expert system dealing with other aspects of paddock design, and provide simple process models of animal water balance and water use rates. Eventually, this module will be linked to a Geographic Information System (GIS) and integrated with erosion prediction models.

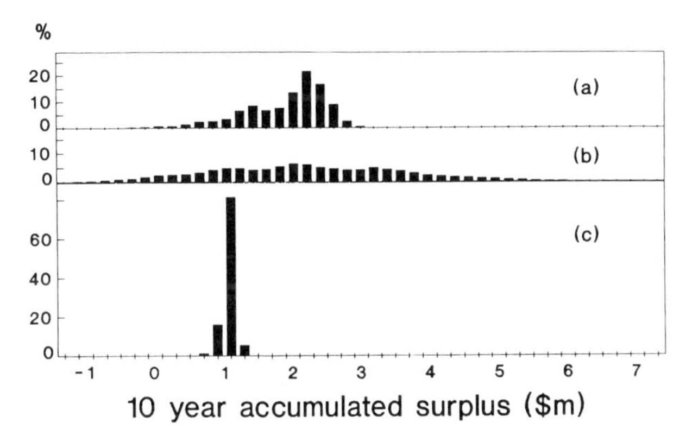

FIG. 4. An example of the use of *HerdEcon* to test three cattle management strategies in central Australia in the face of drought. The three strategies were run for every 10 year combination (3^{10}) of drought, average and good years, and a histogram of the frequency of occurrence of different accumulated cash surpluses over the 10 years is shown for each strategy. (a) is a typical property with average management which does little except wait and hope in the face of drought; (b) is a property which runs more stock on the same area but destocks very quickly in drought times, resulting in higher returns in runs of good years, but a much greater variance overall; (c) is a property which runs many fewer animals on the same area so that there is plenty of feed in dry times but no desire to stock heavily in good times, resulting in a very constant but smaller cash flow. All three strategies are based on real properties.

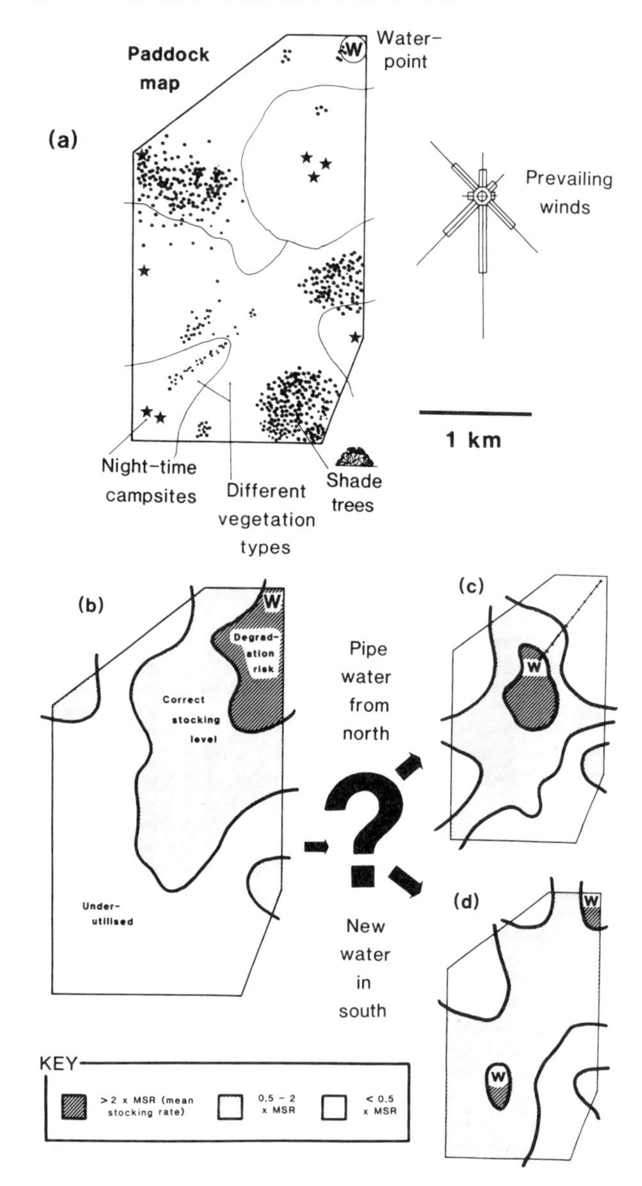

FIG. 5. An example of the use of *Paddock* in a sheep paddock with the features shown in (a); (b) shows the predicted current distribution of sheep grazing pressure, which the manager can compare with his experience to assess the reliability of the model, whilst (c) and (d) predict how the distribution might be improved by two alternative additional waterpoints. (d) results in the more even pattern with a smaller area liable to severe degradation. (*Paddock*'s actual output is more detailed.)

A prototype of the central component of *Paddock* is available, which predicts the distribution of animal grazing impact in relation to the location of paddock features.

Forage and *Animal* are two modules tackling the crucial issue of stocking rates. *Forage* will be developed over the next 2 years by drawing together plant growth models for key vegetation types throughout Australia. Initially, these models will provide a crude level of discrimination, but better regional models can replace earlier ones as they become available. *Animal* will quantify animal growth, and will be largely limited by the quality of the predictions of *Forage*, since good animal production models are now widely available. These two modules will be variously combined in the decision support structures '*MeatPack*' and '*WoolPack*', for the cattle and sheep grazing industries respectively.

Fire will be an extension of the SHRUBKILL program (Ludwig, 1988, 1990) developed for the semi-arid woodlands of eastern Australia, dealing with the planned use of fire for pasture management. *Fire* will advise on burning strategies for the control of undesirable perennial plants, and burning for other reasons, such as promoting green pick in tropical grasslands.

Other modules will be developed with time, in tune with the demands of the day. A special *Drought* module will draw on several modules already mentioned to provide advice and strategies for drought management. *StockingRate* and *LandCare* modules may target specific stocking rate and land restoration and protection decision. These modules will largely be expert systems which use other modules to assist in answering particular questions. Also, an 'umbrella' program will be developed to link these modules together, to facilitate data exchange, and to act as an expert system for a new user. The RANGEPACK framework should evolve indefinitely as new methods, models and understanding become available, and as many different agencies contribute.

KEY ELEMENTS OF DESIGN AND DEVELOPMENT

In this section we review a few key lessons which we have learned during the RANGEPACK project.

Decision support philosophy

Some of the philosophy involved in RANGEPACK is no different to that which should be present in any decision support work, and has been documented variously elsewhere (e.g. Arnold & Bennett, 1975; Davis *et al.*, 1986). For example, it is crucial to identify the decision makers: for RANGEPACK, the primary decision maker is the land manager, but there are other decision 'influencers' of varying significance. For *HerdEcon*, and the economic assessment of management options, loan-assessing bodies are particularly important, but pastoral houses, accountants, stock and station agents and government departments all affect property decisions, and may perceive the risks and profits of alternative strategies differently to managers themselves. Additionally, cash flows are greatly affected by market changes which are beyond the control of the industry.

Once the decision makers are identified, the decisions that they make must be analysed. This analysis results in a vast number of strategic and tactical decisions, with long- or short-term implications. Those which are trivial or too value-laden are unlikely to be worth implementing, and there will be others which are not amenable to a decision support system, or for which a microcomputer-based approach is unsuitable. This still leaves a long list of questions, which must be given priorities on the basis of utility and the existence of helpful information. It is essential to involve potential users in this analysis.

It is important to understand how decisions are made. A

fundamental RANGEPACK tenet is that the decision support system must 'supplement not supplant' the decision-makers' own thinking, or the decision is taken out of their hands. This tenet implies that the user must be able to assess the working reliability of a model on their own terms. This can be particularly difficult to do if large numbers of processes are hidden behind a single answer.

Large and complex process models are obscure to most users. They are difficult for even the expert to validate, and impossible for the inexperienced user to interpret reliably. Outside the research laboratory, our experience with this sort of model is that many users are simply sceptical of it from the start. Others believe too implicitly in what computer models can do and are then sadly disillusioned, since no program can hope to cope perfectly with the idiosyncrasies of specific real systems. Either way, the system is not used.

One way to minimize the problem is to modularize models so that assessment can be made at levels which are natural to the user. Thus a pastoral manager may accept a property model more readily if the pasture growth, animal growth and economics (at least) are performed as separate operations. He can then examine and assess the reliability of the pasture growth results, before these are chained through to animal growth. Overall pasture growth, rather than the seedling density of a particular species or the rate of nitrogen cycling, is likely to be a reasonably natural measure to this user, which can be easily related to his experience. Thus the user can see the process in smaller, more assimilable blocks.

Optimization in complex models is a technique which can reduce the opportunities for assessment by the user. An optimized result may be a useful starting point, but models are not perfect and optimization in a complex model of a real system is liable to hide the limitations of that model. *Paddock*, for example, cannot deal with all the quirks of real paddocks. Here optimization of grazing distributions might hide an unconsidered factor, such as access or poisonous plants, or may remove attention from a solution which is nearly as good as the optimum on the optimization criteria but much better for other reasons. Allowing users to proceed by successive approximation is often important, therefore. Allowing them many opportunities to assess the reliability of model components is essential.

Finally, managers need to be able to apply the models to their own management units – properties or paddocks. Information about the district average or some other hypothetical unit is too easily dismissed, for the good reason that every property and paddock is unique. Decision support systems must be made personal. Thus, for example, *HerdEcon* can be adapted to mimic almost any property in considerable detail, by setting up suitable herd flows and marketing strategies, and entering the relevant biological and economic rates. Similarly, *Paddock* attempts to provide useful information about any paddock design, once the user enters at least a bare minimum of information, including fence and water locations (but preferably also vegetation types, and, for sheep, shade, campsites, and extrinsic variables such prevailing wind directions).

Program development philosophy

Much of the programming philosophy in RANGEPACK is intended to be that of good systems programming anywhere. This includes aims such as clear documentation, good access to help, ease of use, and style consistency between modules.

Ease of use is particularly important because we are targeting an audience who rarely have a good knowledge of computers. Learning can be facilitated by designing modules to produce a result with minimal data, whilst also being capable of using much more information when an answer of greater reliability is required. Thus new users of *HerdEcon* can start by modifying the generalized information of a regional example property, and gradually improve the match with their own property. Alternatively, they can set up their property with yearly biological averages and a single operating cost per head, and only gradually try to detail the biology and economics on a monthly basis. This hierarchical approach is valuable, as long as users recognize the limitations of predictions from simplified or regional data.

Consistency of style requires not only a consistent 'user interface', which in RANGEPACK's case is based on interactive screens and a standard report generator, but also a consistent philosophical approach. This is often harder to achieve, because different modules are tackling very different problems. A significant philosophical thread running through RANGEPACK is the assessment of risk: risky climate, risky markets, risk-taking in decisions. This requires particular attention to the presentation of information, emphasizing interactive risk-based questions, and easy graphical comparison of alternative outcomes (cf. Figs. 3 and 5).

The most important element in RANGEPACK program design is that of modularity and communication between modules. All modules can be run entirely from external instructions, whether typed at a keyboard, provided by another controlling program, or read from a file. Thus, although the user can operate the programs interactively, using cursor keys and easy menu selections on the screen, all menu usage, significant key strokes and value setting on the screens can also be typed as commands. All modules use the same input and output routines (Fig. 6), including a report generator, and screen handling and graphics display routines. The report generator uses externally-defined 'report forms' to format output. These can provide output suitable for entry into other modules, or unrelated packages such as spreadsheets. All these factors have been implemented in existing modules.

The communication protocols allow one program module to control another, so that an expert system can be built to sit over any module. Such a program might lead a user through setting up a property, or perform a series of tedious comparisons without user intervention. This program is 'expert' in the sense that it knows the instructions needed to control the module, and has a suitable defined 'report form' through which to obtain results. RANGEPACK's 'umbrella' superstructure will be developed using this approach.

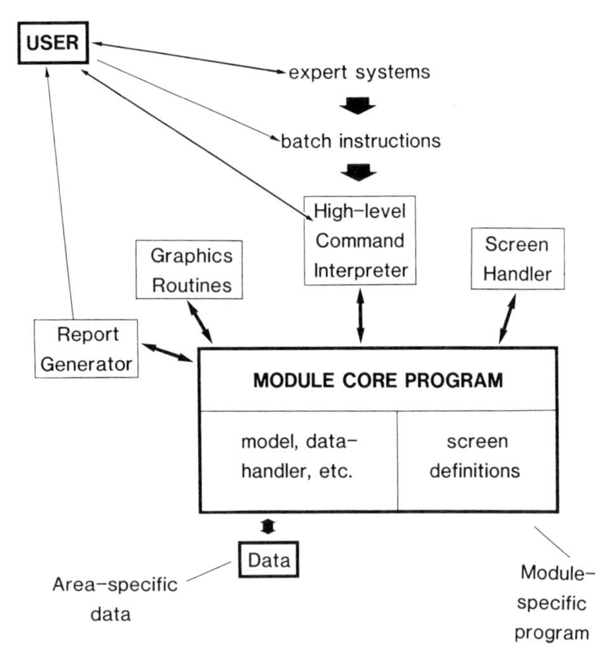

FIG. 6. An outline of the program design of a RANGEPACK module. Note separation of the module core and data from the peripheral utilities, and the several routes for interaction with the user.

LIMITS TO 'HI-TECH' APPROACHES

Projects involving high technology must not be carried away by it. Both the program *HerdEcon* and its underlying decision-making philosophy have benefited enormously from personal feedback from a series of workshops and from early users of the modules. It is also important to recognize when computer-based presentation is less effective than other approaches. For example, the property development options of Fig. 2 (as expanded in Stafford Smith & Foran, 1988) could easily be coded as a simple expert system. However, most users would find this very tedious, and a far better presentation is in the form of a wall chart. In other cases, a supplementary book of standard photographs, a simple written key, or a well-designed manual may be a more useful decision support tool than the overkill of a computer program.

On a wider geographic scale, computer-based approaches which are suitable for land managers in Australia or the United States may be useless to subsistence graziers in Africa or Asia, or even to their advisory or administrative services. Producers in developing countries may be divided between those with private ranches akin to Australia's and subsistence farmers, with respectively much less and no access to computer technology. The concepts underlying systems such as RANGEPACK, translated and simplified where necessary, should still be useful to the former class of managers. It has been suggested that 'generic shells' – simplified programs containing key elements drawn from several decision support systems comparable to RANGEPACK – might facilitate the transfer of the concepts underlying such a system to a new environment. This idea needs further exploration.

For subsistence farmers, or those with centrally-controlled lands, different decision-makers and decisions are important. Although many of the same concepts may still be important, different social, cultural and political limiting factors will necessitate a very different approach, outside the scope of this paper. The development of this approach, drawing on the experience of systems such as RANGEPACK, is a challenge for the future.

ACKNOWLEDGMENTS

D.M.S.S. thanks the RSSD programme for assistance to attend the Workshop. We are grateful to colleagues there for thoughtful discussion, to Oscar Bosman for his work on RANGEPACK, and to John Ludwig, John Ive and Doug Cocks for valuable comments on the manuscript.

REFERENCES

Arnold, G.W. & Bennett, D. (1975) The problem of finding an optimum solution. *Study of agricultural systems* (ed. by G. E. Dalton), pp. 129–173. Applied Science Publishers, London.

Bardsley, J.B. (1981) Farmers' assessment of information and its sources. Unpubl. Ph.D. thesis, University of Melbourne, Australia.

Bennett, J.L. (1983) *Building decision support systems*. Addison-Wesley Publishing Company, Mass.

Crouch, B.R. (1972) Innovation and farm development, a multidimensional model. *Sociologia Ruralis*, **12**, 431–449.

Davis, J.R., Hoare, J.R.L. & Nanninga, P.M. (1986) Developing a fire management expert system for Kakadu National Park, Australia. *J. Environ. Management*, **22**, 215–227.

Donnelly, J.R. (1988) Technology and the rural industries, expert systems and farm management. *Proc. Bicentennial Electric Eng. Cong., Melbourne, April, 1988*, pp. 135–139.

Harmon, P. & King, D. (1985) *Expert systems*. John Wiley & Sons, New York.

Ludwig, J.A. (1988) Expert advice for shrub control. *Austr. Rangel. J.* **10**, 100–105.

Ludwig, J.A. (1990) SHRUBKILL: a decision-support system for management burns in Australian savannas. *J. Biogeogr.* **17**, 547–550.

Mittra, S.S. (1986) *Decision support systems*. John Wiley & Sons, New York.

Norton, G.A. & Mumford, J.D. (1984) Decision-making in pest control. *Adv. appl. Biol.* **8**, 87–119.

Norton, G.A. & Walker, B.H. (1985) A decision-analysis approach to savanna management. *J. Environ. Management*, **21**, 15–31.

Scifres, C.J., Hamilton, W.T., Conner, J.R., Inglis, J.M., Rasmussen, G.A., Smith, R.P., Stuth, J.W. & Welch, T.G. (1985) Development and implementation of integrated brush management systems (IBMS) for south Texas. *Texas. agr. exp. Sta. Bull.* **1493.**

Stafford Smith, D.M. (1988) Modeling: three approaches to predicting how herbivore impact is distributed in rangelands. *New Mex. agr. exp. Sta. Reg. Res. Rep.* ***628***, 1–56.

Stafford Smith, D.M. & Foran, B.D. (1988) Strategic decisions in pastoral management. *Austr. Rangel. J.* **10**, 82–95.

Starfield, A.M. & Bleloch, A.L. (1983) Expert systems: an approach to problems in ecological management that are difficult to quantify. *J. Environ. Management*, **16**, 261–268.

23/SHRUBKILL: a decision support system for management burns in Australian savannas

JOHN A. LUDWIG *CSIRO Division of Wildlife and Ecology, Rangelands Research Centre, Deniliquin, New South Wales 2710, Australia*

Abstract. A microcomputer-based advisory program, SHRUBKILL, has been developed to provide advice on the use of prescribed fire to control dense shrubs. This program is a decision support system (DSS) designed to help managers resolve difficult management decisions – in this example, to decide whether and how to use management burns on pastoral properties in the semi-arid savanna rangelands of northwestern New South Wales and southwestern Queensland, Australia. In this region, the use of fire to improve paddocks by controlling dense shrub is an effective, broad-scale, and economically viable management option for property managers, who have found that the SHRUBKILL program provides valuable expert advice on how to control shrubs with fire without having to directly consult a research expert (often not readily available). Because of its modular design SHRUBKILL can easily be modified for extension to other regions where management burns can be used as an effective control for similar shrub problems.

Key words. Shrub-control, management burns, prescribed fire, decision support systems, savannas, Australia.

INTRODUCTION

A major goal of the CSIRO Division of Wildlife and Ecology is to develop microcomputer-based advisory programs called decision support systems (DSS; also see Stafford Smith & Foran, 1990). A DSS can be designed to help resolve a wide variety of problems involving difficult decision-making. They are particularly well suited for problems dealing with natural resource and land management (Davis *et al.*, 1989; Noble, 1987). Despite this suitability, there are only a few examples of DSS and expert systems applied to resource management problems (Davis & Clark, 1989; Davis, Hoare & Nanninga, 1986). The purpose of this article is to describe certain aspects of SHRUBKILL, a fully operational DSS designed to provide advice on the use of prescribed fire to manage shrub problems on pastoral properties in the savannas of eastern Australia.

The emphasis of this article will be on why SHRUBKILL, as a DSS, provides a valuable management tool, and how SHRUBKILL can be extended to similar problems in other regions. The detailed structure and operation of SHRUBKILL has been described elsewhere (Ludwig, 1988a, b). The philosophy behind the development of a DSS (used in developing SHRUBKILL) has also been well presented elsewhere (Stafford Smith & Foran, 1990).

THE SHRUB MANAGEMENT PROBLEM

The low, semi-arid, open-woodlands or savannas of eastern Australia occupy some 500,000 km^2 (Fig. 1). Mulga (*Acacia aneura* F. Muell. ex Benth.) tends to dominate the tree stratum in the western portion while poplar box (*Eucalyptus populnea* F. Muell.) tends to dominate this stratum in the east (Harrington *et al.*, 1984). The low stratum of these savannas was originally dominated by perennial grasses with few shrubs, but now extensive areas have very high shrub densities with little or no perennial grasses, depending on the history of grazing and fire. This vegetation type is referred to here as savanna (not closed-woodland or forest) because savanna was the original vegetation; some areas remain open and grassy.

Pastoral properties in these savannas are largely managed as wool enterprises. High shrub densities cause numerous management problems, for example, lowered wool-clips, reduced lamb weaning, higher animal mortalities, and difficulties with mustering. The use of prescribed fire to improve paddocks by controlling dense shrubs is argued to be the only effective and economically viable management option for these land managers (Hodgkinson & Harrington, 1985).

DECISION-MAKING

The opportunity to use fire in these savannas occurs infrequently, perhaps only once or twice in a manager's active lifetime, and is risky because of unpredictable follow-up rains. Thus, management decisions about whether (or not) to burn (and if yes, when and how to burn) involve answering many difficult questions. The answers to these questions and advice on the management problem can be obtained from shrub and fire experts (e.g. Burgess, 1988;

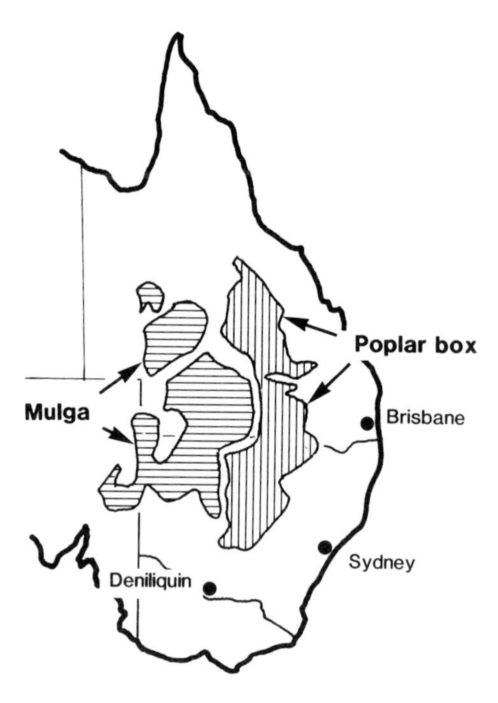

FIG. 1. The distribution of mulga and poplar box dominated semi-arid woodlands and savannas in eastern Australia.

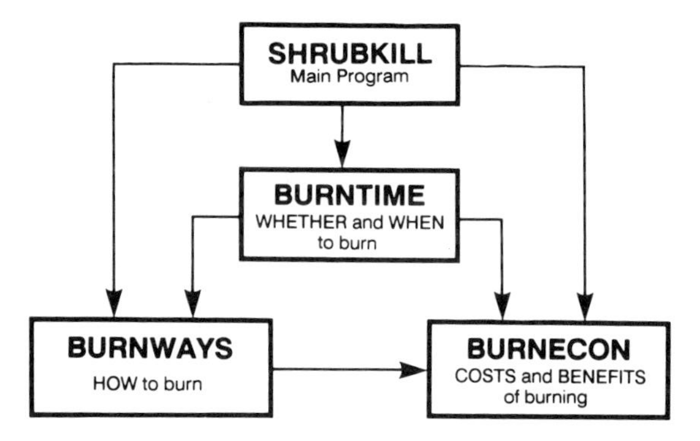

FIG. 2. The structure of SHRUBKILL as a main program with three modules: BURNTIME, BURNWAYS and BURNECON.

Harrington, 1986; Hodgkinson & Harrington, 1985; Hodgkinson *et al.*, 1984). However, obtaining this advice from these experts is difficult for a variety of reasons. Experts undergo career changes and move to new locations (e.g. Burgess is now in the sugar industry, and Harrington now works in tropical forest research). However, if this expertise can be captured within a DSS before these experts shift, then this knowledge can be made readily available through these easy to use or 'user-friendly' microcomputer programs.

SHRUBKILL is a DSS that has captured the knowledge of fire-experts in a user-friendly microcomputer program to provide conversationally-based advice on using fire to manage shrub problems. SHRUBKILL is being used directly by those pastoralists and land managers with personal microcomputers, but most usage is through consultation with agricultural advisors at local offices or in the field using portable microcomputers. The program runs on IBM PCs or compatibles using DOS 2.1 (or higher) and is available upon request from the author.

DESIGN OF SHRUBKILL

SHRUBKILL is designed as a main-program with three modules (Fig. 2). The main-program provides a series of screens explaining the purpose of this DSS, how it prompts the user with questions, and how the user is to respond to these questions using the microcomputer's keyboard. The main-program also asks a set of questions to obtain general information needed later by all three modules, for example, the density or 'thickness' of shrubs (one of several questions designed to define the severity of the shrub problem).

The BURNTIME module is an expert system that provides expert advice on whether a burn is needed, and if so, when is the most effective time to burn; as an expert system it is specifically designed to contain the rationale behind this advice (Starfield & Bleloch, 1986). This involves asking more questions to determine whether the precise nature of the shrub problem is seedlings, mature shrubs, or both (in some cases seedlings are hidden within parent plants). The shrub problem can also be dense shrub resprouts which have established after a previous prescribed fire (or a wildfire).

The BURNWAYS module is an information data-base on how to plan and safely conduct a prescribed fire. It basically focuses a large file of information down to only that information important to the individual manager's shrub problem. It provides specific advice on how to construct adequate firebreaks and how to most effectively ignite the burn based on the amount of grass fuel (and its distribution) and on the time of year (related to air temperature and humidity, and rainfall probability factors). It also provides information on how to prepare other safety precautions.

The BURNECON module is a numerical simulation model which computes a cost–benefit analysis for the prescribed fire. It considers both direct costs (e.g. costs of blading firebreaks), and indirect costs (e.g. loss of income from removing stock from the paddock during and after the burn). For these computations, BURNECON asks numerous questions to set the scenario for costs and benefits (e.g. numbers of stock, expected income from wool production, etc.). It simulates the consequences of *not* burning over a period of 20 years with those of using management burns over the same 20 years. Year-by-year costs and benefits are computed (with and without discounting), along with a 20-year gross-margin. These results, in comparison with other simulation runs using different management scenarios, allow the manager to evaluate the economics of conducting management burns, aiding the final decision-making. BURNECON has been designed to allow the user to easily examine different management scenarios.

SHRUBKILL as a management tool

SHRUBKILL is aimed at pastoral land managers, being used primarily in consultation with their local agricultural advisors. It has been distributed throughout the Western Division of New South Wales as part of a Fire Management

Folder by the Woody Weed Task Force, a Land Care Committee centred around the Cobar and Bourke region; it has also been distributed throughout southwest Queensland. SHRUBKILL has also proven to be an effective tool for disseminating information on management burns at numerous 'field days' held by the New South Wales Department of Agriculture and Fisheries, the New South Wales Soil Conservation Service, the Queensland Department of Primary Industries, and CSIRO.

SHRUBKILL has also been a valuable aid for training new extension officers who arrive in the region and are expected to give advice on management burns; using SHRUBKILL they can learn what advice the 'expert' would have given when asked a particular fire management question. These advisory officers are also learning to link SHRUBKILL to a LANDSAT-based Geographical Information System (GIS) that identifies land areas with statistically significant increases in shrub density (McCloy, 1988).

SHRUBKILL is being used by university educators as a teaching tool in land resources management courses in Australia (e.g. Roseworthy Agricultural College). It has also been a very useful tool for identifying research priorities (where expert knowledge is lacking), hence, provides a useful guide to research planning. The initial development of SHRUBKILL was also useful for learning how to build efficient and 'user-friendly' DSS.

EXTENSION OF SHRUBKILL TO OTHER REGIONS

SHRUBKILL currently provides advice on the use of prescribed fire to control shrubs within paddocks on sheep properties in the semi-arid, open mulga and poplar box woodlands and savannas of Eastern Australia. As described above, SHRUBKILL was designed as a 'modular' microcomputer program; the advantage of this approach is the ease of developing extensions to the program, or simply modifying the existing program to build a new program (other advantages of the modular approach are described by Stafford Smith & Foran, 1990).

For example, SHRUBKILL has been extended to the mallee region of southern Australia (Ludwig, MacLeod & Noble, 1989, 1990); mallee has a tree stratum dominated by low, multi-stemmed *Eucalyptus*. Again, the problem is one of fire management, particularly wildfire mitigation within conservation areas. The BURNTIME module was simply changed to provide expert advice on burning mallee (Noble, 1989a, b); advice quite different from that for the semi-arid savannas. The BURNWAYS module was also modified to include information on the use of aerial ignition, which can effectively be used in mallee (Noble, 1986). The BURNECON module only required minor changes as it provides a relatively general cost–benefit analysis.

A further development will be the extension of SHRUBKILL to provide advice on the use of prescribed fire in arid central Australia, perhaps for the problem of wildfire management for the conservation of Uluru National Park (Saxon, 1984). As with the mallee example, SHRUBKILL (or a program derived from it) provides an ideal DSS framework for this (and other) fire management problems in central Australia because an extensive body of expert knowledge already exists (e.g. Griffin & Friedel, 1984; Griffin & Hodgkinson, 1986). In this case, SHRUBKILL (or its derivative) would be a program within RANGEPACK (see Stafford Smith & Foran, 1990). SHRUBKILL could also be extended to provide advice on the use of fire to manage a number of different types of native pastures in Queensland, where there is also an extensive body of expertise (Anderson *et al.*, 1988).

ACKNOWLEDGMENTS

I thank the RSSD programme for support. The development of SHRUBKILL involved many fire and shrub experts, ecologists and economists, and modellers and computer programmers; all contributions were important. However, I would like to especially thank Graeme Miles, who did most of the computer programming, and Neil MacLeod and David Burgess, who designed the economics module. I also thank Mark Stafford Smith and another reviewer for constructive comments on earlier versions of this paper.

REFERENCES

Anderson, E.R., Pressland, A.J., McLennan, S.R., Clem, R.L. & Rickert, K.G. (1988) The role of fire in native pasture management. *Nature pastures in Queensland – the resources and their management* (ed. by W. H. Burrows, J. C. Scanlan and M. T. Rutherford), pp. 112–124. Information Series Q187023, Queensland Department of Primary Industries, Brisbane.

Burgess, D.M.N. (1988) The economics of prescribed burning for shrub control in the semi-arid woodlands of north-west New South Wales. *Aust. Rangel. J.* **10**, 48–59.

Davis, J.R. & Clark, J.L. (1989) A selective bibliography of expert systems in natural resource management. *AI Applic. in Nat. Res. Management*, **3**, 1–18.

Davis, J.R., Hoare, J.R.L. & Nanninga, P.M. (1986) Developing a fire management expert system for Kakadu National Park, Australia. *J. Environ. Management*, **22**, 215–27.

Davis, J.R., Nanninga, P.M., Hoare, J.R.L. & Press, A.J. (1989) Transferring scientific knowledge to natural resource managers using artificial intelligence concepts. *Ecol. Modelling*, **46**, 73–89.

Griffin, G.F. & Friedel, M.H. (1984) Effects of fire on central Australian rangelands. II. Changes in tree and shrub populations. *Aust. J. Ecol.* **9**, 395–403.

Griffin, G.F. & Hodgkinson, K.C. (1986) The use of fire for the management of the mulga land vegetation in Australia. *The mulga lands* (ed. by P. S. Sattler), pp. 93–97. Royal Society of Queensland, Brisbane.

Harrington, G.N. (1986) Critical factors in shrub dynamics in eastern mulga lands. *The mulga lands* (ed. by P. S. Sattler), pp. 90–92. Royal Society of Queensland, Brisbane.

Harrington, G.N., Mills, D.M.D., Pressland, A.J. & Hodgkinson, K.C. (1984) Semi-arid woodlands. *Management of Australia's rangelands* (ed. by G. N. Harrington, A. D. Wilson and M. D. Young), pp. 189–207. CSIRO Printing Centre, Melbourne.

Hodgkinson, K.C. & Harrington, G.N. (1985) The case for prescribed burning to control shrubs in eastern semi-arid woodlands. *Aust. Rangel. J.* **7**, 64–74.

Hodgkinson, K.C., Harrington, G.N., Griffin, G.F., Noble, J.C. & Young, M.D. (1984) Management of vegetation with fire. *Management of Australia's rangelands* (ed. by G. N. Harrington, A.

D. Wilson and M. D. Young), pp. 141–56. CSIRO Printing Centre, Melbourne.

Ludwig, J.A. (1988a) Expert advice for shrub control. *Aust. Rangel. J.* **10**, 100–5.

Ludwig, J.A. (1988b) SHRUBKILL users guide, CSIRO Div. Wildlife & Ecology, Deniliquin, New South Wales. 14 pp.

Ludwig, J.A., MacLeod, N.D. & Noble, J.C. (1989) MALLEE-FIRES, a decision support system for fire management in mallee conservation parks. *Natural systems management: approaches, methods and applications* (Proceedings of the Simulation Society of Australia), pp. 485–90. Central Printery, Australian National University, Canberra.

Ludwig, J.A., MacLeod, N.D. & Noble, J.C. (1990) An expert system for fire mananagement in mallee reserves. In: *The future of mallee lands: the conservation perspective* (ed. by J. C. Noble and P. J. Joss). CSIRO Printing Centre, Melbourne.

McCloy, K. (1988) NSW Department of Agriculture and Fisheries initiate: an operational GIS. *Remote Sensing Committee Newsletter*, **1**, 6.

Noble, I.R. (1987) The role of expert systems in vegetation science. *Vegetatio*, **69**, 115–21.

Noble, J.C. (1986) Prescribed fire in mallee rangelands and the potential role of aerial ignition. *Aust. Rangel. J.* **8**, 118–30.

Noble, J.C. (1989a) Fire studies in mallee (*Eucalyptus* spp.) communities of western New South Wales: The effects of fires applied in different seasons on herbage productivity and their implications for management. *Aust. J. Ecol.* **14**, 169–87.

Noble, J.C. (1989b) Fire regimes and their influence on herbage and mallee coppice dynamics. *Mediterranean landscapes in Australia: Mallee ecosystems and their management* (ed. by J. C. Noble and R. A. Bradstock), pp. 168–80. CSIRO Printing Centre, Melbourne.

Saxon, E.C. (ed.) (1984) *Anticipating the inevitable: a patch-burn strategy for fire management at Uluru (Ayers Rock – Mt. Olga) National Park.* CSIRO Printing Centre, Melbourne.

Stafford Smith, D.M. & Foran, B.D. (1990) RANGEPACK: the philosophy underlying a microcomputer-based decision support system for pastoral land management. *J. Biogeogr.* **17**, 541–546.

Starfield, A.M. & Bleloch, A.L. (1986) *Building models for conservation and wildlife management.* Collier Macmillan Publ., London.

Conclusions, Changes, and Consequences

Intercontinental comparisons can provide insights to scientists trying to understand the functioning of their own savanna lands. Such comparisons are not perfect by any means – there are always too many uncontrolled variables – yet they can be very useful in developing (or in some cases, testing) hypotheses about how savannas actually work.

Several comparisons of papers in this volume yield interesting ideas about savannas. Firstly, it is important to distinguish comparisons which are derived from percentages from those which use absolute numbers. Statements of *relative* strengths of a particular force are instructive *within* a particular savanna, but can be less useful when applied *among* savannas. Comparisons among savannas need to consider both relative and absolute values to be most useful, although this is not always possible due to lack of data. There are many examples, but two of the major ones which infuse the papers in this volume are the relative roles of vertebrate versus invertebrate herbivory, and the relative roles of herbivory versus fire, especially in Australia compared to other continents.

On the relative roles of vertebrate versus invertebrate herbivory, it is important to note that on all continents the role of herbivory by invertebrates is considered greater than that by vertebrates. Australia is not unique in this regard. However, since Australia has no native ungulates (although it did have many large non-ungulate grazing mammals until recent geological time), ecologists' attention turns to the massive herbivory by invertebrates (e.g. Braithwaite; unless otherwise indicted, references are to papers in this volume, or are referenced in the Preface). Yet, as Anderson and Lonsdale point out, the absolute biomass of invertebrates appears to be just about the same in Australian savannas as in other savannas. When ungulates (and other large mammals) are introduced into the Australian savannas from other continents, they generally proliferate and do very well, often developing population sizes greater than they have on their native savannas of equal plant production (cf. Freeland). Whether they have reduced the Australian abundance of invertebrates through some competitive mechanism for the same food resource is unknown. However, that is unlikely, but rather the vertebrates have been added 'on top of' the herbivore pressure from the invertebrates.

On the relative roles of herbivory and fire, it is commonly observed that in Australia, fire is the most important non-microbial consumer of plant biomass, whereas in Africa herbivory (specifically large mammal herbivory) is more important (e.g. Braithwaite). This is certainly true within each continent. One question is whether herbivory and fire are equally important forces – some sort of ecological analogues. The opposite question is whether there is some sort of a hierarchy of these two factors. It is commonly accepted that they are ecological 'trade-offs', and, indeed, they do seem to be within a given savanna, and within a short time period. However, on a larger scale of space and time, I suggest that there might be a hierarchy, although with significant feedbacks; specifically, that herbivory (large mammal herbivory) is the principal factor, and fire a derivative or secondary factor. Support for this comes from the 'experiments' in additions of large mammals in Australia and the Americas and also subtractions of large mammals in parts of Africa by human activity over the past century. Where large grazing mammals are present in appreciable numbers, they reduce the incidence of fire (via removal of the fuel load); however, where the grazers are not present, then fire becomes the dominant feature – by 'default'.

On the other hand, it is intriguing to speculate that there is indeed a 'trade-off' between fire and microbial decomposition of plant biomass. Assuming that invertebrate grazing is at about the same absolute levels on savannas of equivalent plant productivity (e.g. equivalent plant available moisture and available nutrients, or PAM-AN), then most of the remainder of the plant biomass is consumed by fire, free-living microbes, or animals that have microbial flora in their digestive tracts (ungulates or termites). This makes sense in these tropical savannas dominated by C4 grasslands, as C4 plants are extremely poor quality food for most herbivores, whether they are vertebrate or invertebrate herbivores (Caswell *et al.*, 1973), without a source of cellulase. Of course, both ungulates and termites have such a source of cellulase from symbiotic gut flora; in the absence of these animals (especially ungulates), fire assumes a larger decomposer role than when they are present.

This volume on savannas brought several other scientific issues to the forefront. For example, the several papers on 'patches' of woody plants in savannas, when taken as a whole, are all consistent with the notion that there is some sort of positive feedback mechanism(s) which encourages the spread of woody patches – in some cases authors have identified suppression of grass via shading (Menaut *et al.*, Archer), continual input of seeds by birds and other animals (Archer), nutrient accumulations by activities of humans and/or animals (Blackmore *et al.*, Scholes), reduction in fire (Archer, Menaut *et al.*), etc. Several authors framed the problem in terms of the positive feedbacks versus forces which reduce the size of the woody patches, or eliminate them altogether; it would be interesting to apply this approach to other questions regarding habitat heterogeneity in savannas.

Another scientific issue which can be gleaned from this volume is the extent to which a population approach was proven useful in understanding savanna structure and composition (whether of species or of life forms such as tree–grass ratios). Outstanding examples are those of Menaut *et al.* and of Medina and Silva for plants, and of Freeland for animals. Although it is true that the four major determinants of savannas (Frost *et al.*, 1986; Walker, 1987) are well-thought-out and the appropriate place to begin study to understand savanna ecosystem function, there are instances where they are awkward. These instances involve cases where population phenomena are acutely responsible for structure and composition (e.g. competition, dispersal, etc., especially where it involves a rapidly changing population such as an invading exotic species), but where the relationship of those population parameters to the abiotic factors of PAM and AN is not readily discernible. (In this volume Medina and Silva, and Bilbao and Medina, are examples where the two areas are nicely bridged. In general, however, we scientists have not gone very far over the past 20 years in discovering relationships between physiological and population-level attributes of organisms.) Indeed, the four determinants of savannas we work with are an attempt to get at *function* in terms of the whole ecosystem, but what we are interested in ultimately is *structure* and *composition* of the biota. There are instances where there is an opportunity to understand a rapidly changing structure/composition by a simpler, more direct, route, and these sorts of studies should be encouraged as well.

The volume also deals with the management of savannas, with a special emphasis on rangelands, it being the primary activity in Australian savannas. One theme that emerged was the advocacy of the use of fire as a management tool, not only to reduce undesirable plants and to redistribute stock, but as one of the most effective and economically feasible tools available to rangeland managers (e.g. Winter, Ludwig, Burrows *et al.*) Also, many authors pointed out that the socio-economic factors of decision-making and planning cannot be ignored. To the credit of many of the rangelands scientists working on the economic–ecology interface (e.g. Stafford Smith and Foran, Winter, McKeon *et al.*, Stuth *et al.*, Moog), they have made an concerted attempt to 'understand the wishes of people,' perhaps responding to the concern of Dr Ted Henzell in his summative address at a world symposium on savannas in 1984 (Henzell, 1985). It is not clear whether this has established itself as a new trend, or are isolated cases. Nevertheless, Henzell's admonition must not be forgotten, especially when decisions are made which affect subsistence-level inhabitants of the savannas.

Other points made in the volume are very nicely summarized by the authors (especially those involving abiotic factors and their relationships to biota; cf. Holt and Coventry, McCown and Williams, Medina and Silva), so I shall not attempt to repeat them here. I have concentrated on some of the insights gained by cross-reading groups of papers. In spite of the valuable sessions conducted during the Darwin (1988) symposium and workshop, and the valuable hypotheses, ideas, and principles that have emerged by the individual research programmes and interactions of the scientists represented in this volume, there is still very much that needs to be done in order to understand the ecological determinants of savannas, the biological mosaics and tree/grass ratios, and how to manage savannas for long-term sustainability.

The savannas of the world are currently facing dramatic changes. It is absolutely critical we understand the basic biological processes governing savanna flora and fauna if we are to avert disaster on a massive scale. Decisions regarding savannas will be made on political and economic grounds, mainly; nevertheless, it is crucial that they be enlightened decisions, with scientific knowledge a part of that decision process. What are the major changes ahead, and to what extent do we understand aspects of savanna function which relate to these agents of change?

The changes ahead

Climate change

The climate is predicted to change at a rate unprecedented in geological time, in response to anthropogenic changes in atmospheric composition. Although the exact causes and rate of change are disputed, there is almost universal agreement that the next 50 years will see a warming of the global average temperature and, as a consequence, changes in precipitation patterns. (It is interesting to note that the U.S. Environmental Protection Agency, in a recent study of relative risks to the environment and human welfare, where criteria for the assessment of risk were selected *a priori*, ranked climate change and habitat destruction as the two most important risks facing humankind, ahead of toxic substances and pollution; EPA, 1990.) There are already correlates of global climate events (e.g. ENSO) to regional climate and vegetation cycles, some of which are already being incorporated into management schemes for savannas (see the development of models for management by McKeon *et al.* which explicitly includes ENSO; also Stuth *et al.*).

Indeed, climate change is one of the two most important factors of change affecting savannas, the other being land use changes driven by human and economic pressures. Both will surely challenge our scientific ingenuity in our attempts to understand how savannas function.

Land use changes and related matters

The increasing size of the human populations living in savannas. The growth of human populations living on many of the world's savannas is among the fastest in the world. Especially in savannas of Africa and Asia, the pressure for food and fuel from local sources has been increasing dramatically. (See discussions of Stott, Pemadasa and Yadava.) This has led to greater pressure on plants and soil in traditional agricultural land, as well as a geographic spread in the conversion of land for human settlement and activities. This includes the conversion of both marginal land (dry, unproductive, and with little potential for long-term yields) and forested land for agricultural activities, often at a subsistence level. With further increases in population pressure and climate change anticipated, there are grave concerns for ecological stability of such areas (cf. Stott, Adámoli *et al.*).

Changes in land use as a consequence of changes in economic, social, and technological development. Besides the sheer pressure of more humans, other social pressures affect how land is used; some of these involve changing family structures or increasing expectations in standard of living and quality of life. Others are tied to real changes in technology which affect how daily lives and huge enterprises are conducted; for example, the adoption of new tools to work soil, or the use of genetic lines of cattle, introduced pasture and crop plants, and mineral supplements in livestock production. And, of course, the market place has changed dramatically, leading to high input/high output farming and ranching operations, especially in the Americas and Australia, and also parts of Africa and Asia. (See Winter's insightful discussion on the need to change Australia's current agricultural philosophy in the interest of long-term sustainability.)

Changes in relative values of various land use options. Both increasing human pressure and economic pressure have shifted the relative values of various land use options in savanna regions. The long-term value of forested land, or of ungrazed savanna grassland, for instance, almost always gives way to short-term value of immediate needs for food and/or profit. Ever increasingly, the value of non-agricultural land is being evaluated in economic terms, which, because of the economic assumptions used (e.g. discounting; short-time horizons) comes up the 'loser' to other 'productive' land use options. Ironically, or maybe partly in response to this, there is an ever-increasing awareness of conservation in many areas (cf. Stott, Braithwaite); these opposing forces are part of the rapidly shifting, and sometimes polarizing, values associated with various land use options.

The establishment of introduced plants and animals in native savanna lands

The very common practice of transferring new species or varieties in order to increase production of crops and livestock is not at issue here. The problem comes from either deliberate or accidental introductions that become naturalized (e.g. Bilbao and Medina, Freeland). There are scores, if not hundreds, of examples where exotic plants and animals have wreaked havoc by outcompeting or exerting severe predation or herbivory on desirable native species. Although there is an increasing sensitivity to the problem, and in some cases active campaigns to eliminate exotics and 'screening' processes in place to help prevent new introductions, the reality is that the problem is going to get worse, mainly because many introduced species are in their early stages of exponential growth/spread (e.g. discussion for weeds of Northern Australia; Cowie & Werner, 1990).

Savanna processes affected by the changes ahead

Undoubtedly, climate change will affect all the major determinants of savannas (determinants discussed in Belsky). Using the four major determinants of the RSSD program (cf. Frost, 1986; Walker, 1987), the Plant Available Moisture (PAM) will be affected by changes in precipitation patterns, which will have short-term effects on Available Nutrients (AN) as well as longer-term effects on soil properties. Any changes in plant biomass, temperature and humidity will also influence fire regimes; changes in productivity and fire will both affect herbivory patterns.

Land use changes can certainly effect Available Nutrients (AN) either by additions or subtractions of nutrients via supplements, removal of biomass, etc. Depending on circumstances, land use changes can affect Plant Available Moisture (PAM), especially when the land use change alters plant biomass or organic content of the soil. The direct manipulation of biomass through cutting, culling, or species additions, as well as direct and indirect changes in fire and herbivory patterns, are other major factors which change savanna structure and stability with concomitant changes in land use patterns.

The initial effect of the introduction of exotic species (and the removal of native species by the direct activities of humans) is a change in species composition. This might, in turn, affect fire and/or herbivory patterns, and possibly, PAM and AN. Direct changes in species composition is one of the cases where a physiological/ecosystem approach to understanding the problem is limited, and population dynamics *per se* must be considered. Here, the most useful operative level in the biological hierarchy are the population-level processes such as comparative demography, competition, predation, and/or herbivory. While it is true that these population-level processes operate against a background of changes in available moisture and nutrients, in the first instance they can be studied directly, without first taking the arduous trip down the biological hierarchy to study physiological responses to abiotic factors.

ACKNOWLEDGMENTS

Thanks to A. J. Press and P. Stott for valuable discussions on savannas over the course of the past 2 years.

PATRICIA A. WERNER

REFERENCES

Caswell, H., Reed, R., Stephenson, S.N. & Werner, P.A. (1973) Photosynthetic pathways and selective herbivory: a hypothesis. *American Naturalist*, **107**, 465–480.

Cowie, I. & Werner, P.A. (1988) Weeds in Kakadu National Park: Part II. Final Report to the Australian National Parks and Wildlife Service. CSIRO, Darwin. 87pp.

Environmental Protection Agency (1990) Reducing risk: Setting priorities and strategies for environmental protection. Report of the EPA Science Advisory Board A-101. U.S. Government Printer, Washington, D.C.

Henzell, E.F. (1985) Summative Address. *Ecology and management of the world's savannas* (ed. by J. C. Tothill and J. J. Mott), pp. 367–369. Australian Academy of Sciences, Canberra.

Index

Note: Not all species and taxa named in the volume are listed in this index. Only key ecological, economic, and indicator species are used to allow more space for the referencing of complex processes, interactions, and comparisons.
Index compiled by Professor Patricia A. Werner and Professor David L. Wigston